# Transgenesis Techniques

## SECOND EDITION

# METHODS IN MOLECULAR BIOLOGY™

## John M. Walker, SERIES EDITOR

METHODS IN MOLECULAR BIOLOGY™

# Transgenesis Techniques

## Principles and Protocols

### SECOND EDITION

Edited by

## Alan R. Clarke

*School of Biosciences, Cardiff University*
*Cardiff, UK*

**Humana Press**  **Totowa, New Jersey**

Cover design by Patricia F. Cleary.

Cover illustration: Oocyte injection in the mouse—embryo following injection of DNA (see pp. 63–64.)

For additional copies, pricing for bulk purchases, and/or information about other Humana titles, contact Humana at the above address or at any of the following numbers: Tel: 973-256-1699; Fax: 973-256-8341; E-mail: humana@humanapr.com or visit our website at http://humanapress.com

**Photocopy Authorization Policy:**

Authorization to photocopy items for internal or personal use, or the internal or personal use of specific clients, is granted by Humana Press Inc., provided that the base fee of US $10.00 per copy, plus US $00.25 per page, is paid directly to the Copyright Clearance Center at 222 Rosewood Drive, Danvers, MA 01923. For those organizations that have been granted a photocopy license from the CCC, a separate system of payment has been arranged and is acceptable to Humana Press Inc. The fee code for users of the Transactional Reporting Service is: [0-89603-696-0/02 $10.00 + $00.25].

Printed in the United States of America. 10 9 8 7 6 5 4 3 2 1

Library of Congress Cataloging-in-Publication Data

Transgenesis techniques: principles and protocols/edited by Alan R. Clarke.—2nd ed.
    p. cm. — (Methods in molecular biology; v. 180)
    ISBN 0-89603-696-0 (alk. paper)
        1. Transgenic animals—Laboratory manual. 2. Animal genetic engineering—Laboratory manuals.
I. Clarke, Alan R. II. Methods in molecular biology (Totowa, NJ); v. 180.

QH442.6.T66 2002
576.5'07'24--dc21

2001024458

# Preface

The past decade has witnessed a spectacular explosion in both the development and use of transgenic technologies. Not only have these been used to aid our fundamental understanding of biologic mechanisms, but they have also facilitated the development of a range of disease models that are now truly beginning to impact upon our approach to human disease. Some of the most exciting model systems relate to neurodegenerative disease and cancer, where the availability of appropriate models is at last allowing radically new therapies to be developed and tested. This latter point is of particular significance given the current concerns of the wider public over both the use of animal models and the merits of using genetically modified organisms.

Arguably, advances of the greatest significance have been made using mammalian systems—driven by the advent of embryonic stem-cell–based strategies and, more recently, by cloning through nuclear transfer. For this reason, this new edition of *Transgenesis Techniques* focuses much more heavily on manipulation of the mammalian genome, both in the general discussions and in the provision of specific protocols.

Of all mammalian experimental systems, the laboratory mouse is probably the most widely used, a situation that almost certainly derives from the fact that it is genetically the most tractable. This second edition, therefore, devotes much space to methodologies required for the creation and maintenance of genetically modified murine strains. In addition to protocols for conventional pronuclear injection, chapters have been included covering alternative routes to the germline, by either retroviral or adenoviral infection. Extensive coverage is also given to the generation, maintenance, and manipulation of embryonic stem cell lineages, since this is now widely recognized as an indispensable approach to genotype–phenotype analysis. Part V contains protocols to facilitate gene targeting and so permit both constitutive and conditional gene targeting. The latter approach, reliant on either the Cre-lox or the Flp-frt system, is rapidly gaining favor as a method of choice for the analysis of null mutations because it solves the twin difficulties of embryonic lethality and developmental compensation—two problems that have hampered the analysis of simple "knock-out strains."

The proliferation of newly engineered murine strains has given rise to one problem within the field, namely, that of the long-term storage of lines for which there might be no immediate requirement. Within many laboratories, this is now far from a trivial problem, and, therefore, methodologies are included that detail the cryopreservation of both male and female germlines.

Although the mouse is currently the most genetically tractable system, it is not without its limitations and clearly cannot deliver all appropriate experimental or commercial systems. Transgenic manipulation of the rat germline is now delivering valuable models across a range of fields, perhaps most notably in neurobiology and in the study of vascular diseases. This edition, therefore, also focuses on the generation, maintenance, and cryopreservation of rat transgenic lines.

The mouse and the rat remain essentially laboratory models. However, perhaps the most radical change to occur within the field relates to our emerging ability to genetically engineer livestock. In particular, the advent of cloning as a viable technology has wide ramifications for the scientific and industrial communities as well as for the wider public. Protocols are given for the generation of transgenic sheep by nuclear transfer, and, furthermore, the potential implications and future directions of large animal transgenesis are discussed in some detail.

Finally, this second edition carries a very detailed part relating to the basic analysis of transgenic organisms. Although many of the techniques included are widely used throughout molecular biology, those pertinent to transgenic analysis have been brought together to facilitate the rapid analysis of phenotype. Used in conjunction with the plethora of techniques relating to the generation and maintenance of transgenic strains, the contributors and I anticipate that this new edition of *Transgenic Techniques* will prove an invaluable asset to any laboratory either already engaged in transgenic manipulation or setting out along this route.

*Alan R. Clarke*

# Contents

# Contributors

RICHARD A. BOWEN • *Animal Reproduction and Biotechnology Laboratory, Colorado State University, Fort Collins, CO*

GILLIAN BROOKER • *Molecular Physiology Laboratory, University of Edinburgh Medical School, Edinburgh, UK*

GARY A. J. BROWN • *Transgenic Mouse Core Facility, Shands Cancer Center, University of Florida, Gainesville, FL*

KEITH H. S. CAMPBELL • *Division of Animal Physiology, University of Nottingham, Nr. Loughborough, Leicestershire, UK*

WILLIAM CHIA • *Institute of Molecular and Cell Biology, Singapore*

A. JOHN CLARK • *Division of Gene Expression and Development, Roslin Institute, Roslin, Midlothian, UK*

TIMOTHY J. CORBIN • *Amgen, Inc., Thousand Oaks, CA*

RALF KÜHN • *Institute for Genetics, Cologne, Germany*

JIM MCWHIR • *Division of Molecular Biology, Roslin Institute, Roslin, Midlothian, Scotland*

DAVID W. MELTON • *Molecular Medicine Centre, University of Edinburgh, Edinburgh, UK*

JOHN J. MULLINS • *Molecular Physiology Laboratory, University of Edinburgh Medical School, Edinburgh, UK*

LINDA J. MULLINS • *Molecular Physiology Laboratory, University of Edinburgh Medical School, Edinburgh, UK*

NAOMI NAKAGATA • *Division of Reproductive Engineering, Center for Animal Resources and Development (CARD), Kumamoto University, Kumamoto, Japan*

MARK J. O'CONNOR • *Institute of Molecular and Cell Biology, Singapore*

DOMINIC RANNIE • *Department of Pathology, University of Edinburgh Medical School, Edinburgh, UK*

STEFAN SELBERT • *Mice and More GmbH and Co. KG, Hamburg, Germany*

JILLIAN M. SHAW • *Monash Institute of Reproduction and Development, Monash University, Clayton, Victoria, Australia*

RAUL M. TORRES • *Department of Immunology, University of Colorado Health Sciences Center; National Jewish Medical and Research Center, Denver, CO*

YUTAKA TOYODA • *Department of Reproductive and Developmental Biology, Institute of Medical Science, University of Tokyo, Tokyo, Japan*

A. O. TROUNSON • *Monash Institute of Reproduction and Development, Monash University, Clayton, Victoria, Australia*

TOHRU TSUKUI • *Department of Reproductive and Developmental Biology, Institute of Medical Science, University of Tokyo, Tokyo, Japan*

DAVID WELLS • *Reproductive Technologies Group, AgResearch, Ruakura Research Center, Hamilton, New Zealand*

# I

## TOPICAL REVIEWS IN TRANSGENESIS

# 1

## Biomedical and Agricultural Applications of Animal Transgenesis

### Jim McWhir

## 1. Introduction

In 1980, Gordon et al. *(1)* showed that DNA injected into the pronuclei of single-cell embryos could be incorporated, expressed, and transmitted to the offspring of transgenic mice. Since then, pronuclear injection has become a widely used and invaluable tool for the study of mammalian gene function. The same technique has also been used to generate transgenic livestock *(2)*; however, the proportion of injected and transferred embryos giving rise to transgenic animals is greatly reduced relative to mice (1 to 2% vs 10–25%). Two general disadvantages of pronuclear injection apply equally to all species: unpredictable effects of site of incorporation and transgene copy number on gene expression lead to a requirement for testing multiple lines to ensure appropriate transgene expression, and the technique is restricted to the addition of genetic material.

The disadvantages of pronuclear injection have been partially circumvented in mice with the development of an alternate route to transgenesis through murine embryonic stem (ES) cells *(3,4)*. ES cell lines are isolated from undifferentiated cells of the early embryo and retain in culture their capacity to differentiate into the full range of embryonic tissues. Hence, ES cells can be genetically modified in vitro and returned to the early embryo, where they resume their normal program of development. This procedure leads to the generation of chimeric animals whose tissues, including germ cells, are frequently derived from both host embryo and ES genotypes, and a proportion of chimeras will transmit the ES-derived genetic modification to their offspring. Precise genetic modification can be achieved in ES cells by taking advantage of homologous recombination to target single-copy transgenes to specific sites or

From: *Methods in Molecular Biology, vol. 180: Transgenesis Techniques, 2nd ed.: Principles and Protocols*
Edited by: A. R. Clarke © Humana Press Inc., Totowa, NJ

to modify existing genes *in situ*. A major limitation of this technology is that, at present, germline-competent ES cells are available only in the mouse.

As a consequence of the inefficiency of pronuclear injection in farm animals, the absence of proven ES cells in these species, and the high cost of animal maintenance, the literature describing transgenic livestock has been better served by reviews than by concrete example. Perhaps the single exception has been the use of transgenic livestock to produce a small number of pharmaceutical proteins. This situation may be about to change; the development of techniques for cloning livestock from cultured cells *(5,6)*, of cell-based transgenesis in sheep *(7)*, and the imminent possibility of gene targeting in livestock have dramatically altered the logistic and biologic constraints.

## 2. What Has Changed?

Cell-based methods of transgenesis by nuclear transfer or by ES chimerism have the critical advantage that genetic modification is carried out on cycling cell populations rather than directly on embryos. Hence, mass transfection is followed by selection for expression of a marker transgene in cultured cells and gives rise to hundreds of primary transfectants. These in turn give rise to limitless numbers of clonally derived cells, each with the potential to give rise to a transgenic founder animal. Significantly, DNA from modified cells can be prepared and characterized prior to their use in animal experiments. Subsequent nuclear transfer uses only those cells carrying the desired modification, and 100% of resulting animals will be transgenic. This cell-based approach to transgenesis has recently been exemplified in livestock by the arrival of Polly *(7)*, a transgenic sheep carrying a gene encoding the human blood-clotting factor IX.

In principle, the ability to clone from cultured cells following genetic modification has provided the means to identify rare cells in which DNA has integrated into homologous sequences already present in the genome (gene targeting). Although this has not yet been exemplified in the cell populations proven in nuclear transfer, human somatic cells have been successfully targeted using those same techniques that are now routine for murine ES cells *(8–13)*. It only remains to couple targeting in livestock-derived cells with nuclear transfer. A major application of cloning technology, therefore, will be to generate animals that carry subtle gene modifications generated by gene targeting in cultured somatic cells. The specific advantages to gene targeting that accrue are discussed in later sections. The only caveat at the time of this writing is the formal possibility that the properties of cells necessary to support targeting may be incompatible with those required for nuclear transfer.

## 3. ES Cells in Livestock

Cloning from genetically modified somatic cells may be thought to render ES cells redundant for most applications in livestock. Murine ES cells, however, are particularly well adapted to gene targeting and also provide an in vitro model of differentiation that may offer novel biomedical opportunities in transplantation therapy. It seems likely that interest in livestock ES cells will persist. There are numerous reports of ES-like cells in several species: hamster *(14)*, mink *(15)*, rat *(16)*, chicken *(17)*, sheep *(18–20)*, cattle *(21–25)*, pig *(18,20,26,27)*, rhesus monkey *(28)*, and human *(29)*. None of these reports, however, has yet met the definitive test of germline transmission (there is no intended suggestion that this test should be applied in the special case of human cells); in spite of intensive effort, germline ES cell technology remains restricted to mice.

Even were proven ES cells available, the ES route to transgenesis in farm animals would have the serious disadvantage that it requires an extra chimeric generation to establish transgenic founder animals. By contrast, the cloning option would generate transgenic individuals in the first generation. Perhaps the greatest disadvantage of the chimeric route is that it requires test breeding of all animals generated, including an unknown proportion (possibly 100%) that will be incapable of germline transmission. This contrasts with a cloning experiment in which failure is self-evident at an early stage by the absence of pregnancies.

The aforementioned considerations raise the possibility that one might enjoy the best of both the cloning and ES options by employing targeted ES or ES-like cells in nuclear transfer. Here, there are several unresolved issues. Unlike murine ES cells, the livestock-derived ES-like lines reported to date are poorly adapted to single-cell cloning—a problem that will need to be overcome if these cells are to be used in gene targeting (reviewed in **ref. 30**). In addition, the early results of somatic nuclear transfer suggest that an important ingredient is that the nuclear donor be in a state of quiescence or $G_0$ *(3)*. It seems likely that in addition to issues of cell-cycle compatibility between nucleus and ooplasm, the quiescent nucleus be configured in such a way as to favor reprogramming. Alternatively, it may be important that the somatic cell program of gene expression be shut down before the full developmental program is reinitiated. In either event, ES cells (unlike fibroblasts) do not readily enter quiescence on serum starvation. Several important questions remain unanswered: Can ES cells be entered into quiescence in some novel way? Can the differentiated derivatives of targeted ES cells be entered into quiescence? Do the ES-like cells currently available from livestock support gene targeting?

## 4. A Hierarchy of Complexity

Some transgenic applications in farm animals will involve complex targeting technology and yet in biologic terms may have quite humble goals (simple loss of function mutation). In other instances, the technology may be relatively crude (as with pronuclear injection of growth hormone [GH] genes) whereas the biologic objective (to modify growth rate) is highly ambitious. In mouse transgenic programs, the biologic objective is usually straightforward—to observe and record the effects of ectopic gene expression or, in the case of gene targeting, the effects of loss of gene function. Here, although the phenotypic consequences are often not predictable, the experiment will always be informative.

In contrast to the mouse, many potential livestock applications (particularly in agriculture) will involve intervention in complex metabolic pathways in which the objective is to achieve a predetermined phenotypic change. Here, there is an additional challenge: it is necessary to accurately predict the phenotypic consequences of a single genetic modification. Limited attempts to do this in order to increase the growth rate of transgenic pigs have led to unforeseen consequences on animal health and fertility (31). The most straightforward applications of transgenesis in livestock are those in which the objective is simply to harvest high levels of recombinant protein. In this case, there is no requirement to modify endogenous metabolic pathways, a physiologic response to transgene expression is neither required nor anticipated, and the risks of adverse effects on animal health and welfare are minimized. It is not surprising, therefore, that the most successful transgenic applications to date have involved expression of human therapeutic proteins in the milk of transgenic sheep (32), pigs (33), cows (34), and goats (35).

## 5. Biopharming

Biopharming is the commercial production of pharmaceuticals from the body fluids of transgenic animals. Although most attention has centered on the mammary gland (for reviews see refs. 36–39), other body fluids may have particular benefits for certain applications. For example, human GH has been expressed in mouse bladder epithelium under the control of the mouse uroplakin promoter (40). Advantages of bladder production might include the ability to harvest from all animals at all stages of their lives and the small number of other proteins from which the recombinant protein need to be purified. Transgenic swine have been generated that express human hemoglobin in their blood as a potential cell-free substitute for human plasma (41,42). However, this blood-based approach has been hampered by difficulty in separating human hemoglobin from its porcine counterpart.

By far the most readily harvestable source of recombinant protein is milk. While fermentation technologies and transgenic plant alternatives may be favored for some applications, the mammary gland provides several general advantages. Milk is a less complex fluid than blood, thus enhancing the prospects for rapid purification of recombinant protein. In addition, milk proteins are present in the circulatory system at undetectable or very low levels, thus minimizing potential animal health problems associated with high circulating levels of metabolically active proteins. Unlike fermentation-derived products, recombinant proteins produced in the mammary gland are posttranslationally modified in a manner that closely mimics their modification in humans *(43)*, and are more likely to be stable, have high biologic activity, and be non-immunogenic in patients. While the mammary gland may be the preferred option in this regard, there is still scope for improvement. For example, some proteins purified from milk have a lower than expected molecular mass *(44–46)*; the ovine or bovine mammary gland product does not exactly mimic the human-derived protein. At least one group has addressed this issue by the coinjection of a furin transgene designed to increase the level of posttranslational modification *(47)*.

Production of pharmaceutical proteins in the transgenic mammary gland is rapidly being commercialized to produce products such as $\alpha$-1 antitrypsin for treatment of emphysema and cystic fibrosis *(48)*; the blood-clotting factors antithrombin III *(35)*, factor VIII *(49,50)*, factor IX *(5,51)*, and fibrinogen *(52)* for treatment of bleeding disorders; and protein C *(46,53)* for treatment of blood clots. Recombinant antithrombin III and $\alpha$-1 antitrypsin from transgenic livestock are now undergoing phase III and phase II clinical trials, respectively, and this first generation of transgenic livestock has already spawned a significant biopharming industry.

Most of the achievements in biopharming to date have employed pronuclear injection. How might cell-based techniques and gene targeting be used to improve the rate and direction of progress? As with any biologic system, there are upper limits to the synthetic potential of the mammary gland. One way to boost the production of therapeutic proteins would be to delete nonessential milk protein genes and simultaneously replace them with the desired transgene by gene targeting. This approach would not only introduce the transgene into an active site in the genome, but would simultaneously provide excess synthetic capacity by knocking out the gene for a competing high-volume protein. This method may be essential to providing proteins that are required in very large quantities such as human serum albumin (potentially useful in the treatment of burns).

Gene targeting also can be used to improve the level and repeatability of transgene expression. Microinjection of DNA into the pronucleus usually results in multiple copies of the transgene being integrated in large arrays. In most cases, the level of expression of the transgene is not correlated with the number of copies and is subject to random effects of elements at the site of incorporation. In practice, up to 10 transgenic lines may have to be analyzed to obtain a single line in which the transgene is expressed in the desired temporal and spatial manner. A potential solution to this problem is to introduce the transgene into chosen sites in the genome by homologous recombination in ES cells. In mice this strategy has been used successfully to introduce a *lacZ* cassette into the hypoxanthine phosphoribosyl transferase (*HPRT*) locus *(54)*, to replace the β-globin *(55)* and α-lactalbumin genes *(56)*, and to introduce a *bcl-2* minigene into the *HPRT* locus in ES cells *(57)*. A variation on this theme involves an in vitro prescreen of marked, random sites for appropriate transgene expression followed by targeted replacement to introduce a transgene into the same site *(58)*.

Cloning not only facilitates cell-based transgenesis but also carries innate advantages. Based on averages of the data reported in the "Dolly" article *(5)*, the proportion of nuclear transfer embryos that develop to term is low (approx 1.0%). Fortunately, much of this cost is owing to embryos that fail prior to reimplantation into foster females and is borne in the laboratory rather than on the farm. The proportion of embryos transferred into final recipients that give rise to lambs rises to 6.0%, and since two embryos are generally implanted per recipient, the proportion of recipients that give rise to nuclear transfer lambs rises to about 12%. Although there is still a large requirement for animals to act as embryo donors, cloning compares favorably in efficiency with pronuclear injection. PPL Therapeutics, in collaboration with Roslin Institute, have recently generated the first additive transgenic sheep, Polly, using cell-based transgenesis *(7)*. Even with the present rates of nuclear transfer success, it was estimated that 2.5 times as many sheep would have been required to create Polly by pronuclear injection. The practical consequence of this efficiency gain is that more potential therapeutic products will be tested. Another barrier to the testing of novel milk-derived recombinant proteins is difficulty in obtaining sufficient quantities of purified protein from single founder animals (often male) for preliminary trials. This alone can be a sufficient commercial risk to rule out many potential applications. The cloning option provides a means with which to generate multiple female founders in a single generation.

## 6. Nutraceuticals

Closely related to biopharming is the idea that genetic modification of milk proteins could be used to improve the nutritional or industrial properties of milk. A major nutritional objective is the humanization of bovine milk for the infant formula market. Transgenic cattle have been generated that carry the cDNA for human lactoferrin *(58)*, the major whey protein in human milk (although of low abundance in bovine milk). Lactoferrin may also play roles in iron transport and in protecting against bacteria. Many other strategies toward the humanization of milk have been widely discussed *(60–63)*. In one example, human α-lactalbumin was introduced into mice in order to mimic the balance of whey to casein characteristic of human milk *(54)*. Although several such ideas have been modeled in transgenic mice (for reviews, *see* **refs.** *60–63*), only the lactoferrin approach has been attempted in livestock.

Potential applications of transgenesis to alter the industrial properties of milk include modifying the casein content to alter milk-clotting properties, altering the proportion of hydrophobic residues in β-casein to improve its emulsifying properties *(63)*, and improving the rate of maturation of cheese by introducing an altered αs1-casein transgene *(63)*. According to a 1990 estimate, a 20% increase in the content of αs1-casein would be worth almost $200 million annually in the United States alone. Again, none of these strategies have yet been exemplified in livestock.

A striking aspect of nutraceuticals is that in spite of the broad range of identified opportunity, there are few examples of reduction to practice. This has been attributed in part to the relatively low value of agricultural vs biomedical products, and in part to the inherent conservatism of agricultural industries *(58)*. Consideration of the time and money required to generate sufficient numbers of genetically modified animals to support, e.g., the cheese industry suggests that a certain amount of conservatism may be appropriate. In common with other agricultural applications, the greatest barrier to progress lies in the fact that engineered milks destined for human consumption are still not broadly acceptable to the consumer and in many countries are proscribed by law.

## 7. Animal Models

In mice, the ES system allows us to re-create precisely, genetic lesions that are associated with human genetic disease *(64)*. The production of gene-targeted mouse models has become routine (for a detailed description of targeting, *see* **ref.** *65*). In livestock, the practicality of engineered animal models will be sensitive to the added value of the livestock model over the corresponding mouse model and for broad application will require the development of gene targeting in livestock cell lines. With animal models in general, the resulting phenotype is usually anticipated, although species differences frequently confound this expectation. Mouse models of cystic fibrosis, e.g., fail to present the same lung pathology characteristic of the human disease *(66)*, and mice lacking the gene whose dysfunction in humans is associated with Lesch Nyhan syndrome *(HPRT)* are overtly normal *(67,68)*. While it is clear that the majority of mouse models have been invaluable in the study of human disease, it is equally true that for certain diseases, the mouse models have serious limitations.

Livestock species share similarities with humans in anatomy, size, physiology, and life span, which often renders them better models than rodents. The pig has been particularly useful in the past as a model of kidney dysfunction, ischemic heart disease, hypercholesteremia, and atherosclerosis (reviewed in **refs. 69**). The sheep has been proposed as a potential animal model of the human condition cystic fibrosis, which results from defects in the cystic fibrosis transmembrane conductance regulator (*CFTR*) gene *(70)*. Ovine CFTR protein is 95.3% similar to the human amino acid sequence and has a very similar expression pattern. In addition, the sheep lung epithelium shares anatomic, functional, and electrophysiologic similarities with the human *(68)*. At least one group has embarked on a search for a spontaneous *CFTR* mutant among commercial flocks in New Zealand *(71)*; however, the success of such a large-scale screening program cannot be taken for granted. To date, livestock models have been restricted to spontaneous mutants and to pharmacologic models in which wild-type animals are challenged with disease-causing agents. A single but fortuitous exception to this generalization may be the GH pigs that were originally generated in an attempt to enhance growth rate *(30)*. Although these animals have not proven useful in agriculture, it has been suggested that they may provide a model of the human growth disorder acromegaly *(69)*.

Cell-based transgenesis in livestock and the possibility of gene targeting in these species open new opportunities for engineering large-animal models. The ovine *CFTR* gene, e.g., may be a prime candidate for knockout by gene targeting. Even were a spontaneous ovine mutant available, a significant advantage to the engineered model is that one or two of the commonly occurring point mutations in cystic fibrosis patients could be precisely mimicked. Other candidates include the *prp* genes of sheep and cattle. Misfolding of the *prp* gene product (the prion protein) is associated with the spongiform encephalopathies: scrapie in sheep; BSE in cattle; and CJD, GSS, and Kuru in humans. Although mice carrying inactive *prp* genes show certain subtle alterations in circadian rhythms *(10)*, they are fully viable, developmentally and behaviorally normal, and resistant to scrapie *(73)*. To confirm this circumstantial evidence for the control of scrapie by PrP, it would be invaluable to determine whether sheep carrying inactive *prp* are similarly resistant and to establish whether they can carry and transmit the infective agent.

One of the general limitations of generating knockout animals in livestock is that in most instances only the homozygous knockout is useful. This presents particular problems in disseminating loss-of-function genotypes into commercial populations. As a consequence, PrP-deficient animals are most likely to be restricted to the small numbers required for fundamental research, and possibly to the generation of new cell lines for use as nuclear transfer donors for biomedical applications. If PrP-deficient animals can be shown to be incapable of carrying the infective agent or agents, then animals cloned from such *prp*-deficient cell lines could be declared scrapie/BSE-free.

## 8. Xenotransplantation

At any one time, some 5000 persons await suitable organs for heart transplants in the United Kingdom and about 50,000 in the United States. Many of these patients will die before a suitable donor is available. According to one estimate, only 10% of those patients who could benefit from a heart transplant actually receive one *(74)*. There is, therefore, considerable interest in genetically engineering pigs so that their organs will be acceptable to the human immune system (xenografting).

The major epitope leading to hyperacute rejection of xenografts in humans is a sugar residue produced by the action of the enzyme $\alpha 1,3$ galactosyl transferase. This enzyme is inactive in humans and Old World primates but is functional in all other mammalian species. The binding of xenoreactive antibodies following xenotransplantation activates the classic complement pathway leading to rapid (within minutes) cell lysis (reviewed in **ref. 75**). Hence, two potential transgenic strategies to address the problem of hyperacute rejection are either to block the complement pathway or to reduce levels of the major xenoreactive epitope, Gal $\alpha$ 1,3 Gal.

Two lines of transgenic pig have been produced that carry transgenes encoding two of the three main regulators of human complement activation: human decay accelerating factor (hDAF) *(76)*, and human CD59, respectively *(77)*. In perfusion tests, the genetically modified hearts are protected from the action of human complement *(76,77)* and following transplantation to cynomolgus monkeys, hDAF hearts lead to a significant increase in survival *(75)*. Mice rendered dysfunctional at the *$\alpha 1,3$ galactosyl transferase* locus by gene targeting are fully viable, and several attempts have been made to reduce gal transferase activity in pigs by additive transgenesis. Transgenic mice and pigs have been generated that express human fucosyl transferase *(78,79)*. This gene is not normally expressed in pigs and mice, and its transgenic expression leads to reduced levels of the Gal $\alpha$ 1,3 Gal epitope. Further reduction in Gal $\alpha$ 1,3 Gal levels was obtained by combined expression of $\alpha$-galactosidase and $\alpha 1,3$ fucosyltransferase *(80)*. Perhaps the optimal transgenic strategy would be to use gene targeting to inactivate the *$\alpha 1,3$ galactosyl transferase* gene, although this awaits the development of gene targeting and of somatic cloning in pigs.

Regardless of the method employed, controlling the hyperacute response will not prevent eventual T-cell rejection, and successful xenotransplantation must deal with this downstream problem either by improvements to immunosuppression regimes or by further engineering strategies. If we look to the future and make certain optimistic assumptions, it is possible to imagine the eventual humanization of the porcine major histocompatibility complex, although this sort of strategy would depend greatly on further advances in chromosome engineering.

Safety issues surrounding xenotransplantation have led to intense public debate. Of particular concern is the risk of zoonoses following the demonstration that human cells in vitro can be infected by an endogenous porcine retrovirus *(81)*, although it remains unclear if infection can also occur in the normal in vivo situation. Although porcine pancreatic islets have been transferred to human patients for some time with no evidence of viral infection, xenotransplantation involves extra factors associated with viral activation such as heavy immunosuppression. At present time, most countries have imposed a moratorium on human transplantation of xenografts. The heavy demand for organs may nonetheless make it likely that clinical trials will proceed in the near future. Guidelines to minimize risk are presently being prepared by the appropriate regulatory bodies.

## 9. Agriculture

The first application of additive transgenic technology to improving the performance of livestock was the introduction of extra genes for GH in an attempt to improve the growth rate and feed efficiency of pigs *(30)*. The resulting pigs did show improvements in feed efficiency and fat content, but they also suffered from a variety of debilitating defects associated with poor control of transgene expression. Since then, potential applications of transgenesis in agriculture have been widely reviewed *(56,82,83)*, but have seldom been reduced to practice. Notable exceptions include sheep engineered for improved wool production either by transfer of bacterial genes for cysteine synthesis *(84)*; by addition of genes for wool keratin proteins, which improve wool fiber ultrastructure *(85)*; or by expression of insulin-like growth factor-1 in hair follicles *(86)*. A second promising area for agricultural transgenesis is disease resistance. Pigs have been generated that express low levels of the murine *Mx1* gene associated with resistance to influenza *(86)*. It has been suggested that high expression of *Mx1* may be developmentally lethal *(87)*. Hence, as with GH pigs, it is again apparent that successful transgenic programs often require extremely tight control of transgene regulation. In general, agricultural applications of transgenesis have been hindered by inefficiencies in the production of transgenic founders, in the proportion of founders with appropriate transgene expression, and in dissemination of transgenic stock to commercial populations.

Two major sources of inefficiency in the production of transgenic founders have been the large numbers of embryos required for injection and the large requirement for recipient females to bring nontransgenic embryos to term. The production of in vitro matured and fertilized oocytes taken from ovarian follicles has dramatically reduced the cost of transgenic programs in cattle *(88,89)*. Attempts to reduce the number of transfers of nontransgenic embryos by polymerase chain reaction screening prior to transfer have been hampered by difficulty in distinguishing between integrated and nonintegrated transgenes; however, such techniques have proven useful in increasing the percentage of transmission from transgenic founders *(89)*. Identification of transgenic embryos immediately following microinjection has been achieved in mice by inclusion of a fluorescent marker, green fluorescent protein, whose product can be visualized prior to transfer without harming the embryo *(90)*. It remains to be seen if this technique can be adapted to large-animal transgenesis. An alternative technique based on *in situ* hybridization of metaphase spreads obtained from biopsied material also shows promise *(91)*.

Our understanding of gene expression has improved in recent years, and it now seems possible that in future applications many of the problems associated with inappropriate transgene expression can be avoided. Tissue-specific and copy number–independent expression of transgenes can be improved by the inclusion of a locus control region in the transgenic construct *(92,93)*. The complementary approach is to introduce the transgene into chosen sites in the genome by homologous recombination. This latter strategy has been used successfully in mice *(51–55)*. Other approaches are the cointroduction of the transgene with fragments that "rescue" genes from positional silencing *(94)* and the flanking of transgenes with inverted terminal repeat sequences from adenoassociated virus *(95)*; however, the latter method is not yet exemplified in mammalian cells. ES cells also can be modified by the introduction of a *loxP* site that is recognized by the bacterial site-specific recombinase Cre *(96,97)*. In a subsequent step, and in the presence of Cre recombinase, this allows site-specific insertion of transgenes that carry flanking *loxP* sites *(97–99)*.

Improvements in transgenic technology *per se* will not affect the downstream economics of animal breeding. Incorporating transgenes into commercial populations by conventional breeding will take many years and will reduce the rate of genetic progress by conventional selection. In one study simulating the effect of transgenic strategies to increase the male:female sex ratio in beef cattle, it was concluded that reduction in genetic progress in other characters could offset any gains in average growth rate *(100)*. Clearly, any transgenic program in agriculture that required the establishment of a nucleus herd would need to offer a very large potential benefit (reviewed in **ref. *101***). However, if we assume substantial improvements in the efficiency of embryo cloning and cell-based transgenesis, then it is possible to imagine the provision of cloned, transgenic embryos directly to the producer. In this scenario, transgenic strategies offering even modest improvements could be introduced to the most productive genotypes and disseminated to commercial herds in a single generation. Even without transgenesis, producers could raise the average performance of their herd to that of the best within a single generation by purchasing cloned embryos. An added advantage is that all embryos could be of chosen sex. The development of coherent strategies for the application of cloning and transgenesis in agriculture will require careful consideration of selection schemes to accommodate continued genetic improvement by traditional breeding.

In addition to the efficiency advantages of cell-based transgenesis, this technique opens the possibility of applying gene-targeting techniques to livestock to generate novel phenotypes. One example of a candidate "knockout" gene is *myostatin*, which has been identified as the gene whose dysfunction results in the double muscling condition in cattle *(102)*. Other candidate genes may well be identified in the search for quantitative trait loci (QTL) in livestock.

## 10. Impact of Gene Mapping and Functional Genomics

Genetic linkage maps are now being developed for many species. In cattle, sheep, and pigs, this has allowed investigators to map genes that have a large effect on quantitative traits—the so-called QTL. Many genes on the human and mouse maps have resulted from mapping expressed sequence tags (ESTs), short sequences obtained from cDNA libraries that are often highly conserved across species (for reviews, *see* refs. *103* and *104*). How the information obtained from mapping programs can be used to establish gene function is the new and still poorly defined area of "functional genomics." At least one company, Lexicon, now offers a database of sequence tags obtained from randomly "knocked out" genes in murine ES cells. Hence, in principal, candidate QTL genes obtained from EST databases can be cross-referenced to an ES "knockout" in the Lexicon database and the corresponding mouse generated and analyzed for phenotype.

So far only a few QTL genes have been identified. Two of these are the melatonin receptor *(105)* and the KIT gene *(106)*, which are associated with coat color. Arguably, these are not strictly QTLs because coat color is not a quantitative trait. However, their identification has arisen directly from the QTL effort, so it is perhaps appropriate in this case to stretch the definition. Others include genes associated with stress susceptibility in pigs *(107)* and hyperplasia in cattle *(101)*. Transgenesis, both in model species and in livestock, is likely to play a key role in functional genomics by providing the information that either proves or disproves an association with phenotype.

Transgenesis also could be used to introgress transgenes into commercial populations. Although a discussion of the economics of transgenesis vs traditional allele introgression is beyond the scope of this review, it seems likely that the key advantage to the transgenic route would be the opportunity to introduce the transgene on a variety of genetic backgrounds.

## 11. The Future: Gene and Cell Therapies

ES cells can be induced to differentiate in vitro into a variety of lineages with potential therapeutic value such as hematopoietic stem cells *(108)*, neural stem cells *(109)*, and myoblasts *(110)*. In principle, one can imagine the application of nuclear transfer to generate cloned embryos for the subsequent isolation of isogenic ES cell lines tailored to individual patients. Isogenic cell lines could then be genetically modified to repair oncogenes or to replace defective alleles and returned to the patient following in vitro differentiation. The simplest example of such an approach would be the complete replacement of the patient's hematopoietic system following oncogene repair by gene targeting. Clearly, applying such a strategy in medicine raises serious ethical issues. The most troublesome of these arises from the requirement for recipient oocytes. So, is the oocyte really necessary?

That nuclear transfer in embryos may be simply a special instance of a more general nuclear reprogramming phenomenon is an exciting prospect. As we learn more about the processes involved in nuclear reprogramming in the embryo, it seems at least possible that we may identify factors that can be used to achieve the direct transformation of somatic cells to ES cells without the requirement for an embryo's intermediate. One possibility is that the ES cytoplasm itself may have reprogramming activity. It is intriguing that in fusions of embryonic germ cells with thymocytes, the ES phenotype is dominant *(111)*.

## 12. Conclusion

This discussion has been rather broad in scope and, of necessity, has been less than exhaustive in its treatment of individual areas of application. For more comprehensive information, the reader is referred to the many excellent reviews cited throughout. In addition to the areas of application discussed, it seems likely that unusual uses will arise that are neither biomedical nor agricultural. For example, Nexia Pharmaceuticals (Canada) has generated transgenic goats expressing spider silk proteins in their milk and anticipate wide use of the recombinant product in the manufacture of highly resilient industrial materials. Other omissions in this limited review include avian transgenesis and detailed discussion of the ethical, public acceptability of and economic constraints to the further uptake of transgenic technology.

Until now, livestock transgenesis has been largely restricted to the production of high-value biomedical products. The advent of cell-based technology has greatly broadened the scope of potential future applications, and the ethical debate has broadened, in turn, to meet this challenge. Agricultural and animal model applications of transgenesis offer relatively low financial returns. As a consequence, these were previously constrained by the limitations of the pronuclear injection procedure. The impact of the recent developments in cell-based transgenesis and control of transgene expression is that these applications are now constrained more by issues of public acceptability. It now seems likely that many of the once futuristic ideas discussed herein may find reduction to practice.

## References

1. Gordon, J. W., Scargos, G. A., Plotkin, D. J., Barbosa, J. A., and Ruddle, F. R. (1980) Genetic transformation of mouse embryos by microinjection of purified DNA. *Proc. Natl. Acad. Sci. USA* **77,** 7380–7384.
2. Hammer, R. E., Pursel, V. G., Rexroad, C. E., Wall, R., Bolt, J., Ebert, D. J., Palmiter, R. D., and Brinster, R. L. (1985) Production of transgenic rabbits, sheep and pigs by microinjection. *Nature* **315,** 680–683.
3. Evans, M. J. and Kaufman, M. H. (1981) Establishment in culture of pluripotential cells from mouse embryos. *Nature* **292,** 154–156.
4. Martin, G. (1981) Isolation of a pluripotential cell line from early mouse embryos cultured in medium conditioned by teratocarcinoma stem cells. *PNAS* **78,** 7634–7638.
5. Wilmut, I., Schnelke, A. E., McWhir, J., Kind, A. J., and Campbell, K. H. S. (1997) Viable offspring derived from fetal and adult mammalian cells. *Nature* **385,** 810–813.
6. Campbell, K. H. S., McWhir, J., Ritchie, W. A., and Wilmut, I. (1996) Sheep cloned by nuclear transfer from a cultured cell line. *Nature* **380,** 64–66.
7. Schneike, A. E., Kind, A. J., Ritchie, W. A., Mycock, K., Scott, A. R., Ritchie, M., Wilmut, I., Colman, A., and Campbell, K. H. S. (1998) Human factor IX transgenic sheep produced by transfer of nuclei from transfected fetal fibroblasts. *Science* **278,** 2130–2133.
8. Smithies, O., Gregg, R. G., Boggs, S. S., Koralewski, M. A., and Kucherlapati, R. S. (1985) Insertion of DNA sequences into the human chromosomal beta-globin locus by homologous recombination. *Nature* **317,** 230–234.
9. Shesely, E. G., Kim, H.-S., Shehee, W. R., Papayannopoulou, T., Smithies, O., and Popovich, B. W. (1991) Correction of a human bs-globin gene by gene targeting. *PNAS* **88,** 4294–4298.
10. Williams, S. R., Ousley, F. C., Vitez, L. J., and DuBridge, R. B. (1994) Rapid detection of homologous recombinants in nontransformed human cells. *PNAS* **91,** 11,943–11,947.

11. Arbones, M. L., Austin, H. A., Capon, D. J., and Greenburg, G. (1994) Gene targeting in normal somatic cells: inactivation of the interferon-gamma receptor in myoblasts. *Nat. Genet.* **6,** 90–97.
12. Scheerer, J. B. and Adair, G. M. (1994) Homology dependence of targeted recombination at the Chinese hamster APRT locus. *Mol. Cell. Biol.* **14,** 6663–6673.
13. Itzhaki, J. E., Gilbert, C. S., and Porter, A. C. (1997) Construction by gene targeting in human cells of a "conditional" CDC2 mutant that re-replicates its DNA. *Nat. Genet.* **15,** 258–265.
14. Doetschman, T. C., Williams, P., and Maeda, N. (1988) Establishment of hamster blastocyst-derived embryonic stem (ES) cells. *Dev. Biol.* **127,** 224–227.
15. Sukoyan, M. A., Vatolin, S. Y., Golubitsa, A. N., Zhelezova, A. N., Zemenova, L. A., and Serov, O. L. (1993) Embryonic stem cells derived from morula, inner cell mass and blastocysts of mink: comparison of their pluripotencies. *Mol. Reprod. Dev.* **36,** 148–158.
16. Iannaccone, P. M., Taborn, G. U., Garton, R. L., Caplice, M. D., and Brenin, D. R. (1994) Pluripotent embryonic stem cells from the rat are capable of producing chimaeras. *Dev. Biol.* **163,** 288–292.
17. Pain, B., Clark, M. E., Shen, M., Nakazawa, H., Sakurai, M., Samurut, J., and Etches, R. J. (1996) Long-term in vitro culture and characterisation of avian embryonic stem cells with multiple morphogenetic properties. *Development* **122,** 2339–2348.
18. Piedrahita, J. A., Anderson, G. B., and Bon Durant, R. H. (1990) On the isolation of embryonic stem cells: comparative behaviour of murine, porcine and ovine embryos. *Theriogenology* **34,** 879–901.
19. Tsuchiya, Y., Raasch, G. A., Brandes, T. L., Mizoshita, K., and Youngs, C. R. (1994) Isolation of ICM-derived cell colonies from sheep blastocysts. *Theriogenology* **41,** 321.
20. Notorianni, E., Galli, C., Laurie, S., Moor, R. M., and Evans, M. J. (1991) Derivation of pluripotent embryonic cell lines from the pig and sheep. *J. Reprod. Fertil.* **43,** 255–260.
21. Sims, M. M. and First, N. L. (1993) Production of fetuses from totipotent cultured bovine inner cell mass cells. *Theriogenology* **39,** 313.
22. Tsuchiya, Y., Raasch, G. A., Brandes, T. L., Mizoshita, K., and Youngs, C. R. (1994) Isolation of ICM-derived cell colonies from sheep blastocysts. *Theriogenology* **41,** 321.
23. Stice, S., Strelchenko, N. S., Betthauser, J., Scott, B., Jurgella, G., Jackson, J., David, V., Keefer, C., and Matthews, L. (1994) Bovine pluripotent embryonic cells contribute to nuclear transfer and chimaeric fetuses. *Theriogenology* **41,** 304.
24. Stice, S. L., Strelchenko, N. S., Keefer, C. L., and Matthew, L. (1996) Pluripotent bovine embryonic cell lines direct embryonic development following nuclear transfer. *Biol. Reprod.* **54,** 100–110.
25. Strelchenko, N. and Stice, S. (1994) Bovine embryonic pluripotent cell lines derived from morula stage embryos. *Theriogenology* **41,** 304.

26. Talbot, N. C., Powell, A. M., Nel, N. D., Pursel, V. G., and Rexroad, C. E. Jr. (1993) Culturing the epiblast cells of the pig blastocyst. *In Vitro Cell Dev. Biol.* **29,** 543–554.

27. Gerfen, R. W. and Wheeler, M. B. (1995) Isolation of embryonic cell lines from porcine blastocysts. *Anim. Biotech.* **6,** 1–14.

28. Thomson, J. A., Kalishman, J., Golos, T. G., Durning, M., Harris, C. P., Becker, R. A., and Hearn, J. P. (1995) Isolation of a primate embryonic stem cell line. *Proc. Natl. Acad. Sci. USA* **92,** 7844–7848.

29. Thomson, J. A., Itskovitz-Eldor, J., Shapiro, S. S., Waknitz, M. A., Swiergiel, J. J., Marshall, V. S., and Jones J. M. (1998) Embryonic cell lines derived from human blastocysts. *Science* **282(5391),** 1145–1147.

30. Stice, S. L. (1998) Opportunities and challenges in domestic animal embryonic stem cell research, in *Animal Breeding Technology for the 21st Century* (Clark, A. J., ed.), Harwood Academic, Amsterdam, pp. 63–73.

31. Pursel, V. G., Hammer, R. E., Bolt, D. J., Palmiter, R. D., and Brinster, R. L. (1990) Genetic engineering of swine: integration, expression and germline transmission of growth related genes. *J. Reprod. Fertil. Suppl.* **41,** 77–87.

32. Simons, J. P., Wilmut, I., Clark, A. J., Archibald, A. L., and Bishop, J. O. (1988) Gene transfer into sheep. *Bio/Technology* **6,** 179.

33. Lee, T. K., Bangalore, N., Velander, W., Drohan, W. N., and Lubon, H. (1996) Activation of recombinant human protein C. *Thrombos. Res.* **82(3),** 225–234.

34. Colman, A. (1997) Transgenic production of specialty nutritionals, in *IBC Conference Proceedings on Transgenic Nutraceuticals.*

35. Young, M. W., Okita, W. B., Brown, E. M., and Curling, J. M. (1997) Production of biopharmaceutical proteins in the milk of transgenic dairy animals. *Biopharm—Technol. Business Biopharm.* **10(6),** 34–38.

36. Wilmut, I. and Whitelaw, C. B. A. (1994) Strategies for production of pharmaceutical proteins in milk. *Reprod. Fertil. Dev.* **6,** 625–630.

37. Ziomek, C. A. (1998) Commercialisation of proteins produced in the mammary gland. *Theriogenology* **49,** 139–144.

38. Houdebine, L. M. (1994) Production of pharmaceutical proteins from transgenic animals. *J. Biotechnol.* **34,** 269.

39. Kerr, D. E., Liang, F. X., Bondioli, K. R., Zhao, H. P., Kreibich, G., Wall, R. J., and Sun, T. T. (1998) The bladder as a bioreactor: urothelium production and secretion of growth hormone into urine. *Nat. Bio/Technol.* **16(1),** 75–79.

40. Sharma, A., Martin, M. J., Okabe, J. F., Truglio, R. A., Dhanjal, N. P., Logan, J. S., and Kumar, R. (1994) An isologous porcine promoter permits high level expression of human hemoglobin in transgenic swine. *Bio/Technology* **12,** 55–59.

41. Swanson, M. E., Martin, M. J., O'Donnell, J. K., Hoover, K., Lago, W., Huntress, V., Parsons, C. T., Pinkert, C. A., and Logan, J. S. (1992) Production of functional human haemoglobin in transgenic swine. *Bio/Technology* **10,** 557–559.

42. Archibald, A. L., McClenaghan, M., Hornsey, V., Simons, J. P., and Clark, A. J. (1990) High-level expression of biologically active human $\alpha 1$ antitrypsin in the milk of transgenic mice. *PNAS* **87,** 5178–5182.

43. Clark, A. J. (1998) The mammary gland as a bioreactor: expression processing and production of recombinant proteins. *J. Mammary Gland Biol. Neoplas.* **3(3)**, 337–350.

44. Denman, J., Hayes, M., O'Day, C., Edmunds, T., Bartlett, C., Hirani, S., Ebert, K. M., Gordon, K., and McPherson, J. M. (1991) Transgenic expression of a variant of human tissue-type plasminogen activator in goat milk: purification and characterisation of the recombinant enzyme. *Bio/Technology (NY)* **9**, 839–843.

45. Velander, W. H., Johnson, J. L., Page, R. L., et al. (1992) High level expression of a heterologous protein in the milk of transgenic swine using the cDNA encoding human protein C. *PNAS* **89**, 12,003–12,007.

46. Drews, R., Paleyanda, R. K., Lee, T. K., Chang, R. R., Rehemtulla, A., Kaufman, R. J., Drohan, W. N., and Lubon, H. (1995) Proteolytic maturation of protein C upon engineering the mouse mammary gland to express furin. *PNAS* **92**, 10,462.

47. Carver, A., Wright, G., Cottom, D., et al. (1992) Expression of human α1 antitrypsin in transgenic sheep. *Cytotechnology* **9**, 77–84.

48. Niemann, H., Halter, R., Espanion, G., Wrenzycki, C., Herrmann, D., Lemme, E., Carnwath, J. W., and Paul, D. (1996) Expression of human blood clotting factor VIII (FVIII) constructs in the mammary gland of transgenic mice and sheep. *J. Anim. Breed. Genet.* **113(4–5)**, 437–444.

49. Paleyanda, R. K., Velander, W. H., Lee, T. K., Scandella, D. H., Gwazdauskas, F. C., Knight, J. W., Hoyer, L. W., Drohan, W. N., and Lubon, H. (1997) Transgenic pigs produce functional human factor VIII in milk. *Nat. Biotech.* **15(10)**, 971–975.

50. Clark, A. J., Bessos, H., Bishop, J. O., Brown, P., Harris, S., Lathe, R., McClenaghan, M., Prowsae, C., Simons, J. P., Whitelaw, C. B. A., and Wilmut, I. (1989) Expression of human anti-hemophilic factor IX in the milk of transgenic sheep. *Bio/Technology (NY)* **7**, 487–492.

51. Prunkard, D., Cottingham, I., Garner, I., Lasser, G., Bishop, P., and Foster, D. (1997) Expression of recombinant human fibrinogen in the milk of transgenic sheep. *Thromb. Haemost.* PS158.

52. VanCott, K. E., Lubon, H., Russell, C. G., Butler, S. P., Gwazdauskas, F. C., Knight, J., Drohan, W. N., Velander, W. H. (1997) Phenotypic and genotypic stability of multiple lines of transgenic pigs expressing recombinant human protein C. *Transgen. Res.* **6(3)**, 203–212.

53. Shaw-White, J. R., Denko, N., Albers, L., Doetschman, T. C., and Stringer, J. R. (1993) Expression of the *lacZ* gene targeted to the HPRT locus in embryonic stem cells and their derivatives. *Transgen. Res.* **2**, 1–13.

54. Detloff, P. J., Lewis, J., John, S. W. M., Shehee, W. R., Lengenbach, R., Maeda, N., and Smithies, O. (1994) Deletion and replacement of the mouse β-globin genes by a "plug and socket" repeated targeting strategy. *Mol. Cell. Biol.* **14**, 6936–6943.

55. Stacey, A., Schnieke, A., McWhir, J., Cooper, J., Colman, A., and Melton, D. W. (1994) Use of double replacement gene targeting to replace the murine alpha lactalbumin gene with its human counterpart in embryonic stem cells and mice. *Mol. Cell. Biol.* **14(2)**, 1009–1016.

56. Bronson, S. K., Plaehn, E. G., Kluckman, K. D., Hagaman, J. R., Maeda, N., and Smithies, O. (1996) Single copy transgenic mice with chosen site integration. *PNAS* **93,** 9067–9072.

57. Wallace, H., Ansell, R., Clark, A. J., and McWhir, J. (2000) Pre-selection of integration sites imparts repeatable transgene expression. *Nucleic Acids Res.* **28,** 1455–1464.

58. Krimpenfort, P., Rademakers, A., Eyestone, W., van der Schans, A., van den Broek, S., Kooiman, P., Kootwijk, E., Platenburg, G., Pieper, F., Srijker, R., and De Boer, H. (1991) Generation of transgenic dairy cattle using "in vitro" embryo production. *Bio/Technology* **9,** 844–847.

59. Wall, R. J., Kerr, D. E., and Bondoli, K. R. (1997) Transgenic dairy cattle: genetic engineering on a grand scale. *J. Dairy Sci.* **80,** 2213–2224.

60. Clark, A. J. (1996) Genetic modification of milk proteins. *Am. J. Clin. Nutr.* **63,** 633S–638S.

61. Garner, I. and Colman, A. (1998) Therapeutic protein from livestock, in *Animal Breeding: Technology for the 21st Century* (Clark, A. J., ed.), Harwood Academic, Amsterdam, pp. 215–227.

62. Karatzas, C. N. and Turner, J. D. (1997) Toward altering milk composition by genetic manipulation: current status and challenges. *J. Dairy Sci.* **80,** 2225–2232.

63. Jiminez-Flores, R. and Richardson, T. (1988) Genetic engineering of the caseins to modify the behaviour of milk during processing: a review. *J. Dairy Sci.* **71,** 2640–2654.

64. Clarke, A. R. and McWhir, J. (1995) Transgenic models of disease, in *Progress in Pathology, Volume 2* (Kirkham, N. and Lemoine, N. R., eds.), Churchill Livingstone, Edinburgh, pp. 227–246.

65. Joyner, A. (1993) *Gene Targeting: A Practical Approach*, Oxford University Press, NY.

66. Snouwaert, J. N., Brigman, K. K., Latour, A. M., Malouf, N. N., Boucher, R. C., Smithies, O., and Koller, B. H. (1992) An animal model for cystic fibrosis made by gene targeting. *Science* **257,** 1083–1088.

67. Dunnett, S. B., Sirinathsinghihji, D. J. S., Heaven, R., Rogers, D. C., and Kuehn, M. R. (1989) Monoamine deficiency in a transgenic (HPRT-) mouse model of Lesch-Nyhan syndrome. *Brain Res.* **501,** 401–416.

68. Finger, S., Heaven, R. P., Sirinathsinghji, D. J. S., Kuehn, M. R., and Dunnett, S. B. (1988) Behavioural and neurochemical evaluation of a transgenic mouse model of Lesch-Nyhan syndrome.

69. Petters, R. L. (1994) Transgenic livestock as genetic models of human disease. *Reprod. Fertil. Dev.* **6,** 643–645.

70. Harris, A. (1997) Towards an ovine model of cystic fibrosis. *Hum. Mol. Genet.* **6(13),** 2191–2193.

71. Tebbutt, S. J., Harris, A., and Hill, D. F. (1996) An ovine CFTR variant as a putative cystic fibrosis causing mutation. *J. Med. Genet.* **33(7),** 623, 624.

72. Tobler, I., Gaus, S. E., Deboer, T., Achermann, P., Fischer, M., Rulicke, T., Moser, M., Oesch, B., McBride, P. A., and Manson, J. C. (1996) Altered circa-

dian activity rhythms and sleep in mice devoid of prion protein. *Nature* **380(18),** 639–642.

73. Bueler, H., Fischer, M., Lang, Y., Bluethmann, H., Lipp, H.-P., DeArmond, S. J., Prusiner, S. B., Aguet, M., and Weissmann, C. (1992) Normal development and behaviour of mice lacking the neuronal cell-surface PrP protein. *Nature* **356,** 577–582.

74. Evans, R. W., Orians, C. E., and Asher, N. L. (1992) The potential supply of donor organs: an assessment of the efficiency of organ procurement efforts in the United States. *JAMA* **267,** 239–246.

75. White, D. and Langford, D. (1998) Xenografts from livestock, in *Animal Breeding Technology for the 21st Century* (Clark, A. J., ed.), Harwood Academic, Amsterdam, pp. 229–242.

76. Langford, G. A., Yannoutsos, N., Cozzi, N., Lancaster, E., Elsome, R., Chen, K., Richards, P., and White, D. J. (1994) Production of pigs transgenic for human decay accelerating factor. *Transplant. Proc.* **26,** 1400–1401.

77. Fodor, W. L., Williams, B. L., Matis, L. A., Madri, J. A., Rollins, S. A., Knoght, J. W., Velander, W., and Squinto, S. P. (1994) Expression of a functional human complement inhibitor in a transgenic pig as a model for the prevention of xenogeneic hyperacute rejection. *PNAS* **91,** 11,153–11,157.

78. Sandrin, M. S., Fodor, W. I., Mouhtours, E., Osman, N., Cohney, S., and Rollins, S. A. (1995) Enzymatic remodelling of the carbohydrate surface of a xenogenic cell substantially reduces human antibody binding and complement-mediated cytolysis. *Nat. Med.* **12,** 1261–1265.

79. Koike, C., Kannagi, T., Muramatsu, T., Yokoyama, I., and Takagi, H. (1996) Converting ($\alpha$) Gal epitope of pig into H antigen. *Transplant. Proc.* **28,** 553.

80. Osman, N., McKenzie, I. F. C., Ostenreid, K., Ioannou, Y. A., Desnick, R. J., and Sandrin, M. S. (1997) Combined transgenic expression of $\alpha$-galactosidase and $\alpha$-1,2-fucosyltransferase leads to optimal reduction in the major zenoepitope Gal$\alpha$(1,3)Gal. *PNAS* **94,** 14,677–14,682.

81. Patience, C., Takeuchi, Y., and Weiss, R. A. (1997) Infection of human cells by an endogenous retrovirus of pigs. *Nat. Med.* **3,** 282–286.

82. Clark, A. J., Simons, J. P., and Wilmut, I. (1992) Germline manipulation: applications in agriculture and biotechnology, in *Transgenic Mice in Biology and Medicine* (Grosveld, F. and Kollias, G., eds.), Academic, London, pp. 247–270.

83. Pursel, V. G. (1998) Modification of production traits, in *Animal breeding: Technology for the 21st Century* (Clark, A. J., ed.), Harwood Academic, Amsterdam, pp. 183–200.

84. Bawden, C. S., Sivaprasad, A. V., Verma, P. J., Walker, S. K., and Rogers, G. E. (1995) Expression of bacterial cysteine biosynthesis genes in transgenic mice and sheep: toward a new in vivo amino acid biosynthesis pathway and improved wool growth. *Transgenic Res.* **4,** 87–104.

85. Powell, B. C., Walker, S. K., Bawden, C. S., Sivaprasad, A. V., and Roger, G. E. (1994) Transgenic sheep and wool growth: possibilities and current status. *Reprod. Fertil. Dev.* **6,** 615–623.

86. Damak, S., Su, H.-Y., Jay, N. P., and Bullock, D. W. (1996) Improved wool production in transgenic sheep expressing insulin-like growth factor 1. *Bio/Technology* **14,** 185–188.

87. Muller, M., Brenig, B., Winnacker, E.-L., and Brem, G. (1992) Transgenic pigs carrying cDNA copies encoding the murine Mx1 protein which confers resistance to influenza virus infection. *Gene* **121,** 263–270.

88. Eyestone, W. H. (1994) Challenges and progress in the production of transgenic cattle. *Reprod. Fertil. Dev.* **6,** 647–652.

89. Eyestone, W. H. (1999) Production and breeding of transgenic cattle using in vitro embryo production technology. *Theriogenology* **51(2),** 509–517.

90. Takada, T. T., Lida, K., Awaji, T., Itoh, K., Takahashi, R., Shibui, A., Yoshida, K., Sugano, S., and Tsujimoto, G. (1997) Selective production of transgenic mice using green fluorescent protein as a marker. *Nat. Biotechnol.* **15,** 458–461.

91. Lewis-Williams, J., Sun, Y., Han, Y., Ziomek, C., Denniston, R. S., Echelard, Y., and Godke, R. A. (1997) Birth of successfully identified transgenic goats using preimplantation stage embryos biopsied for FISH. *Theriogenology* **47,** 226.

92. Bonifer, C., Yannoutsos, N., Kruger, G., Grosveld, F., and Sippel, A. E. (1994) Dissection of the locus control function located on the chicken lysozyme gene domain in transgenic mice. *Nucleic Acids Res.* **22,** 4202–4210.

93. Grosveld, F., Van Assendelft, G. B., Greaves, D. R., and Kollias, G. (1987) Position independent, high-level expression of the human β-globin gene in transgenic mice. *Cell* **51,** 975–985.

94. Clark, A. J., Couper, A., Wallace, R., Wright, G., and Simons, J. P. (1992) Rescuing transgene expression by co-integration. *Bio/Technology* **10,** 1450–1454.

95. Fu, Y., Wang, Y., and Evans, S. M. (1998) Viral sequences enable efficient and tissue-specific expression of transgenes in Xenopus. *Nat. Biotechnol.* **16,** 253–257.

96. Fukushige, S. and Sauer, B. (1992) Genomic targeting with a positive-selection lox integration vector allows highly reproducible gene expression in mammalian cells. *PNAS* **89,** 7905–7909.

97. Kolb, A. F. and Siddel, S. G. (1997) Genomic targeting of a bicistronic DNA fragment by Cre-mediated site-specific recombination. *Gene* **209,** 209–216.

98. Rucker, E. B. and Piedrahita, J. A. (1997) Cre-mediated recombination at the murine whey acidic protein (nWAP) locus. *Mol. Reprod. Dev.* **48,** 324–331.

99. Kolb, A. F., Ansell, R., McWhir, J., and Siddel, S. G. (1999) Insertion of a foreign gene into the β-casein locus by Cre-mediated site-specific recombination. *Gene* **227,** 21–31.

100. Bishop, S. C. and Woolliams, J. A. (1991) Utilization of the sex-determining region Y gene in beef cattle breeding schemes. *Anim. Prod.* **53,** 157–164.

101. Gibson, J. (1998) Breeding genetically manipulated traits, in *Animal Breeding: Technology for the 21st Century* (Clark, A. J., ed.), Harwood Academic, Amsterdam.

102. Smith, T. P. L., Lopez-Corrales, N. L., Kappes, S. M., and Sonstegard, T. S. (1997) Myostatin maps to the interval containing the bos mh locus. *Mamm. Genome* **8,** 742–744.

103. Kappes, S. M. (1999) Utilization of gene mapping information in livestock animals. *Theriogenology* **51,** 135–147.
104. Archibald, A. (1998) Comparative gene mapping—the livestock perspective, in *Animal Breeding: Technology for the 21st Century* (Clark, A. J., ed.), Harwood Academic, Amsterdam, pp. 137–164.
105. Klungland, H., Vage, D. I., Gomex-Raya, L., Adalsteinsson, S., and Lien, S. (1995) The role of melanocyte-stimulating hormone (MSH) receptor in bovine coat colour determination. *Mamm. Genome* **6,** 636–639.
106. Johansson Moller, M., Chaudhary, R., Hellmen, E., Chowdahary, B., and Andersson, L. (1996) Pigs with the dominant white coat colour phenotype carry a duplication of the KIT gene encoding the mast/stem cell growth factor receptor. *Mamm. Genome* **7,** 822–830.
107. Fujii, J., Otsu, K., Zorzato, F., De Leon, S., Khanna, V. K., Weiler, J. E., O'Brien, P. J., and MacLennan, D. H. (1991) Identification of a mutation in porcine ryanodine receptor associated with malignant hyperthermia. *Science* **253,** 448–451.
108. Potcnik, A. J., Nielson, P. J., and Eichmann, K. (1994) In vitro generation of lymphoid precursors from embryonic stem cells. *EMBO J.* **13(22),** 5274–5283.
109. Bain, G., Kitchens, D., Yao, M., Huettner, J. E., and Gottlieb, D. I. (1995) Embryonic stem cells express neuronal properties in vitro. *Dev. Biol.* **168,** 342–357.
110. Rohwedel, J., Maltsev, V., Bober, E., Arnold, H.-H., Hescheler, J., and Wobus, A. M. (1994) Muscle cell differentiation of embryonic stem cells reflects myogenesis in vivo: developmentally regulated expression of myogenic determination genes and functional expression of ionic currents. *Dev. Biol.* **164,** 87–101.
111. Tada, M., Tada, T., Lefebvre, L., Barton, S. C., and Surani, M. A. (1997) Embryonic germ cells induce epigenetic reprogramming of somatic nucleus in hybrid cells. *EMBO J.* **16(21),** 6510–6520.

# II

## Transgenesis in Invertebrate and Lower Vertebrate Species

# 2

# Gene Transfer in *Drosophila*

## Mark J. O'Connor and William Chia

## 1. Introduction

The generation of germline transformants in *Drosophila melanogaster* has relied on the utilization of transposable elements to effect the chromosomal integration of injected DNA *(1,2)*. The success of this approach has depended largely on our understanding of the biology of P elements and the syncytial nature of the early Drosophila embryo. The first 13 embryonic divisions following fertilization are nuclear, resulting in the formation of a syncytium. Consequently, if microinjection into the posterior end of the embryo is carried out prior to cellularization, a proportion of the microinjected DNA will be present in the cytoplasm of the pole cells, the progenitor cells of the germline.

In practice, the DNA to be injected comprises two components. The first consists of a helper plasmid containing a defective P element that, although capable of producing the P transposase, which can act in *trans* to mobilize P transposons, is itself immobile (*see* **Note 1**). The second component consists of a transposon construct in which the sequence to be integrated as a transgene is situated between the 31-bp P element inverted terminal repeats along with a suitable marker (*see* **Note 2**). The transposase produced by the helper plasmid will act on the inverted repeats of the transposon construct and facilitate the integration of the transposon into essentially random chromosomal sites of the recipient's germline. Both P element biology and the characteristics of P element–mediated transformation have been reviewed extensively (e.g., *see* **ref. 3**). In this chapter, we deal primarily with the technical details necessary for obtaining germline transformants.

### 1.1. Outline of Events Involved in Generation of Germline Transformants

1. Construct the desired plasmid containing the transgene, marker, and necessary P element sequences for transposition.

From: *Methods in Molecular Biology, vol. 180: Transgenesis Techniques, 2nd ed.: Principles and Protocols*
Edited by: A. R. Clarke © Humana Press Inc., Totowa, NJ

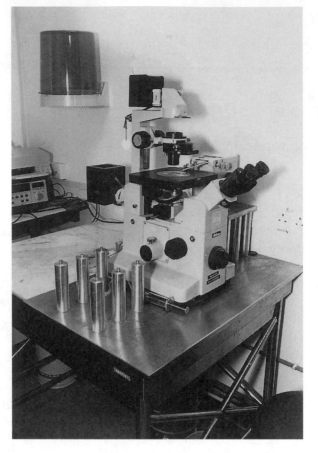

Fig. 1. Typical arrangement of the apparatus used for injection of *Drosophila* embryos.

2. Coinject the transposon along with a defective helper plasmid supplying the P element transposase.
3. Mate the survivors (Go) to an appropriate strain that will allow for the scoring of the marker carried on the transposon construct.
4. Select for transformed progeny that have acquired the marker carried on the transposon and balance the transformants.
5. Test the structure and copy number of the transgene(s) in the transformant lines.
6. Choose unrearranged single insert lines for phenotypic analysis.

## 2. Materials

### 2.1. Microinjection System

**Figure 1** shows the injection apparatus we use. This system consists of the following:

1. Leitz micromanipulator.
2. Nikon inverted phase-contrast microscope.
3. Vibration-free table, on which the microscope is mounted.
4. Loaded needle, containing the DNA to be injected.
5. Collar (Narishige, Tokyo) into which the needle is placed, which, in turn, is attached to the micromanipulator.

Although the micromanipulator is used to position the needle, injection is carried out by moving the microscope stage with the embryos on it. We use an air-filled system to deliver the DNA into the embryos. This consists of a 60-mL glass syringe attached to the collar by a piece of rubber tubing (Narishige Teflon™ tubing also may be used).

This system may appear very basic, but we find that the syringe imparts adequate control of DNA delivery without producing the problems often encountered when using a fluid-filled transmission system, and the system has the advantage of being much cheaper. Injection needles are prepared from borosilicate capillaries (e.g., Clark Electromedical [Reading, UK] GC100TF-15 capillaries, which contain an internal filament) using a pipet puller. A relatively inexpensive two-stage vertical needle puller can be used, such as the PB-7 model from Narishige.

## 2.2. Fly Requirements

In general, a large number of embryos (in the region of 500–1000) need to be injected for each construct in order to produce several independent transformants. In our hands, between 25 and 75% of injected embryos will hatch as larvae. Approximately 50% of the larvae will survive as adults, and between 50 and 80% of the surviving adults will be fertile. Each surviving adult will be individually mated, and approx 200 progeny from each mating will be scored for the marker present on the transposon construct. Although the frequency with which germline transformants are produced varies depending on the construct injected *(4)*, in general, on the order of 10% of the surviving adults will produce at least one germline transformant among its progeny. Therefore, it is reasonable to aim at obtaining about 100 adult survivors for any given construct injected. We usually collect only one transformant from the progeny derived from each surviving adult with which to establish stocks. This ensures that different transformants originated from independent events.

Since the injections must be performed prior to pole cell formation, 1-h embryo collections are used (*see* **Subheading 3.3.**). Therefore, the fly strain used for embryo collections must be robust enough to provide sufficient eggs (at least 100) during a 1-h interval. One further consideration is that the presence of defective P elements in the injected host strain can affect the frequency of transformation. Consequently, care should be taken to ensure that such elements are not present in the chosen host strain.

### 2.3. Miscellaneous

### 2.3.1. Preparation of DNA

1. Qiagen anion-exchange columns.
2. Injection buffer: 5 m$M$ KCl, 0.1 m$M$ Na phosphate, pH 7.8.
3. Millipore filters (0.45-μm).

### 2.3.2. Egg Collection and Egg Processing

1. Egg collection chamber. This can be made from open-ended plastic cylinders of any sort large enough to contain a few hundred flies. The chambers should have fine gauze placed over one end for ventilation, and once the flies have been placed into the chambers, small Petri dishes containing yeast-glucose food and smeared with moist, live yeast are taped to the other end.
2. Glass or plastic tube with a nitex gauze over one end.
3. Freshly diluted 50% household bleach.
4. 0.02% (v/v) Triton X-100.
5. Black nitrocellulose filters.
6. Fine paint brush.
7. Cover slips (22 × 40 mm).
8. Solution of Sellotape in $n$-heptane.
9. Voltalef oil.

## 3. Methods

### 3.1. Preparation of DNA

Plasmid DNA for microinjection may be prepared either by the cesium chloride–ethidium bromide centrifugation method, or by the more convenient Qiagen anion-exchange columns produced by Qiagen (Chatsworth, CA). The latter method produces clean DNA and is not only quicker but also avoids the use of ethidium bromide and organic solvents, such as phenol and chloroform, which could potentially reduce embryonic survival rates.

The concentration of DNA for microinjection needs to be quite high (between 400 and 600 μg/mL) with "helper" plasmid, if used, at a concentration of 200 μg/mL. The DNA to be injected should be ethanol precipitated and given an 80% ethanol wash before being redissolved in injection buffer. Aliquots of 20 μL can then be stored at –20°C.

Prior to loading the DNA into injection needles, the aliquots should be heated to 65°C for 10 min to ensure that the DNA is fully dissolved and then spun through 0.45-μm Millipore filters for a couple of minutes to remove any dust or particles, which could potentially block the needle.

## 3.2. Preparation of Needle

To obtain a needle that possesses the appropriate shape, the first-stage pull should generate a stretch with a length of about 8 mm and a diameter of approx 200 µm. The heating filament should then be moved to the center of this stretch so that the second pull produces a very fine tip of approx 2 mm in length with an end of between 1 and 5 µm in diameter. The heater settings for the first and second pull will need to be determined empirically in order to produce a good-quality needle.

Once a needle has been prepared, the simplest way to load it with the DNA solution is to add 1 to 2 µL of the injection DNA at the back of the capillary with a micropipet. The internal filament that runs along the length of the capillary draws the DNA solution to the front of the needle, which can then be placed into the collar of the microinjection system.

The survival of injected embryos is affected to a large extent by the sharpness of the needle. To obtain a sharp point, the needle can be broken at an angle against a cover slip mounted onto a glass slide. This process is visualized using the inverted-phase microscope and is made easier by placing a drop of Voltalef halocarbon oil on the junction between the slide and the cover slip where the needle is to be broken. When the needle breaks, a small amount of the oil can usually be seen to enter the tip. The flow of DNA can then be tested by applying a little pressure to the syringe. The needle is now ready to use for microinjection.

In between injecting embryos, the needle can be lowered into a small (5-cm) Petri dish lid containing Voltalef halocarbon oil. This helps prevent evaporation of the DNA solution and the concomitant clogging of the needle that can otherwise occur.

## 3.3. Egg Collection

Synchronous and abundant batches of eggs are required for injections. In general, 300–600 adults will produce enough eggs for a few days of microinjections. The flies should be transferred into collection chambers. To optimize egg laying, the flies should be kept at 25°C for a further 2 d in the chambers before starting egg collections for injection, and the Petri dishes containing the food should be changed every day. At the end of the second day, and every subsequent day, the flies should be transferred to 18°C overnight and then returned to 25°C on the morning of collection. The first hour's collection should be discarded because female flies tend to retain eggs until fresh food is supplied. Thereafter, at 60-min intervals, the collection plates can be removed and replaced with new ones.

The eggs to be injected are washed off the collection plates with distilled water and passed down a glass or plastic tube containing a nitex gauze over one end to retain the embryos. The eggs are then ready for dechorionation.

## 3.4. Preparation of Embryos for Microinjection

1. The first step in preparing the eggs for microinjection requires the removal of the tough outer chorion (*see* **Note 3**). To achieve chemical dechorionation, place the tube with nitex gauze and embryos into a beaker containing 10 mL of a 50% solution of household bleach. Gently shake the beaker and tube and, after 2–2.5 min of dechorionation, dilute the bleach by adding an equal volume of a 0.02% Triton X-100 solution. Then remove the tube from the beaker and wash the eggs thoroughly with distilled water.
2. Transfer the embryos onto a black nitrocellulose filter with a fine paint brush and line up along one of the ruled lines on the filter in such a way that the micropile is nearest to you. It is important to keep the filter damp to prevent the eggs from drying out.
3. When 50–60 embryos have been lined up, transfer them to a 22 × 40 mm cover slip; the cover slip can be made adhesive by the prior application of a solution of Sellotape in *n*-heptane. Stick the cover slip with attached embryos onto a microscope slide using a small drop of Voltalef oil and a little pressure. Place the whole slide inside an airtight box containing silica gel in order to desiccate the embryos (*see* **Note 4**).
4. At the end of the desiccation period, take the eggs out of the box containing the silica gel and cover with a layer of Voltalef oil. This oil, although being oxygen permeable, is water impermeable and therefore prevents any further desiccation of the embryos. The embryos are now ready to be injected.

## 3.5. Microinjection of Drosophila Embryos

1. Once the needle is lifted safely out of the way, place the slide containing the embryos on the microscope stage so that the eggs have their posterior facing the needle. Use the micromanipulator to bring the needle into the same plane as the line of eggs.
2. Bring the tip of the needle level with the center of the first egg; this is gaged by running the very end of the needle up and down the edge of the embryo. This method ensures that the needle will not slide over the surface of the egg and will also help decrease the amount of damage to the embryo. Then move the embryo toward the needle with a purposeful motion so that the vitelline membrane is just penetrated. Draw back the needle so that the tip is only just within the cytoplasm. Most of the embryos to be injected will be in the early cleavage stage (15 min to 1 h 20 min) and will have a space between the posterior pole and the vitelline membrane. It is important that the needle be inserted through the space and that the DNA be deposited in the posterior pole of the embryo proper. It is here, at the posterior pole, that the germline will be formed. Next, inject the embryo with a quantity of DNA solution equivalent to approx 1% of the egg's total volume and

remove the needle. Repeat the procedure until all the embryos have been injected (*see* **Note 5**).

3. Kill any embryos in which pole cell formation has already taken place running them through with the needle. Do not count these among those eggs that have been successfully injected.

4. Remove the cover slip containing the injected embryos from the slide and place onto a flat yeast-glucose-charcoal plate. Apply a further thin layer of Voltalef oil to the line of embryos and place the plate into a box kept humid by damp tissues. Then place the box on a level surface in an 18°C incubator for 48 h. If the plate is not kept level, the Voltalef oil will run off, and the embryos will overdesiccate and die.

5. After this time, count the hatched larvae, transfer into vials containing fly food, and return to the 18°C incubator to develop. The percentage survival to first instar larvae can be determined by dividing the number of survivors by the number of successfully injected embryos.

## 4. Notes

1. There exist a number of plasmids that, when injected, can provide the P element transposase necessary to mobilize the coinjected transposon. Two of the most widely used sources are pp25.7wc (wings clipped; **ref. 5**) and pUChsΔ2-3 *(6)*. The wings-clipped transposase source contains a complete 2.9-kb P element in which the last 22 bp has been deleted so that the element is no longer mobile. The pUChsΔ2-3 transposase source comprises the engineered transposase gene (Δ2-3) in which the intron separating the second and third exons (normally only spliced in the germline) has been removed *(6,7)*. This modified transposase gene is placed under the control of the *HSP70* promoter, although the constitutive expression of this promoter is of a sufficiently high level such that heat shock is not necessary. Injecting this construct will result in the transient expression of a functional transposase in both germline and somatic tissues. An alternative approach to coinjecting a plasmid that provides a transposase source is to inject embryos that possess a chromosomal source of the Δ2-3 transposase *(3)*.

2. Many vectors suitable for constructing transposons have been described. We consider here three of the more widely used ones. The transformation vectors based on *rosy* (*ry*) as a scorable marker were the first to be used. One of the most versatile versions of the *ry*-based vectors is pDM30 *(8)*. The major advantage of using *ry*-based vectors is that since 1% of wild-type *ry* expression is sufficient to yield *ry*⁺ eye color, insertions into positions that result in a low level of expression can still be recovered. However, the *ry* gene is large (usually a 7.2-kb *Hin*dIII fragment carrying *ry* is used), and this results in a less-than-optimal vector size. For example, the largeness of *ry*-based vectors can make the construction of transposons more difficult and can also contribute to a decreased transformation frequency.

   Another popular series of transformation vectors use the *white* (*w*) gene as a marker *(9)*. In the most widely used *w* vectors, a mini-*white* gene *(10)* with a subthreshold of $w^+$ activity is used. There are several advantages associated with these mini-$w$-based vectors. First, the gene is small, ~4 kb, compared with $ry$. Second, since mini-$w$ has subthreshold activity, for most insertions, flies that are heterozygous for mini-$w$ can be distinguished from flies that are homozygous on the basis of eye color. Finally, $w$ is easier to score than $ry$ when large numbers of flies are involved. The latest versions of these vectors (the *Casper* series) may be requested from the Thummel or Pirrotta laboratories.

A third series of vectors are those based on G418 antibiotic selection *(11)*. In these vectors, the bacterial neomycin resistance gene is used as a selectable marker in place of visible markers such as $ry$ and $w$. The advantage of using such vectors is that transformants can be selected on *Drosophila* food containing G418 (usually 500–1000 mg/mL), eliminating the chore of screening many flies for a visible marker. However, the major disadvantage is that the window of G418 concentration that will allow true transformants to survive, but that will reduce the leakage of nonresistant animals to an acceptable level, is narrow. Consequently, transformants owing to insertions into chromosomal sites resulting in a low level of expression will not be recovered.

Other transformation vectors, such as those based on *Adh*, which allow for selection on media containing alcohol, have also been described. In addition, a transformation vector (pCaWc) in which both the transposon and the transposase are carried on the same plasmid molecule (with the transposase located outside the P element 31-bp repeats) has been successfully employed for obtaining transformants *(12)*. There are also "shuttle vectors" that greatly facilitate the construction of complex transposons. These vectors (e.g., pHSX, referred to in **ref.** *12*) contain large polylinkers flanked by restriction enzyme sites such as *Not*I (which occurs only very rarely) and enable several DNA fragments to be assembled and then excised as one contiguous piece. The construct can then be inserted into the single *Not*I site of transformation vectors such as pDM30 or the *Casper* series. Finally, transformation vectors designed for placing genes under the control of *HSP70* and actin promoters have been described *(13)*, as have transformation vectors designed to facilitate the insertion of desired sequences upstream of a *LacZ* reporter gene to drive its expression *(10,13)*.

3. Two methods of dechorionation can be employed: chemical and mechanical. However, we favor the chemical method because it is far easier and less time-consuming.

4. This stage is of vital importance if the embryos are to withstand being punctured and accommodate the volume of DNA being introduced. Moreover, this step of the procedure is probably the most crucial, in terms of survival rates, because there is only a narrow margin between a sufficient reduction in egg turgor and excessive drying, which kills the embryos. If possible, embryos should be prepared in an environment with constant temperature and humidity conditions, because this will facilitate the determination of the optimum desiccation time. However, if this is not possible, the experimenter will have to determine the desiccation time empirically, since this will tend to fluctuate depending on the cli-

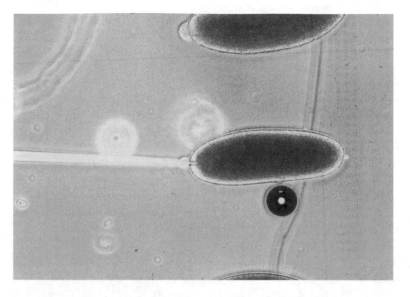

Fig. 2. Microinjection of *Drosophila* embryos illustrating the region of the embryo targeted for injection. Note also the "bubble" of cytoplasmic material leaking from the embryo, which should be removed.

matic conditions. As a starting point, we generally have used desiccation times of between 10 and 15 min.

5. If the embryo has not been desiccated enough, or if too much DNA solution has been injected, cytoplasm may leak out of the egg, reducing its chances of survival (*see* **Fig. 2**). We have found that increased survival rates can be achieved by removing the "bubbles" of cytoplasm. This is easily achieved by having a constant flow of DNA coming out of the needle, which is then brushed passed the line of embryos.

## References

1. Spradling, A. and Rubin, G. (1982) Transposition of cloned P elements into Drosophila germ line chromosomes. *Science* **218,** 341–347.
2. Rubin, G. and Spradling, A. (1982) Genetic transformation of *Drosophila* with transposable element vectors. *Science* **218,** 348–353.
3. Robertson, H., Preston, C., Phillis, R., Johnson-Schlitz, D., Benz, W., and Engels, W. (1988) A stable genomic source of P element transposase in *Drosophila melanogaster. Genetics* **118,** 461–470.
4. Spradling, A. (1986) P element mediated transformation, in *Drosophila: A Practical Approach* (Roberts, D. B., ed.), IRL, Oxford, pp. 175–197.
5. Karess, R. and Rubin, G. (1984) Analysis of P transposable element functions in *Drosophila. Cell* **38,** 135–146.

6. Rio, D., Laski, F., and Rubin, G. (1986) Identification and immunochemical analysis of biologically active *Drosophila* P transposase. *Cell* **44,** 21–32.

7. Laski, F., Rio, D., and Rubin, G. (1986) Tissue-specificity of P element transposition is regulated at the level of mRNA splicing. *Cell* **44,** 7–19.

8. Mismer, D. and Rubin, G. (1987) Analysis of the promoter of the nina E opsin gene in Drosophila melanogaster. *Genetics* 116, 565–578.

9. Klemenz, R., Weber, U., and Gehring, W. (1987) The *white* gene as a marker in a new P element vector for gene transfer in *Drosophila. Nucleic Acids Res.* **15,** 3947–3959.

10. Pirrotta, V. (1988) Vectors for P element mediated transformation in *Drosophila,* in *Vectors: A Survey of Molecular Cloning Vectors and Their Uses* (Rodrigues, R. L. and Denhardt, D. T., eds.), Butterworths, Boston, MA, pp. 436–457.

11. Steller, H. and Pirrotta, V. (1985) A selectable P vector that confers selectable G418 resistance to Drosophila larvae. *EMBO J.* **4,** 167–171.

12. Moses, K., Ellis, M., and Rubin, G. (1989) The Glass gene encodes a zinc finger protein required by *Drosophila* photoreceptor cells. *Nature* **340,** 531–536.

13. Thummel, C., Boulet, A., and Lipshitz, A. (1988) Vectors for *Drosophila* P element-mediated transformation and tissue culture transfection. *Gene* **74,** 445–456.

# III

## TRANSGENESIS IN THE MOUSE: *OOCYTE INJECTION*

# 3

# Oocyte Injection in the Mouse

**Gary A. J. Brown and Timothy J. Corbin**

## 1. Introduction
### 1.1. Mouse Production Colony

To provide fertilized eggs for microinjection, a production colony needs to be established. This should be carefully planned in order to provide enough material for your requirements. There are several items for consideration in this regard, detailed in the following sections.

### 1.1.1. Mouse Strain

Consideration of the chosen strain is important because of the differences in genetic makeup, parental suitability, fecundity, response to administered gonadotropins, or future experiments involving breeding against a specific genetic background. Popular strains for use in transgenic mouse production as embryo donors are C57/BL6 x SJL F1 animals, FVB/N *(1)*, or C57BL6 x C3H F1s, whereas those used for embryo recipients include Swiss Webster or C57BL/6 x DBA/2 F1s.

### 1.1.2. Colony Size

It is necessary to plan colony size according to expected frequency of injections. A typical day of microinjection at most institutions involves the injection of upward of 200 embryos. To provide this number of fertile embryos, and to constrain operational expenses by limiting cage per diem costs and the number of animals to be used, the technique of superovulation *(2)* is frequently employed. Typically between 7 and 10 animals are superovulated to yield sufficient embryos for one day's microinjection. These animals should be used such that the administered gonadotropins initiate the animals' first estrous cycle (3.5–4 wk of age). Additionally, the same number of male stud animals to mate

From: *Methods in Molecular Biology, vol. 180: Transgenesis Techniques, 2nd ed.: Principles and Protocols*
Edited by: A. R. Clarke © Humana Press Inc., Totowa, NJ

to the superovulated females will be required. Male stud animals should be used no more than twice weekly and tracked as to their ability to plug females.

For the embryo recipient mice, a colony large enough to consistently produce sufficient animals in estrus for each day's injection effort is required. Planning involves knowing the length of estrous cycle for the strain used for embryo recipients. To vaginally plug these animals, a stud colony of vasectomized males is required. These should be monitored for their ability to plug estrous females. It is recommended that a strain with a different coat color be used for embryo recipient animals and vasectomized studs than used for the embryo donors for microinjection, to allow easy detection of any vasectomized male that may have re-ligated a vas deferens and regained fertility.

### 1.1.3. Efficiency of Superovulation

The ability of the intended embryo donor strain to be superovulated should be considered, because some strains do not react well to gonadotropin treatment. Hybrid strains tend to respond well and to produce substantially higher egg yields than by natural matings, although some inbred strains also have a consistently good performance.

### 1.1.4. Parental Suitability

Some strains exhibit poor parenting and cannibalize their pups with a high frequency, and these should be avoided, if possible. Often outbred animals such as ICR or Swiss Webster are suitable for use as embryo recipients, although any strain with a high mean litter size and size (for best postoperative recovery) will be effective. Outbred strains are usually less expensive and better suited for this element of transgenic production than inbred or hybrid strains.

### 1.1.5. Pseudopregnancy

It is necessary to generate animals that will have a receptive environment to implanted embryos. This is carried out by inducing animals to exhibit pseudopregnancy through sham fertilization by vasectomized males. The intended recipient females ideally should be between 8 and 16 wk old when sham fertilized and a strain with proven parental abilities chosen for this purpose.

## 1.2. Collection of Fertilized Eggs

The fertilized embryos used for microinjection should be approx 0.5 d postcoitum (p.c.), typically obtained from female mice that have been mated to stud male mice in the afternoon of the previous day. The goal is to time the mating of the mice such that the sperm from the stud males has enough time to complete fertilization, and that the pronuclei from both gametes will be visible for several hours after embryo isolation has occurred. This will provide a time window in which the embryos may be successfully injected with prepared DNA, before the pronuclei fuse and the membranes are no longer visible. To minimize the number of animals to be used for embryo donors, and to boost the efficiency of embryos recovered per donor, the technique of superovulation *(7)* is frequently employed.

### 1.2.1. Superovulation

Superovulation is achieved by the injection of gonadotropins to stimulate and increase natural ovulation. This is most commonly done by administering pregnant mare's serum gonadotropin (PMSG) to mimic the endogenous effects of follicle-stimulating hormone, followed by human chorionic gonadotropin (hCG) to mimic the effect of luteinizing hormone (LH). The importance of this hormone treatment is twofold: to increase the number of ovulations for each female and to control the timing of ovulation independent from the natural estrous cycle. It has been documented, however, that the administration of hormones to elicit superovulation does increase the rate at which there are chromosomal errors in the embryos obtained through this process *(8)*.

PMSG is most commonly supplied as a lyophilized powder and, for best results, should be stored at −80°C until ready for use. PMSG will remain stable as a powder but is very unstable once reconstituted and should only be resuspended immediately before administration. For convenience, aliquots of PMSG can be stored at −20°C, but this can greatly decrease its efficacy over time. Most commonly, PMSG is administered intraperitoneally at a dose of 5 IU (international units)/mouse. For administration, PMSG is resuspended in sterile water or 0.9% NaCl.

The second gonadotropin administered is hCG. It is given to induce the rupture of the mature follicle and is typically administered 42–48 h following administration of PMSG. hCG is also typically supplied as a lyophilized powder and is far more stable than PMSG when reconstituted. For administration, hCG is resuspended at 500 IU/mL in sterile water or 0.9% NaCl and is divided into 100-µL aliquots. These can be stored for approx 1 mo at −20°C. For administration of hCG, these aliquots may be diluted in 1 mL of sterile water or 0.9% NaCl for a final concentration of 50 IU, and then 0.1 mL is injected into each animal at a dose of 5 IU/mouse.

### 1.2.2. Mating Mice

The mice to be used for the generation of fertilized embryos for microinjection and those for use as pseudopregnant embryo recipients should be mated on the day before microinjection is to occur. This will give rise to fertilized embryos that are 0.5 dp.c., and recipients that are timed appropriately to receive those that have been successfully injected.

The mice should be mated in the late afternoon following the administration of hCG, to ensure that the majority of the mice mate during the dark period of the room's light cycle. Mice that mate earlier may yield a much higher number of fertilized embryos that have already fused their pronuclei and lost any visible pronuclear membrane, or have advanced past the first cellular division to the two-cell stage of development. Neither of these embryos can be successfully microinjected, because the placement of the injection needle to deliver the DNA construct cannot be determined. Any mice to be used as pseudopregnant embryo recipients for these injected eggs should be mated in synchrony with these mice also.

The stud males' ability to plug female mice should be carefully monitored and tracked to ensure maximum mating efficiency and a high yield of fertilized eggs harvested per superovulated female.

### 1.2.3. Light Cycle

Several factors control the reproductive performance of superovulated females and stud males. Breeding conditions such as light cycle and timing must be carefully controlled and regulated. The time of administration of PMSG and hCG and the light–dark cycle of the animal facility is critical to the synchronous development of the eggs and the number harvested from the superovulated female. If female mice are ordered from an outside supplier, they should be allowed approx 1 wk to adjust to the animal room's light–dark cycle before administering superovulatory hormones. This will also synchronize the endogenous release of LH stimulated by the PMSG injection. Controlling the release of endogenous LH is important because the hCG must be administered prior to the natural LH surge in order to precisely control ovulation. The endogenous LH release is controlled by the light–dark cycle and generally occurs approx 18 h following the second dark period after administration of PMSG.

A typical injection and mating schedule is as follows: For a 12-h light–dark cycle set at 6 AM lights on and 6 PM lights off, PMSG is given at approx 12:00 noon. hCG should then be given 46–48 h later at approx 11:00 AM to 12:00 noon. This will allow several hours before endogenous release of LH. Following the administration of hCG, mice should be monogamously mated one female to one stud male. The female is then carefully checked the following morning for the presence of a copulatory plug.

### 1.3. Microinjection of DNA

It is critically important that the DNA to be used for microinjection is as clean as and of the highest quality possible. Vector sequences can significantly alter the expression of transgenes (mechanism unclear) and must be separated from the insert by restriction digest. DNA that contains particulate matter will clog the injection needle and slow the operator and if fine enough will pass into the cell with possibly deleterious effects. Additionally, purification reagents that contaminate the sample to be microinjected are frequently toxic to the cell, and nuclear condensation or lysis will occur. Several protocols exist to purify injection-quality DNA, in print and also on the Internet. One such protocol is detailed in **Subheading 2.3.** *(9)*.

## 1.4. Oviduct Transfer

Following microinjection of foreign DNA, the manipulated embryos must be transferred to pseudopregnant recipient females. Embryos from the one-cell through to the morula stage (0.5–2.5 d p.c.) are transferred into the ampulla by oviduct transfer. Generally, microinjected eggs are transferred at the one- to two-cell stage. One-cell embryo transfer is best performed after allowing injected embryos a period in which to recover (1–3 h) in culture media, such as M16 (Sigma, St. Louis, MO) Z (*see* **Note 5**). This allows better evaluation of the cells' survival and easier identification of viable cells for transfer. Eggs also can be incubated in culture overnight and allowed to develop to the two-cell stage. This gives an even better indication of cell viability because only healthy, undamaged cells will divide to the two-cell stage. However, it is best to minimize the time in culture, so the increased confidence in transferring embryos at the two-cell stage must be balanced against the increased in vitro exposure. The number of eggs transferred to a recipient female also depends on whether the transfer is performed at the one- or two-cell stage. In general, approx 20–30 one-cell eggs can be delivered into each recipient mouse. This number can be reduced to approx 15–20 for transfer of two-cell embryos. These numbers generally will produce litters of approx 5–10 pups, considering that 50–75% of transferred embryos will develop to term for one- and two-cell embryos, respectively.

## 1.5. Alternate Technologies and Strategies

In recent years, new methods of transgenic production have been developed. These alternate strategies may allow some benefits over the established "standard" technique of transgenic production outlined in previous chapters. Although none of the techniques described yield higher efficiencies (indeed they often have lower yields per treated embryos), these do afford other tangible benefits that may prove valuable.

Injection into fertilized embryos normally requires placement of the DNA insert into the pronuclear envelope, a skill that may take many months to acquire and gain proficiency. A technique has been described in which the DNA for injection is complexed with a polylysine mixture *(13)*, enabling the generation of transgenic mice by injection into the cytoplasm. Clearly, such a technique has value in the training of a new microinjection operator, whereby even unsuccessful injections where the pronucleus is not injected within the membrane but beside it may yield viable founder animals. In such a manner, training periods would still be able to contribute meaningfully to the injection projects at hand.

A methodology has been described whereby previously vitrified embryos can be injected *(14)* into the pronucleus by standard means and give rise to transgenic founders at a similar rate to nonvitrified embryos. This has the advantage that numbers of embryos can be obtained either over time or when it is convenient to do so and be maintained in a cryopreserved state. When microinjection needs to be performed, sufficient embryos for a day's injection are thawed and injected as normal. This has the advantage of limiting the costs of maintaining a large production colony of mice, although it lends itself more to facilities where microinjection is a relatively infrequent activity.

Additionally, the use of adenoviral vector delivery has been used to generate transgenic mice *(15)*. These viruses are replication defective and are used to infect one-cell fertile embryos that have had the zona pellucida removed. This strategy is particularly relevant to many researchers because it eliminates the requirement for a considerable component of the standard equipment used in transgenic animal production, thus representing a considerable saving in startup costs. An additional benefit is that this system delivers a single copy of the gene of interest rather than the multiple copies that the "standard" technique often imparts by concatomer formation *(16)*. This allows the evaluation of the insert without the complication of gene dosage.

## 2. Materials

### 2.1. Mouse Production Colony

#### 2.1.1. Vasectomy

1. One pair of curved serrated forceps.
2. One pair of straight serrated forceps.
3. Two pairs of watchmaker's forceps.
4. One pair of 4-in. sharp/sharp dissecting scissors.
5. One pair of Autoclip applicators.
6. Autoclips.
7. 4/0 Silk suture, with curved needle swagged on.
8. One-half of a plastic 10-cm Petri dish half full of 70% EtOH (ethanol).

#### 2.1.2. Anesthetic for Surgery

2,2,2 Tribromoethanol (avertin) is found to be quite effective. For the method of preparation, *see* **ref. 3**. This agent should be stored wrapped in tin foil (or another opaque wrapper) at 4°C, because it is light sensitive, and tested after a new batch is made. To anesthetize the mouse, the avertin is administered intraperitoneally, and the mouse is returned to its original cage. The mouse will be more relaxed when placed in a familiar environment, and the anesthetic will act more quickly than it would on a distressed mouse. To check that the mouse is fully anesthetized, press or squeeze the pads of the feet. If the mouse has not reached the depth or plane of anesthesia required for surgery, it will try to withdraw its leg from your grasp (pedal reflex). Do not commence surgery until there is no reflex reaction to this test. Use of mice that weigh <25 g is not recommended.

Note that there has been a reported incidence of peritoneal irritation or inflammation associated with the use of avertin, and that some institutions prohibit its use for this reason. Most incidents of this nature are owing to repeat administration of this agent, which would not normally be permitted under the regulations pertaining to the number of survival procedures allowed by Institutional Animal Care and Use Committee boards. Other anesthetic agents may be used; ketamine/xylazine is popular in veterinary circles, and ketamine/medetomidine *(5)* (with atipamizole as an agonist) is becoming more widely used, particularly in the United Kingdom.

## 2.2. Collection of Fertilized Eggs

### 2.2.1. Harvesting Fertilized Eggs

1. 70% Ethanol.
2. Surgical scissors.
3. One pair of 4-in. sharp/sharp dissecting scissors.
4. Surgical forceps.
5. Working media, (M2, Sigma, St. Louis, MO) (*see* **Note 5**).
6. Three-well watchglass.

### 2.2.2. Collection of Fertilized Embryos from Isolated Oviducts

1. $CO_2$ incubator to maintain manipulated embryos at 37°C in an atmosphere of 5% $CO_2$.
2. Isolated oviducts in three-well watchglass.
3. Fine forceps.
4. Watchmaker's no. 5 forceps or Pattern no. 3c forceps.
5. Sterile 30-mm tissue culture dish.
6. Under-stage illumination dissection stereomicroscope.
7. Hyaluronidase (cat. no. H 3884; Sigma, St. Louis, MO).
8. Heat-pulled transfer pipets.
9. Working media.
10. Culture media, (M16, Sigma, St. Louis, MO) (*see* **Note 5**).
11. 3-mL Syringe with 30-gage, short-bevel, hypodermic needle (Popper & Sons).

## 2.3. Microinjection of DNA

### 2.3.1. Collecting and Moving Embryos

1. Soda-lime glass microcapillary pipets (50 µL; Kimble Glass) or Pasteur pipets.
2. Bunsen burner.
3. Mouth pipet tubing and mouthpieces.
4. Diamond pencil.

### 2.3.2. Slide Preparation

1. One standard depression slide, siliconized.
2. Light mineral oil, embryo tested (cat. no. M8410; Sigma).
3. Working media.

### 2.3.3. Micrometer Setup

1. One micrometer/hydraulic drive unit (Bunton).
2. One Hamilton Gas-Tite™ syringe, 250 µL (Thomas Scientific).
3. Two female luer lock tubing connectors (Popper & Sons).
4. Intramedic™ tubing, 1.27 mm od, 0.86 mm id (Thomas Scientific).
5. Three-way stopcock tubing connector (Thomas Scientific).
6. One disposable plastic 60-mL syringe with male luer connector (e.g., VWR, Sigma).
7. One instrument collar (Leica).
8. Light mineral oil (Sigma).

### 2.3.4. Microinjection of Fertilized Eggs

1. One slide, prepared for injection.
2. One inverted stereomicroscope with phase contrast (Hoffman) or Nomarski optics.
3. One pair of micromanipulators.
4. One micrometer with attached, oil-filled, gastight Hamilton 250-µL syringe.
5. One gas pressure source (50-mL gastight syringe, Eppendorf Transjector automated pressure system, or other similar automated product).
6. One holding pipet.
7. One injection needle preloaded with DNA for injection (preparation described in **Subheading 3.3.1.**).

## 2.4. Oviduct Transfer

### 2.4.1. Surgical Equipment

All surgical instruments should be clean and sterilized with 70% ethanol or heat sterilized prior to use.

1. Anesthetic (avertin or ketamine/xylazine) (*see* **Note 6**).
2. 70% Ethanol.
3. One blunt pair of serrated forceps.
4. Two no. 5 sharp watchmaker's forceps.

5. One pair of 4-in. sharp/sharp dissecting scissors.
6. Autoclip metal wound clipper.
7. One serrefine clip (Dieffenbach).
8. Surgical suture (5.0).
9. Dissecting stereomicroscope.
10. Fiberoptic illuminator.
11. Syringe (1 cc) with needle (26 gage).
12. Several Kimwipe tissues or surgical gauze.
13. Lid of 9-cm plastic Petri dish.
14. Working media.
15. One mouth pipettor and hand-pulled transfer pipet.

## 2.4.2. Recipient Females

Recipient females used for oviduct transfer should be ideally 4–6 wk of age and weigh approx 20–25 g. Do not use a mouse that appears lighter than 20 g; underweight mice tend to reabsorb the embryos because they are not physically ready to support a pregnancy. Overweight mice can make surgery difficult because the anesthetic is absorbed into the fat, reducing its potency. In addition, the presence of fat means the presence of blood vessels, and cutting through all the extra fat causes much unnecessary bleeding. This makes it difficult to see what one is doing and may also clog up the tip of the transfer pipet. Select a mouse that has been plugged by a vasectomized male the night prior to the transfer. The plug should be visualized in the early morning of the injection day. Outbred females such as Swiss Webster mice make excellent mothers, although they do become overweight quickly and exhibit bad planes of anesthesia when heavy. These mice also are very inexpensive. Another very successful strain is B6D2F1. These mice are hardy and display hybrid vigor.

## 2.4.3. Anesthetic

Avertin is found to be an effective mouse anesthetic for oviduct transfer. For method of preparation, *see* **ref. 3**. The pseudopregnant recipient mouse should be anesthetized intraperitoneally as previously described (*see* **Subheading 2.1.2.**).

# 3. Methods
## 3.1. Mouse Production Colony
### 3.1.1. Detection of Estrus

The ability to detect females in estrus (4) is a highly useful skill throughout the process of developing a transgenic model, whether it is for generating suitable recipient animals for embryo implantation or for examining differing developmental time points. The estrous cycle in the mouse has four main phases:

1. Metestrus: The coloration of the vaginal tissue is a dusky white, with little swelling around the opening.
2. Diestrus: The vaginal opening is small, and tissue around the perimeter is gray/blue.
3. Proestrus: The vaginal opening has increased markedly, and surrounding tissues are pink to red.
4. Estrus: Coloration of the tissues is less intensely pink, with striations visible on the dorsal and ventral vaginal lips with pronounced swelling.

By selecting animals at random, 20–25% may be in estrus at any given time. However, the synchronization of estrous cycle by several individuals that are group housed may cause considerable variation in the numbers obtained at any given time, because the mean duration of the estrous cycle for most murine strains is 4 to 5 d. Selecting mice in estrus requires that two variables be noted in their evaluation: coloration and degree of tissue distension. An animal coming into estrus will exhibit a highly pink coloration, yet not so pronounced as to appear inflamed. Additionally, the dorsal and ventral tissue around the vaginal opening will be swollen and glossy, with striations visible primarily on the ventral side. Animals may be observed with a high coloration yet little distension or alternately somewhat distended yet grayish—these would be pre- or postestrus and should not be selected for mating.

### 3.1.2. Confirmation of Mating: Vaginal Plugs

After the animals are mated, it is necessary to determine whether or not the male has successfully impregnated the female. This is done by checking the females for the presence or absence of a vaginal plug. This plug is formed soon after mating has occurred, as the ejaculatory residue from the male solidifies into a waxy obstruction to the vaginal tract. Plugs are easily visualized by eye, only occasionally necessitating the use of a small probe to detect those that are deeper within the vagina. Plugs should always be checked in the early part of the morning, because they will be expelled with increasing frequency 10–12 h p.c.

### 3.1.3. Vasectomy

Vasectomized mice are used for mating to sham fertilize females used for embryo transfer. Vasectomy is performed on mice at 6–8 wk of age. The CD-1, B6D2F1 (C57BL6 x DBA/2), or Swiss Webster strains are all popular choices. The anesthetic ip dose of avertin is 240 mg/kg and xylazine/ketamine is 80–100/(10 mg·kg).

1. Weigh the mouse and anesthetize accordingly. Lie the mouse on its back and swab the intended incision site with 70% EtOH. The surgical site can be shaved close to the skin using clippers to avoid hair contamination of the incision.

2. Using a pair of curved forceps and dissecting scissors, cut across the lower abdomen approx 0.5 in. anterior to the genitals, making the incision about 1 cm wide. Wipe the incision site with a lintfree tissue dampened with 70% EtOH to remove excess cut hair.
3. Incise the body wall, reach in, and locate the bladder. Both tubes of the vas deferens will be visible lying on either side. Grasp the left vas deferens gently with forceps and lift a section clear of the incision.
4. Tuck the curved forceps underneath the vas deferens and allow them to spread to their natural unsprung state (*see* **Fig. 1, top panel**; *[6]*), with the ends pointing vertically. While in this position, use the suture to make two firm knots in the vas deferens, about 4 to 5 mm apart. Cut out a section of vas deferens from in between the knots and place on a tissue to be sure that one side has been done (*see* **Fig. 2, lower panel**; *[6]*).
5. Gently ease the two cut ends of the vas deferens back inside the abdominal cavity, and repeat the procedure on the right-hand side. When both sides are done, sew up the incision in the body wall with separate stitches; two or three should suffice. The suture should be kept in the EtOH in the Petri dish when not being used; this keeps the silk from drying and sticking to tissue and fat when another suture is required. Close the skin using two to three Autoclips.
6. Wrap the mouse in a tissue to keep it warm (loss of body heat is common in abdominal surgery), or, alternatively, place the mouse on a heat pad and allow it to recover. Animals that are placed under anesthetic should always be supervised and monitored until fully awake. Following the operation, allow the mice to recover for 2 wk before test breeding them to confirm sterility.

### 3.1.4. Test Breeding

One or two female mice are placed with the vasectomized male and are checked for plugs the following morning. Females that plug are sacrificed, and their oviducts are flushed 24 h later with phosphate-buffered saline or pH-buffered embryo medium. The eggs should be at the one-cell stage or unfertilized. If they are at the two-cell stage, the vasectomy was not performed correctly and the male should be culled.

## 3.2. Collection of Fertilized Eggs

### 3.2.1. Harvesting of Fertilized Eggs

Fertilized eggs for pronuclear microinjection are best harvested in the earlier part of the morning following mating, several hours prior to injecting. This will allow a greater window in which the pronuclei are visible for microinjection. Additionally, there is less chance that any of the mice set up to mate overnight will have lost their copulatory plugs, the indicator that mating has occurred. Assessing the number of plugs visualized in a superovulated, mated group of females allows for the possibility of troubleshooting should this number be low. A low percentage of plugged females may be an indicator that one

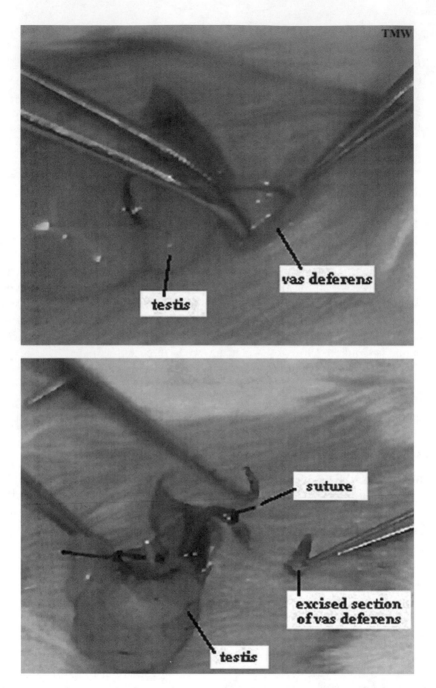

Fig. 1. **Top panel**, isolation of the vas deferens. **Lower panel**, proper suture placement and excision of ligated portion of vas deferens. The removal of a section of vas deferens greatly reduces the potential for subsequent re-ligation and return to fertility.

of the gonadotropins has lost its effectiveness, or that the stud male group used either has been overutilized or are becoming too elderly to be effective. To harvest fertilized eggs, the abdominal cavity must be opened and the oviducts carefully dissected out. Either flushing from the oviduct or dissection from the ampulla can be used as the preferred method for embryo collection.

### 3.2.2. Dissection of Oviducts from Abdominal Cavity

1. Humanely euthanize pregnant donor females by cervical dislocation or terminal $CO_2$ inhalation.
2. Place each animal on its back on a clean dry piece of absorbent paper and wet the surgical area completely with 70% ethanol. This procedure will prevent loose hair from contaminating the work area.
3. Grasp the skin at the midline of the lower abdomen and make a small lateral incision with the surgical scissors. Then, using the scissors, cut up and laterally toward the shoulders of each mouse making a V-shaped flap. Pull the flap of skin upward to reveal the entire abdominal cavity. The abdomen also can be opened by grasping the skin firmly on both sides of the abdominal incision and pulling the skin outward toward the head and tail to expose the peritoneum. Next use the fine forceps and scissors to cut open the peritoneum and expose the abdominal cavity. The scissors may be used in their closed position to brush the viscera upward toward the thoracic cavity; this motion is then duplicated in a downward motion toward the tail. This will move most of the organ systems, viscera, and fat away from the uterine horns, which should now be clearly visible. The uterus is a Y-shaped, muscular organ initiating from the pelvic region behind the bladder that branches off to the horns of the uterus, which travel upward and laterally into the body cavity.
4. Using the forceps, grasp each uterine horn in turn approx 6 to 7 mm (¼-in.) from the oviduct and ovary and gently pull away from the body cavity. This will expose the thin, transparent mesometrium, a vascular membrane connecting the reproductive tract to the body wall. Use a single blade of the forceps to pierce the mesentery, close to and underneath the uterine horn beneath the forceps' point of contact. Sweep this blade down and toward the oviduct and run it carefully along the bottom edge of the uterine horn. Care must be taken to avoid placing stress on the uterine horn/oviduct junction and causing this to break. This motion will clear the mesentery tissue away from the uterine horn and oviduct.
5. Using fine forceps, pull the fat pad, ovary, oviduct, and uterus taut. A very thin membrane surrounds the ovary and oviduct that holds them together. Carefully cut this membrane between the ovary and oviduct, using a slight sawing action with a single blade of the dissecting scissors. Subsequently use the scissors to cut through the uterine horn behind the forceps, approx 5 to 6 mm (¼-in.) posterior to the uterotubule junction (*see* **Fig. 2**). Place the isolated oviduct with a small section of uterine horn in a three-well watchglass containing working media. Repeat this process for the other oviduct, and for each of the embryo donors.

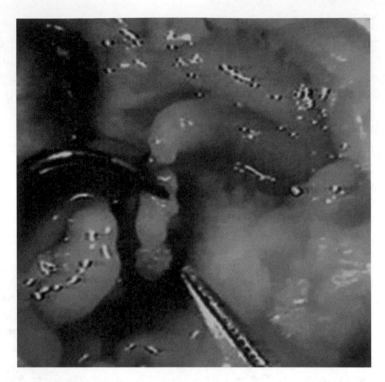

Fig. 2. Dissection and excision of oviduct and small portion of uterus showing separation of ovary and oviduct and proper location of cut at the top of the uterine horn.

### 3.2.3. Collection of Fertilized Embryos from Isolated Oviducts

When examined through a dissection microscope, a swollen area is apparent at the upper part of the oviduct near the infundibulum. This area is the ampulla, which contains clusters of newly ovulated eggs surrounded by sticky follicular cumulus cells. The cumulus cells will need to be removed in a hyaluronidase solution of 300 mg/mL in working media as discussed subsequently.

1. Transfer one oviduct to the second well of the watchglass containing the hyaluronidase solution. Using a pair of fine forceps, grasp the oviduct near the swollen ampulla where the eggs are located.
2. Using a second pair of watchmaker's forceps, gently tear open the oviduct to release the fertilized eggs. The eggs should flow out of the oviduct quite easily. Forceps also can be used to push the eggs out through the tear by gently squeezing the oviduct.

As an alternative method to tearing the ampulla, the oviducts may be flushed with working media. This technique expels the embryos out through the infundibulum. Once all the oviducts have been collected, the embryos can be flushed out. To do this, a finer style of forceps is required (pattern 3c suggested), as well as a 30-gage hypodermic needle on a 3-mL syringe. The use of a short bevel needle is recommended, since the goal for this procedure is to insert the needle into the lumen of the last oviduct loop prior to the uterine horn junction. A shorter bevel will allow the sharp needle tip less opportunity to pierce through the other side of the loop before the entire needle aperture resides within this structure.

1. With the forceps in one hand, and the syringe filled with working media in the other, use the forceps and needle point to turn the oviduct to be flushed to the preferred orientation under the dissecting microscope. The oviduct will be at the top and the section of uterine horn at the bottom, with the last oviduct loop uppermost.
2. Using the forceps, grasp firmly this last or terminal loop at the uterine junction, with the points of the forceps riding underneath this loop and elevating it slightly. Place the hypodermic needle on the syringe bevel up in the groove made by the closed blades of the forceps, and use this to guide the point downward, directly into the terminal loop. This action requires a slow and steady hand, and care should be exercised to avoid running the needle tip out through the far side of the loop (*see* **Note 1**).
3. Once the entire bevel is inside the loop, lightly depress the syringe plunger to flush media throughout the loop structure of the oviduct, expelling the embryos out into the media. Then use the forceps to pick up the empty oviduct and discard.

### 3.2.4. Hyaluronidase Treatment and Embryo Cleaning

After all the embryos have been isolated into working media, some may be observed to be in clumps of much smaller cells known as cumulus. For embryos that will be used for microinjection, it is necessary that they be free of any adherent cells or cellular debris. To clean such cells, they are put through a number of washes in working media to remove cellular debris. All cells having no adherent cumulus can be removed to another Petri dish or watchglass containing fresh working media using a finely drawn out transfer pipet, controlled by mouth suction. Those that remain should be treated with hyaluronidase, an enzyme found in the head portion of sperm that breaks down intercell adhesion between cumulus cells. Hyaluronidase dissolved in working media can be applied to the existing media containing embryos with associated cumulus, or a crystal of this compound can be placed into the media and allowed to dissolve. The embryos in media after hyaluronidase treatment should be watched carefully and removed to fresh working media as soon as the cumulus cells have dissociated. This care is necessary because the hyaluronidase over time also will begin to break down the zona pellucida protecting the embryos.

After two to three washes in working media to clean any associated debris and remove any hyaluronidase carried over, the embryos may be transferred to a culture media microdrop culture using a fresh transfer pipet and placed in a 37°C, 5% $CO_2$ incubator to await injection. This culture microdrop is assembled using culture media covered by a layer of autoclaved mineral oil. The oil is used to prevent contamination, as well as evaporation leading to pH changes that could be harmful to the eggs.

### 3.3. Microinjection of DNA

### 3.3.1. Protocol for DNA Purification for Pronuclear Microinjection

The injection-quality DNA must be of a specific concentration. Studies have shown that DNA has to be injected at a threshold concentration near 1 ng/mL *(10)*. However, increasing the DNA concentration much above this level proves to be no more efficient as toxicity increases substantially past 5 ng/mL.

1. Separate the insert from the vector on an agarose gel run in Tris/acetate/EDTA (not Tris/borate/EDTA) buffer. Use 5 mg/mL of ethidium bromide for gel staining.
2. Visualize the DNA with long-wavelength ultraviolet light to avoid damaging the ethidium bromide–intercalated DNA.
3. Excise the gel slice containing the gene fragment of interest and electroelute the DNA, or process using a Qiaex gel extraction kit (Qiagen) according to the manufacturer's instructions.
4. Ethanol precipitate the DNA. For ethanol precipitation of sample, add 1/10 vol of 3 *M* Na acetate, mix, and then add 2–2.5 vol of sterile 100% ethanol.
5. Incubate at –20°C overnight, and then spin 5 min in a microcentrifuge at 10,000 rpm to pellet the DNA. Resuspend in Elutip buffer (Schleicher & Schuell).
6. Pass the DNA through an Elutip-D minicolumn (cat. no. 27370; Schleicher & Schuell).
7. Ethanol-precipitate the DNA as in **step 4**. Wash the pellet several times with 70% ethanol and dry the pellet under vacuum; the washing and drying steps are extremely important, because residual salt and ethanol are lethal to the developing embryo.
8. Resuspend in injection buffer (10 m*M* Tris-HCl/0.1 m*M* EDTA, pH 7.5) prepared-HCl with Milli-Q-quality water.
9. Estimate the DNA concentration by using a fluorimeter or by comparing ethidium bromide staining of a sample run on an agarose gel to a serial dilution of a standard of known concentration.
10. Adjust the DNA concentration to 5–10 ng/μL using injection buffer (filtered, autoclaved 0.1 × Tris-EDTA [TE]).

## 3.3.2. Preparation of Glassware

### 3.3.2.1. INJECTION NEEDLES

Injection needles are produced using thin-walled glass capillary tubes containing an internal glass filament. This type of tube is necessary because it allows the injection needle to be filled by capillary action through the open end of the needle. Injection needles can be pulled with either a horizontal or vertical pipet puller. The capillary tube is secured into the pulling apparatus so that the heating element is approximately at the center of the capillary tube. The setting on the pipet puller should be selected such that the filament temperature and pulling force are best for injection needle production. This may require making several test needles to determine optimum settings. One capillary tube should produce two usable needles. A further technique involving micropipet beveling *(11)* has been described to modify further microinjection needles to increase transgenic embryo yield. A slurry of grit 120 silicon carbide is used in a beaker on a stir plate to provide a gentle yet abrasive force. Micropipets drawn out previously on a vertical or horizontal puller can be further modified by immersing their tips into the rotating slurry at approx 45° for 15–30 s. This imparts an aperture that is less prone to clogging and also reduces the shearing force applied to the DNA as it is expelled through the glass needle. Needles prepared in this manner should be carefully washed in distilled water prior to use to remove any residual particles of carbide.

### 3.3.2.2. HOLDING PIPETS

The pipets used for holding embryos in a fixed position to allow microinjection have a blunt end with a restricted aperture. This prevents mechanical damage to the embryos, and the small aperture allows them to nestle into this depression while held under negative pressure. This high degree of contact resists breakage of the seal that this seating provides and makes the embryo far less likely to rotate while being injected.

The holding pipet starts as a finely drawn micropipet, somewhat similar to the injection needle. This is pulled using nonfilamented glass. This pipet must be modified using a microforge, which delivers controllable voltage (and thus heat owing to resistance) across a platinum or platinum-iridium filament. This filament has a small glass bead melted onto it to provide a contact point for the manipulations required. This glass bead also heats up after the voltage is applied across the filament. The drawn-out pipet is held by a holding arm horizontally on the microforge and slowly brought into gentle contact with the glass bead in the same focal plane until a slight deflection in the glass is observed. Use of a scale in an ocular on the microscope is highly recommended to correctly size the outer diameter of the holding pipet. This should be 90–

100 μm wide. The filament and associated glass bead are slowly heated until the two glass surfaces fuse; care should be taken not to overheat and melt the capillary. The filament is then quickly turned off. As the filament cools, it will contract slightly and the pipet will break cleanly at a right angle to the point of contact, resulting in a flush end. The holding arm on the microforge is rotated to the vertical position and the pipet adjusted so that the end is slightly above the glass bead on the filament. The broken end of the pipet is heat-polished by slowly increasing the temperature until the glass begins to melt. The tip of the pipet is allowed to melt until the id constricts to 15–20 μ across. The opening of the pipet should be smooth, straight, and perpendicular to the long axis of the pipet.

The holding pipet can be further modified by introducing a slight bend (15°) at the end. This is done by positioning the pipet horizontally in the microforge so that it almost touches the glass bead with the end of the pipet slightly past the glass bead (2 to 3 mm). The temperature of the filament is slowly increased until the end of the pipet starts to bend under its own weight to the desired angle. The filament is quickly turned off to stop the bending. The bend in the end of the holding pipet may make it easier to position it horizontally in the injection chamber.

### 3.3.2.3. Transfer Pipet for Collecting and Moving Embryos

1. Light a bunsen burner and adjust to a fine flame. Rotate a glass capillary tube or Pasteur pipet back and forth between the thumb and forefinger in the upper part of the flame.
2. When the glass begins to soften, quickly withdraw the tube from the heat and pull sharply outward stretching and narrowing the softened area of the glass. This will produce a tube with an id of approx 200 μ. It will take some practice to determine the appropriate time to remove the pipet from the flame and the necessary force needed to pull the pipet to achieve the desired id. However, with time this technique will become quite easy and pipets will be uniformly produced.
3. Score the pipet and snap gently for a clean break. Alternatively, pull the pipet or simply bend it until the two halves are broken apart.
4. Examine the pipet under a dissection microscope and modify the end by recracking to achieve the desired id and a clean, smooth opening.

### 3.3.2.4. Pipets for Transferring Embryos to Recipient Females

The pipets used to deliver the injected eggs to a pseudopregnant female are fashioned exactly as described for the transfer pipets. These pipets, however, must have a slightly smaller id of approx 150 μ. This diameter is slightly larger than a single embryo and will allow the eggs to be nicely packed inside for transfer into the oviduct (as described later). The oviduct transfer pipet should also be further modified by heat-polishing the end to provide a smooth finish so as not to damage the oviduct during insertion. This can be done using the microforge as described in **Subheading 3.3.2.2.** for heat-polishing the broken end of the holding pipet or by rapidly touching the tip of the pipet in the gas flame. Care must be taken during this step not to leave the tip in the flame too long, because it will quickly melt and close.

### 3.3.3. Preparation of Slide

A slide suitable for presenting the eggs for microinjection has to be prepared. This should have eggs in a pH-buffered working medium such as M2 (*see* **Note 5**), enabling them to be protected outside the incubator environment for a period of 30–40 min. This drop of medium should be protected from dehydration and subsequent concentration by overlaying with light mineral oil.

Using a handheld pipettor, working medium is added to the bottom of the slide depression to form a brimming drop approx ¼-in. in diameter. Medium is then removed from the drop, allowing it to flatten. This allows a wider coverage of the bottom of the depression than placement of the final volume would give. The medium is immediately overlaid with embryo-tested light mineral oil; this should not overfill but should cover the top of the drop of medium (*see* **Note 2**). The eggs are removed from the incubator, and the desired number of eggs to be injected at one sitting are placed into the medium on the slide. Their placement should be such that there is ample space to easily keep injected eggs separate.

### 3.3.4. Microinjection Apparatus

The basic elements for any injection platform can be found in any institution where transgenic technology is applied, although the degree of complexity of each setup is contingent on the budget available and the individual preferences of the operators involved (*see* **Note 3**) (*see* **Fig. 3**). The elements necessary for any platform are as follows:

1. A sturdy base (*H*) that protects the microscope, manipulators, and associated equipment from the effects of vibration, which on the microscopic scale have the capacity for disruption of sensitive membranes while the injection needle is entering or inside the cell in the process of DNA delivery. The systems used to provide such vibration protection include the following:
   a. Specialized vibration-free tables, of which the working surface is cushioned from vibration pneumatically by the application of gas suspension. This is usually provided by a nitrogen gas cylinder.
   b. A large concrete platform (for mass and stability) resting on neoprene rubber stoppers or even squash balls that are suitably immobilized.
2. An inverted microscope (*F*), with either Hoffman phase contrast or Nomarski optics and at least two selectable objectives ranging from ×10 to ×32. Specialized

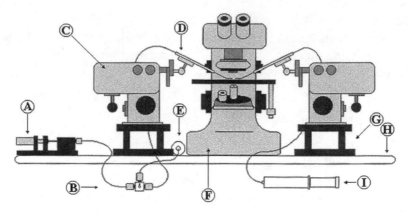

Fig. 3. Standard microinjection platform. The most frequently used equipment suppliers are as follows: A, micrometer/hydraulic drive unit (Bunton, Narishige); B, three-way stopcock valve (Thomas Scientific); C, micromanipulator (Leica, Narishige, Eppendorf); D, needle holder/instrument collar (Leica, Bunton, Eppendorf); E, 60-mL luer end syringe (Becton Dickinson, Sigma, VWR, Fisher); F, inverted light stereomicroscope (Nikon, Olympus, Leitz); G, micromanipulator stands (Leica); H, "breadboard" instrument base (Leica); I, glass 60-mL syringe (Becton Dickinson, Sigma, VWR, Fisher).

optics are a necessary component of any microscope to be used in the injection of single-cell-fertilized mouse eggs, because visualization of the pronucleus without their aid is extremely difficult. The use of several objective powers is highly recommended, because it is important to be able to segregate embryo populations on the injection slide into injected and uninjected groups.

3. Micromanipulators (*C*) to maneuver the glass needles used to hold and inject the embryos. These are essential to this technique's success, because the operator's hands cannot move these needles as precisely as required without some assistance. The manipulators first used in this technique were originally designed for the positioning of microelectrodes in neuroscience applications and were awkward for use in transgenic animal production. As this field has become more commonplace in academia and industry, more advanced models have become available. Undoubtedly, the most commonly used manipulator is a mechanical model (Leitz) that utilizes a system of pinions resting on a brass globe within the unit. Their sensitivity may be adjusted by their relative position in one of three dimensions to the equatorial plane of the globe. The base of the globe has an attached joystick arm that descends out from the bottom of the unit, allowing ease of use. More recently, companies such as Eppendorf have produced electronically driven joystick units that are also fully programmable in all three dimensions.

4. Needle pressure delivery systems. These are a standard part of all microinjection platforms, and similar to the tables or work surfaces used in order to dampen or

eliminate vibration, such systems vary greatly in their degree of complexity and cost. Also, the emphasis on complexity and the degree of control required tends to skew heavily toward the injection needle requirements rather than those of the holding pipet. Described next are several approaches that may be employed:

a. The most basic pressure delivery used is that provided by the operator's use of a mouth pipet coupled to the instrument collar retaining the holding pipet. The application of slight suction or positive pressure can be used to pick up or set down embryos, as well as to hold them to the needle surface during injection. This is never used to drive the DNA out from the injection needle; this action requires more positive pressure than can be applied in this manner.

b. Affectionately known in the transgenic field as the "bicycle (bike) pump" method, a syringe can be used to deliver the requisite pressure to either the holding or injection needle. Since the holding pipet requires far less pressure to be applied, a smaller syringe (10-mL disposable or smaller) is usually used. For the injection needle, a 50- to 60-mL glass syringe (*1*) is frequently used, since significantly higher pressures must be delivered to the needle tip to drive the DNA into the cell because the aperture is so much smaller than that of the holding pipet. Additionally, if the aperture should be further restricted through external or internal clogging, the application of sufficient pressure to purge the blockage can be applied.

c. Usually reserved for holding pipet control, but also used in (*A*) the injection of embryonic stem cells, a micrometer-driven gastight syringe coupled to a light mineral oil hydraulic system may be used (*see* **Fig. 4**). This can provide precise control of the pressure applied to the holding pipet in order to pick up, hold, and set down cells for injection. Other micrometer systems may use air or dimethylpolysiloxane (DMPS), viscosity 200 centistrokes as the hydraulic agent for a similar degree of control.

d. An enterprising adaptation of the syringe pressure delivery method, designed for control of injection needle pressure, uses a standard air gas supply. This may be provided through the use of a regulated compressed air gas cylinder connection, or through an "in-house" compressed air supply. A line using Tygon™ tubing that can be connected to the injection needle instrument collar is run from the gas supply provided. Close to the instrument collar, a T-connector is placed in the line, such that the line runs through the top part of the "T," leaving the arm exposed. The gas pressure is set such that there is a slight positive residual pressure applied to the injection needle tip, even though most of the compressed air escapes through the exposed T arm. This in-line connector can be held while the operator's hand also holds the micromanipulator joystick, and the thumb or other finger (as preferred) may be used to constrict or block the T arm opening. This will cause a controllable increase in line pressure when the DNA is to be expelled into the pronucleus.

e. The expanding use of transgenic technology has seen the arrival of electronically mediated pressure delivery "injection machines." These, at first satisfying the need for precise and consistent injection pressures and volumes, have evolved into systems that are capable of applying independently controllable injection and holding pipet pressures. Earlier models used coupling to nitrogen gas cylinders to provide the pressure source and regulated the application of this pressure internally. More recently, internal compressors have eliminated the need for bulky gas cylinders. Using the pressure provided by the unit, a standardized baseline positive pressure could be established at the injection needle tip. This pressure, although low, would resist the adhesion of external blockages and allow the operator to customize this parameter to the slight variations in any injection needle. An injection pressure could be set in a similar fashion and regulated through the use of a foot pedal. The inherent advantages of these systems are that the amounts of DNA delivered per cell can be controlled more consistently than through more manual means, and that the use of a foot pedal control for such a unit allows the operator's hands to remain on the micromanipulator controls throughout the process. This allows for a more efficient throughput of the available embryos for injection. The more advanced microinjector machines are programmable, having set keypads allowing for "one-touch" embryo pick up or set down, and defined injection time periods at a set pressure.

### 3.3.5. Micrometer Setup

A micrometer-driven pressure system for the control of a holding pipet can be assembled using the following equipment. (Note that only the exact units/equipment and suppliers are suggested). **Figure 4** presents an assembly diagram for this system.

#### 3.3.5.1. ASSEMBLY OF EQUIPMENT

Assembly should begin with the attachment of a female luer tubing connector to the male luer end of the 60-mL syringe. Intramedic tubing should be used to connect the syringe, which will become the oil reservoir for the system, to the three-way stopcock valve (*see B* in **Fig. 3**). To establish this connection, the tubing should be passed through the cap end section of the female luer connector, and a flare end made. This, when made correctly, will form a funnel-shaped end onto the cone end of the other female luer connector section. The flare end is made by using a standard butane gas lighter (or equivalent) and heating the plastic end close to a metal or stone vertical flat surface, such as that on the edge of a laboratory bench or sink. When the end is softened (but not molten), it should be rapidly pressed against this edge such that the end is evenly spread outward, resembling a "T" shape when viewed from the side. Ideally, this end should be two to three times wider than the tubing itself, with the internal space exiting as an unobstructed hole in the center of the flare or flange. This should be unable to be drawn backward through the hole in the luer connector cap section. With the tubing drawn against the inside of this

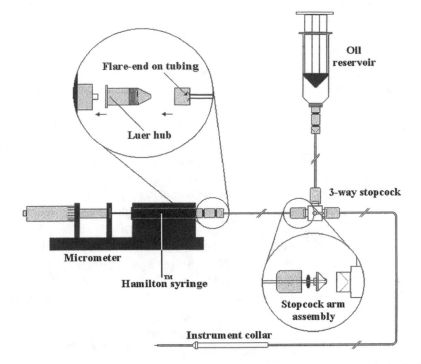

Fig. 4. Set up for oil-based hydraulic system for use in conjunction with the holding needle. This configuration can be used to provide traction as required during propulsion of the egg.

hole, the two luer sections should be joined together such that the cone section is brought against the flare end and molds it into a gastight funnel shape by the application of pressure through the tightening of this junction.

The length of tubing used between the syringe reservoir (*E* in **Fig. 3**) and the stopcock is normally about 90 cm (3 ft). A smaller flare or flange connection will be necessary to establish a gastight seal between the tubing and one arm of the three-way stopcock valve in a similar manner as described previously. Using one of the in-line arms of the three-way valve, a flare end gastight connection is established, and at least 90 cm (3 ft) of tubing is needed before another flare end connection is made to a female luer connector. This connector is twisted into the male luer acceptor end of the Hamilton Gas-Tite 250-μL syringe. This glass syringe can be loaded into and controlled by the hydraulic drive unit micrometer. The opposing and last arm of the three-way valve has tubing attached by way of another flare end connection and is routed to the holding pipet instrument collar. The length of tubing here should be at least 60 cm (2 ft) but will vary depending on the variability of the platform being used. As a rule of thumb, the lengths of tubing should be kept to a minimum to reduce the effects of the hydraulic fluid's compressibility.

The hydraulic fluid, in this case light mineral oil, has to be introduced into the system. As with any hydraulic system using liquids, air needs to be eliminated. This is done by first filling the reservoir syringe with oil while detached from its tubing connector. The connector is then reconnected. The three-way stopcock valve is aligned in a manner such that the arm leading to the instrument collar/holding pipet is blocked, allowing flow from the reservoir through to the Hamilton Gas-Tite syringe. The Hamilton syringe should be removed from the hydraulic drive unit, the plunger removed, and held in an upright position. The plunger on the reservoir syringe is then depressed to drive oil through the tubing, three-way connector, and up into the body of the Hamilton glass syringe. Oil should be expelled through the top of this syringe until all air bubbles have been evacuated successfully, after which the plunger is replaced carefully in order not to trap any air. The Hamilton syringe is then fastened into the hydraulic drive unit and is ready for use. Excess amounts of oil may be flushed out through the instrument collar section of the system. The three-way stopcock valve should then be adjusted to prevent flow to or from the Hamilton syringe, and the reservoir syringe used in a similar fashion to drive oil through (and eliminate air from) the tubing leading to the end of the holding pipet instrument collar.

This system can be used by affixing the holding pipet to the instrument collar or associated tubing and using the reservoir syringe to drive oil to the neck of the needle. In this manner, little oil inside the Hamilton syringe needs to be expended when the reservoir arm is sealed off. The micrometer control can be used to draw a little buffered media from a slide once the oil is at the needle tip, ensuring that no air is trapped at the oil/media boundary. Micrometer adjustments may be used to pick up, hold, and set down embryos using this system.

Both the holding pipet and injection needle must be mounted onto the microinjection platform's micromanipulators for injection. These are held using either a hollow instrument collar, where the hydraulic fluid descends into the back of the needle, or a stem clamp holding the glass needle itself with the oil- or air-filled tubing hanging freely from the posterior portion. The needles may be modified with an angled bend, such that the angle of the main needle section being held by the micromanipulator is offset in the injection slide. The holding pipet and injection needle should be within 30° from the horizontal plane for injection, such that there is a negligible change in the height of the needle tip as it moves from outside the embryo to inside the pronuclear membrane. This will ensure that the DNA is expelled within, rather than above or below, the pronucleus.

## 3.3.6. Loading DNA into Injection Needle

The injection needle is filled by dipping the blunt end of the needle into the tube containing the DNA solution. The DNA will flow into the tip of the needle by way of capillary pressure. The use of needles containing an internal filament (as previously discussed) will facilitate this process. The end of the needle should be left in the DNA solution until a small bubble forms from the blunt end. This is usually a good indication that the DNA has filled the needle. By carefully examining the tip of the needle, a small meniscus should be seen a few millimeters from the tip of the needle. The filled pipet can then be inserted into the needle holder/instrument collar.

## 3.3.7. Microinjection of Fertilized Eggs

The process by which fertilized embryos are injected is relatively simple; however, the large number of coordinated manipulations that this process entails does involve a substantial amount of practice for proficiency in the technique to be attained (*see* **Note 4**).

1. Place the slide on the microscope and focus at low power on the embryos.
2. Lower the attached holding and injection needles to be in similar focus as the embryos and slightly below them, before switching to high power (×32).
3. Draw the injection needle to the edge of the media/oil boundary, slightly within the oil. Here, the presence of DNA flow may be established prior to injection by increasing the pressure delivered to the injection needle tip and visualizing the bubble of DNA solution encapsulated in the oil.
4. If no DNA flow can be established, gently break open the injection needle tip by rubbing it lightly on the blunt face of the holding pipet. Withdraw into the oil to confirm success in gaining DNA flow.
5. Move the needles back to the position beneath the embryos. Apply a gentle vacuum to the holding pipet by using the micrometer-driven hydraulic control, and use this to pick up an embryo. This should be held such that the bottom of the embryo is in light contact with the bottom of the injection slide.
6. Use soft adjustments in the vacuum applied to the holding pipet to allow the embryo to be rotated to allow positioning of a pronucleus directly opposite the holding pipet aperture and close to the zona pellucida (**Fig. 5, top panel**).
7. Maintaining the holding pipet steady, use the injection needle to puncture through the zona pellucida and outer cell membrane and through the pronuclear membrane. The pronuclear membrane is quite elastic and may necessitate pushing the injection needle most of the way through the entire structure. Care should be exercised to avoid contact with the nucleoli.
8. With the needles in this position, apply pressure to the injection needle tip to expel the DNA solution into the pronucleus, such that it swells to twice its size. Then remove the injection needle directly and deliberately (*see* **Fig. 5, lower panel**).

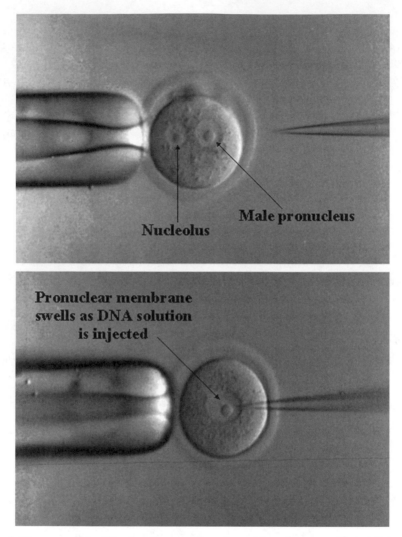

Fig. 5. **Top panel**, fertilized embryo with male and female pronucleus visible. Holding pipet and injection needle are also shown. **Lower panel**, embryo following injection of DNA. Note the swelling of the injected pronucleus.

9. Using the holding pipet, move the embryo to a separate area on the slide, distinct from the uninjected embryos. Return the holding pipet to begin this process.

## 3.4. Oviduct Transfer

### 3.4.1. Oviduct Injection

1. Place the anesthetized mouse on a Petri dish lid, taking care to keep the airway clear by resting the teeth on the edge of the dish. This makes it easier to move the

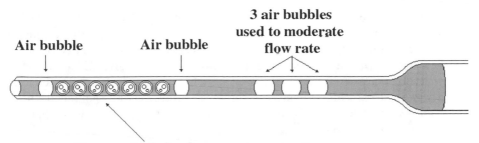

Fig. 6. Diagram of transfer pipet showing proper loading pattern of injected embryos for transplantation into recipient female.

mouse around without having to touch it. Swab the incision area with 70% ethanol. The surgical area of the lower back also can be shaved, if desired.

2. Load the transfer pipet with embryos. The embryos should be moved from culture media to working media, because they will be outside the incubator during the transfer procedure.

3. Proper loading of the transfer pipet is essential for successful completion of the embryo transfer. Take up an amount of working media in the tip of the transfer pipet, and then make a small bubble by taking up a little air. Then take up some more media—roughly the same volume as the air bubble followed by another air bubble the same size as before. Then take up about 2 to 3 cm of media and then a tiny air bubble once more. The air bubbles will help regulate pressure and provide better control when moving the eggs. Collect the embryos in the smallest possible volume of medium, lining them up side by side in the transfer pipet. Introduce another tiny air bubble when all the eggs are loaded, followed by a final short column of media (*see* **Fig. 6**). An alternative method is the use of mineral oil instead of or in conjunction with air bubbles for pressure control. This is a perfectly acceptable practice; however care must be taken to avoid introducing mineral oil into the oviducts because this can dramatically reduce litter size by interfering with the ciliary-driven egg transport in this structure, as well as predispose the animal to uterine infection. Great care should be taken in storing the loaded pipet until the eggs are transferred into the oviduct. The loaded pipet can be affixed to the surgical microscope by using a piece of plasticene.

4. Use a pair of sharp scissors and one pair of blunt serrated forceps to cut the skin. Make a small transverse incision approx 0.5 cm away from the midline and between the natural hump of the back and the point where the rear leg joins the abdomen, at the level of the last rib. Carefully wipe the incision site using a (lintfree) tissue dampened with 70% ethanol sweeping away any cut hair. Grasping one side of the incision at a time, carefully introduce the blades of the scissors (while closed) between the inside of the pelt and the body wall for approx 1 cm.

Open and close the blades to clear the connective tissue in this area. Move the skin around until the nerve (a white line, usually seen with associated capillary) can be seen running across the body wall. The ovary (red) or light color of the ovarian fat pad can be seen under the body wall in this area. Using a pair of watchmaker's forceps, pinch the body wall and make a small incision approx 0.5 cm across. Reach in with the blunt, straight serrated forceps (while holding the body wall with another pair of forceps) and grasp the ovarian fat pad. Gently pull the fat pad and associated ovary, oviduct, and uterus and remove to the exterior (**Fig. 7**). Clip the fat pad with a serrefine clip and rest across the mouse's back to hold the uterus in place. If the uterus or uterine horn continually slip back into the cavity, it may be necessary to gently lie the mouse on its side being careful not to block the airway.

5. Gently move the Petri dish lid supporting the mouse to the dissecting microscope and illuminate with a fiberoptic light source. Adjust lights and focus, and properly position the mouse with the coils of the oviduct clearly in view.

6. Use clean, sharp watchmaker's forceps to gently tear open a small hole in the transparent bursa membrane at the point between the ovary and oviduct where the infundibulum is located. Take care to avoid rupturing the small capillaries that run across the bursa, because these will obscure the view of the infundibulum. An application of epinephrine to the bursa membrane before tearing the hole can help to reduce any bleeding that may occur. Excess blood also can be mopped up using small pieces of tissue.

7. Once the infundibulum can be clearly visualized, grasp the edge with one pair of forceps while inserting one blade of the other set into the tube itself. This will ensure that the mouth of the infundibulum will be open and accessible to the transfer pipet. Insert the transfer pipet into the opening of the infundibulum far enough to ensure delivery of the eggs into the ampulla or until the natural curve of the tube allows no more forward progress without risk of damage to the ampulla.

8. Gently blow and expel the eggs into the ampulla followed by the air bubbles used for pressure regulation. These bubbles will prevent backflow of the eggs and easily drive them forward into the ampulla region of the oviduct. If excessive pressure is required to expel the eggs, the tip of the pipet is probably against the wall of the oviduct. If this occurs, withdraw the pipet slightly and try to expel the eggs again. It is also possible that the pipet tip has become clogged with clotted blood. If this is the case, it may be necessary to return the cell to the culture dish and reload the pipet.

9. With the transfer complete, withdraw the transfer pipet from the infundibulum. Remove the serafine clip, and using blunt forceps, gently pick up the fat pad and push the uterus, oviduct, ovary, and fat pad back inside the body wall.

10. Suture the incision in the body wall with one or two stitches. Close the skin with Autoclips; two per incision is usually sufficient. Autoclip wound clips are used on the skin instead of suture because the mice frequently will chew at the suture thread and effectively open their wounds.

Fig. 7. Mouse in proper orientation for oviduct transfer, showing location of incision and exposed oviduct complex with associated fat pad, ovary, oviduct, infundibulum, and uterus. Illustration by Scott Baswell, Amgen, Inc.

11. If doing a bilateral transfer, repeat this procedure on the other uterine horn.
12. Once surgery is complete, place the mouse in a fresh, clean cage. Under anesthetic, small mammals are unable to retain heat as effectively as when conscious. For this reason, the mouse should be wrapped in a tissue to help keep it warm. A heating pad can be used to care for the animal postoperatively until it regains consciousness. All animals should have recovered sufficiently from anesthetic

**Table 1**
**Analgesic Agents and Oral Dose**

| Agent | Oral dose (mg/kg) |
|-------|-------------------|
| Acetylsalicylic acid | 100 |
| Acetominophen (Tylenol) | 300 |
| Ibuprofen | 7 |

(20–30 min) before being returned to the animal room and left unattended. Recipient mice should be handled with care because pregnant mice become easily stressed, sometimes leading to abortion, or even cannibalism of pups.

### 3.4.2. Postinjection Monitoring

It important to monitor the condition of the recipient animals postoperatively, ensuring that no complications arise. Small rodents are prone to hypothermia induced by anesthesia and should be observed minimally for at least 2 h postoperatively. The use of a heat pad or a slide warmer with the cover on is suggested for this purpose.

The anesthetized mouse should be rolled gently in lint-free tissue (i.e., *K*imwipes, Kimberly-Clark, Roswell, GA). This will help retain heat. Then the mouse should be placed into a cage with bedding and torn nesting materials and the cage placed on a heat source on medium heat. The maintenance of normal body temperature will reduce the length of anesthesia.

Any mouse that has undergone surgery should be observed routinely daily for 4 or 5 d after surgery to ensure that full recuperation is taking place. The mouse should be observed to confirm that the conscious animal is able to move around with no impairment. An animal that has had surgery into the abdominal cavity has a slight risk of hernia postoperatively. Keeping the size of the incision into the abdominal cavity to a minimum and closing the body wall by suture or the use of tissue glue may markedly reduce the incidence of such a problem. The skin must be closed by using sutures or stainless steel skin staples. These may be removed 1 to 2 wk after surgery. If an animal appears to be in discomfort, usually indicated by failure to groom, dehydration, or a pronounced hunched appearance, Tylenol™ or a similar analgesic (**Table 1**) may be administered by adding to the animal's water supply. If dehydration is indicated, 0.9% sterile saline or lactated Ringer's solution may be administered by ip injection, as required. If there is no improvement, the surgical site may need to be reopened following anesthesia to ensure that no hernia has occurred. The interests of the animal should, however, be paramount, and the elimination or minimization of pain should be the primary goal of the researcher. Should there be no marked improvement in the condition of any such animal, the final option of humane euthanasia must be applied.

## 4. Notes

1. Should the terminal loop be torn in this process, or the needle tip be pushed through the far side of the loop, the needle tip also may be used to tear open the visible ampulla area of the oviduct to release the embryos as described in **Subheading 3.2.3.**
2. The most common problems are overfilling the depression with oil or, alternatively, not covering the top of the medium, allowing medium concentration with resulting plasmolysis. A small quantity of phenol red is sometimes added to assess pH change over time, to avoid the eggs having prolonged exposure to the HEPES-buffered environment, which is mildly toxic.
3. The degree of sophistication that needs to be employed generally depends on the environmental conditions relating to local vibration at the injection site, and before setting up such a platform this must be considered along with the setup budget for the project.
4. Care should be exercised to avoid injecting too much DNA into the pronucleus, because this may cause the membrane to rupture. The regulation of DNA flow is critical here. If using an air-filled gastight 50-cc syringe, adjust the manual administration of the pressure. If using an automated pressure delivery device, decrease the injection pressure if necessary or the time of pressure delivery. If the flow is constantly heavy, the aperture on the injection needle may be too wide, and, thus, a new needle is needed. Additionally, care should be exercised to avoid contact with the nucleoli within the pronucleus. These condensed bodies are sticky and may clog the injection needle and will also cause the cell to die if removed with the injection needle. Studies have indicated that injection into the male pronucleus (the larger of the two) may prove slightly more effective with respect to integration efficiency *(10)*. An additional study has indicated that higher transgenic yields are possible following injection of both pronuclei *(12)*.
5. An alternate medium to M16 and M2 is Brinster's BMOC-3 and HEPES-buffered BMOC-3, respectively. Both media work equally well as a culture medium and working medium. Choice of medium is entirely up to the individual researcher.
6. The ip xylazine/ketamine anesthetic dose is 80–100/(10 mg·kg). The equivalent avertin anesthetic ip dose is 270 mg/kg.

## References

1. Taketo, M., Schroeder, A. C., Mobraaten, L. E., et al. (1991) FVB/N: an inbred mouse strain preferable for transgenic analyses. *Proc. Natl. Acad. Sci. USA* **88,** 2065–2069.
2. Runner, M. N. and Gates A. H. (1954) Conception in prepubertal mice following artificially induced ovulation and mating. *Nature* **174,** 222–223.

3. Hogan, B., Beddington, R., Constantini, F., and Lacy, E. (1994) *Manipulating the Mouse Embryo*, 2nd ed., Cold Spring Harbor Laboratory Press, Cold Spring Harbor, NY.

4. Champlin, A. K., Dorr, D. L., and Gates, A. H. (1973) Determining the stage of the oestrous cycle in the mouse by the appearance of the vagina. *Biol. Reprod.* **8,** 491–494.

5. Flecknell, P. (1997) Medetomidine and Atipamezole: potential uses in laboratory animals. *Lab. Anim.* **26.**

6. Brown, G. A. J., *The Microinjection Workshop*, <http://ourworld.compuserve.com/homepages/thebroons/>

7. Gates, A. H. (1971) Maximizing yield and developmental uniformity of eggs, in *Methods in Mammalian Embryology* (Daniel, J. C., ed.), W.H. Freeman, San Francisco, CA, pp. 64–76.

8. Takagi, N. and Sasaki, M. (1976) *Nature* **264,** 278–281.

9. Pearson-White, S. University of Virginia Transgenic Mouse Core Facility Web page <http://www.med.virginia.edu/~sp3i/tghome.html>.

10. Brinster, R. L., Chen, H. Y., Trumbauer, M. E., Yagle, M. K., and Palmiter, R. D. Factors affecting the efficiency of introducing foreign DNA into mice by microinjecting eggs. *Proc. Natl. Acad. Sci. USA* **82,** 4438–4442.

11. Gundersen, K., Hanley, T., and Merlie, J. (1993) Transgenic embryo yield is increased by a simple, inexpensive micropipette treatment. *Biotechniques* **14(3),** 412–414.

12. Kupriyanov, S., Zeh, K., and Baribault, H. (1998) Double pronuclei injection of DNA into zygotes increases yields of transgenic mouse lines. *Transgen. Res.* **7,** 223–226.

13. Page, R. L., Butler, S. P., Subramanian, A., Gwazdauskas, F. C., Johnson, J. L., and Velander, W. H. (1995) Transgenesis in mice by cytoplasmic injection of polylysine/DNA mixtures. *Transgen. Res.* **4,** 353–360.

14. Tada, M., Sato, M., Kasai, K., and Ogawa S. (1995) Production of transgenic mice by microinjection of DNA into vitrified pronucleate stage eggs. *Transgen. Res.* **4,** 208–213.

15. Tohru Tsukui, T., Kanegae, Y., Saito, I., and Toyoda, Y. (1996) Transgenesis by adenovirus-mediated gene transfer into zona-free eggs. *Nat. Biotechnol.* **14,** 982–985.

16. Palmiter, R. D. and Brinster, R. L. (1986) Germ-line transformation of mice. *Ann. Rev. Genet.* **20,** 465–499.

# IV

## ALTERNATIVE ROUTES TO THE GERMLINE

# 4

## Adenoviral Infection

### Tohru Tsukui and Yutaka Toyoda

### 1. Introduction

Replication-defective adenoviruses (Ads), in which the E1A and E1B genes essential for replication are deleted, have several advantages as the vectors for introducing foreign genes into host cells. They are able to infect a large range of host cells and to transfer genes into nonproliferating cells without the replication of viral genome. However, there had been no reports of generating transgenic animals using this type of vector until we demonstrated that Ad-mediated transgenesis is possible and has distinct advantages over conventional microinjection and retrovirus vectors *(1)*. The major advantage compared to the microinjection method is that the methodology is relatively simple so that sophisticated skill and an apparatus for micromanipulation are not needed and a large number of eggs can be handled simultaneously. Instead, you will need some expertise for constructing a recombinant Ad (rAd) and also for handling zona-free eggs. The former aspect is described herein only briefly; for details *see* **refs. 2–4**.

Another characteristic of this methodology is that the transgenic mice so far produced have a single transgene of almost the entire viral genome that can be stably transmitted to the next generation. This is in sharp contrast to the case of microinjection, in which multiple copies of the injected genes are frequently integrated in tandem fashion, although the precise mechanism of this single-copy integration is not yet well understood. The advantage as compared to the retroviral method is that the one-cell pronuclear stage eggs can be infected with Ad quite efficiently, resulting in the nonmosaic transgenic mice that have integrated foreign DNA in every cell. One of the disadvantages of this method is that the size of the inserted expression unit is limited, at about 7.5 kb. Although the Ad-mediated transgenesis is possible, it must be emphasized that

From: *Methods in Molecular Biology, vol. 180: Transgenesis Techniques, 2nd ed.: Principles and Protocols*
Edited by: A. R. Clarke © Humana Press Inc., Totowa, NJ

the conditions for introducing Ad into the pronuclear mouse eggs in vitro, characterized by the removal of egg coat (zona pellucida) and a high concentration of the virus around the eggs, are quite different from the situation for in vivo gene therapy using recombinant Ad. Because the Ad-mediated transgenesis has so far been demonstrated only in the mouse by using a recombinant vector (AxCANLacZ), as described in the text and in **Fig. 1**, further studies are needed before this technique can be established as a new tool for transgenesis in a range of constructs and species. Recently, other applications have become available for using rAd in transgenesis, and for investigating molecular function of a gene in vivo (*see* **Subheading 4.**).

## 1.1. Generation of rAd

Recombinant adenoviruses have several benefits for in vitro and in vivo gene transfer. They are able to infect both nondividing and dividing cells (such as zygotic eggs and neurons). Adenoviruses can be generated as a high-titer viral stock ($-10^{9-10}$ pfu/mL) and are able to accommodate up to 7.5 kb of foreign DNA (*see* **Note 1**). Although it takes at least 1 mo to generate an rAd, these advantages make adenoviruses a powerful tool for molecular biologists. Different methods are available for generating rAd *(3,5,6)*; however the COS-TPC method is more efficient than the other methods *(3)*. An outline of the procedure we use for generating recombinant rAd is shown in **Fig. 2**.

## 1.2. Choice of Cassette Cosmids

Two types of cassette cosmids are available for constructing rAds:

1. pAxcwt *(4)*: This type of cosmid cassette is useful for site or stage specific expression. Lack of promoter in this cosmid requires that the inserted DNA include both promoter and cDNA.
2. pAxCAwt *(4)*: This type of cosmid cassette is useful for ubiquitous expression through its own CAG promoter (cytomegalovirus IE enhancer, chicken β-actin promoter, and rabbit β-globin poly [A] signal) *(7)*; therefore, only cDNA is inserted.

## 2. Materials

1. F1 hybrid strains (C57BL/6JxC3H/HeJ) at 6–10 wk of age.
2. Female recipient mouse (7–10 wk of age) mated to a sterile male 2.5 d before transfer.
3. High-titer rAd (>1 × $10^8$ pfu/mL).
4. Pregnant mare's serum gonadotropin (PMSG; Sankyo Zoki, Japan).
5. Human chorionic gonadotropin (hCG; Sankyo Zoki, Japan).
6. Hyaluronidase (Sigma, St. Louis, MO).

## Experimental Procedures

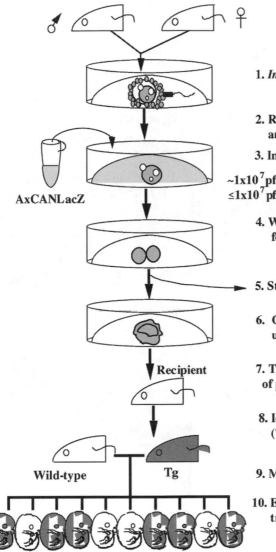

1. *In vitro* fertilization

2. Remove cumulus cells and zona pellucida

3. Infect the eggs with rAd

~1x10$^7$pfu/ml ➤ Transient expression
≤1x10$^7$pfu/ml ➤ Transgenesis

AxCANLacZ

4. Wash with culture medium for 4 times

5. Stain the eggs with X-gal

6. Culture the infected eggs up to the blastocyst stage

Recipient

7. Transfer eggs into the uterus of pseudopregnant foster mother

8. Identify the transgenic mice (Tg) by Southern analysis

Wild-type    Tg

9. Mate Tg with wild-type mice

10. Examine the germline transmission to F1 progeny

Fig. 1. Outline of generating transgenics by rAd.

7. TYH medium: 697.6 mg/100 mL NaCl, 35.6 mg/100 mL KCl, 25.1 mg/100 mL CaCl$_2$ · 2H$_2$O, 16.1 mg/100 mL KH$_2$PO$_4$, 29.3 mg/100 mL MgSO$_4$ · 7H$_2$O, 210.6 mg/mL NaHCO$_3$, 100 mg/100 mL glucose, 5.5 mg/100 mL Na-pyruvate, 7.5 mg/100 mL penicillin G · potassium salt, 5 mg/100 mL streptomycin sulfate, 50 mg/100 mL 1% phenol red, and 400 mg/100 mL BSA.

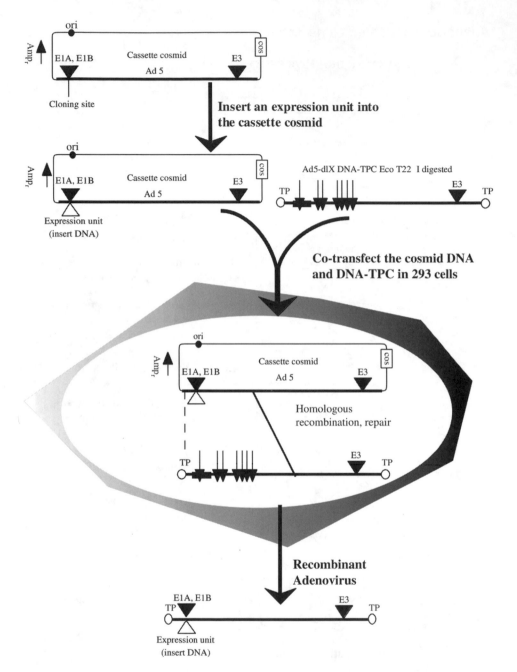

Fig. 2. Outline of generating recombinant adenovirus (COS-TPC methods). Ad5, human-Ad type 5; E1A and E1B, deletions of E1A and E1B regions in the adenovirus; open circle, Ad TP (terminal protein); filled circle, pBR322 replication origin, Amp$^r$, Ampicilline resistant; COS, COS site of λ phage; Ad5-dlX DNA-TPC, a part of adenoviral DNA-terminal protein complex; vertical dashed line, a possible second homologous recombination.

8. Whitten's medium supplemented with 0.1 mM EDTA (WM-EDTA): 514 mg/ 100 mL NaCl, 36 mg/100 mL KCl, 16 mg/100 mL $KH_2PO_4$, 29 mg/100 mL $MgSO_4 \cdot 7H_2O$, 190 mg/100 mL $NaHCO_3$, 100 mg/100 mL glucose, 242 mg/100 mL Na-lactate (or 0.37 mL of 60% syrup), 53 mg/100 mL Ca-lactate $\cdot$ $5H_2O$, 3.5 g/100 mL Na-pyruvate, 8 mg/100 mL penicillin G $\cdot$ potassium salt, 5 mg/100 mL streptomycin sulfate, 50 mg/100 mL 1% phenol red, and 300 mg/100 mL BSA.
9. 2.5% Avertin, anesthetic (Sigma).
10. Mineral oil (Squibb).
11. Mouth pipet.
12. Needle (26-gauge, 1.3-in).
13. Syringe (1 mL)
14. Transfer pipet prepared from hard glass capillary pulling it to approximately 100-μm internal diameter (for egg transfer) and 200-μm internal diameter (embryo transfer for implantation). Tip of the capillary is fire polished.
15. Tissue culture dishes.
16. Carbon dioxide (5% [v/v]) incubator set at 37°C.
17. Two stereomicroscopes.
18. Fine dissection scissors.
19. Watchmaker's forceps.
20. Blunt forceps.
21. Fiberoptic illuminator.
22. X-gal staining solution for preimplantation embryos: 0.04% 5-bromo-4-chloro-3-indolyl-β-D-galactoside (X-Gal), 1 m$M$ $MgCl_2$, 10 m$M$ potassium ferricyanide, 10 m$M$ potassium ferricyanide in phosphate buffer (pH 7.5).
23. X-gal staining solution for embryos at d 11–13.5 of gestation: 0.5 mg/mL 5-bromo-4-chloro-3-indolyl-β-D-galactoside (X-Gal), 0.05% NP-40, 1 m$M$ $MgCl_2$, 5 m$M$ potassium ferricyanide, 5 m$M$ potassium ferricyanide in phosphate buffer, pH 7.5.

## 3. Methods

### 3.1. In Vitro Fertilization in TYH Medium

1. Supcrovulate female F1 mice (C57BL/6xC3H or C57BL/6 strain) by an ip injection of 7.5 IU of PMSG followed 48 h later by an injection of 7.5 IU hCG.
2. Collect spermatozoa from the cauda epididymis of mature F1 male mice.
3. Collect unfertilized eggs from the oviducts 16 h after the hCG injection and transfer 50–150 cumulus-enclosed oocytes to 0.2 mL droplets of TYH medium *(8)* under mineral oil.
4. After 2 h of preincubation, inseminate with a small amount of sperm suspension into the TYH medium containing the oocytes. Final sperm concentration should be 150 spermatozoa/μL.
5. Incubate at 37°C in a 5% $CO_2$ incubator for fertilization.

## 3.2. Removal of Zona Pellucida in Acidic Ringer's Solution

1. Two hours after insemination, remove the cumulus cells by treatment with hyaluronidase (300 µg/mL in TYH) for 5–10 min. Ensure complete removal of the cumulus cells.
2. To remove the zona pellucida, place the cumulus-free eggs in acidic tyrode solution *(9)* for a few seconds.
3. Immediately rinse the zona-free eggs by transferring through four droplets of WM-EDTA *(10)*.
4. Place the zona-free eggs in a 100 µL droplet of WM-EDTA under mineral oil.
5. Add the virus stock (*see* **Subheading 3.3.1.** for details) solution to the droplets (each containing 40–80 eggs), and incubate at 37°C in a 5% $CO_2$ atmosphere.

## 3.3. Adenoviral Infection of the Mouse Eggs

We have tried to assess adenoviral infection of the sperm and eggs at different stages. First, we asked, can the spermatozoa be infected with an adenovirus? We could not detect transgene expression in eggs fertilized with spermatozoa that had been incubated with rAd, suggesting that Ads fail to infect the spermatozoa. Second, we tested whether the oocytes can be infected by rAd before fertilization. Although oocytes were infected by the rAd, it was difficult to generate embryos by in vitro fertilization. Most eggs showed abnormal development. Finally, we tested the zona-free, fertilized eggs for infection with an rAd. The zona-free fertilized eggs, morulae, and blastocysts could be infected with rAd at high efficiency, and normal embryos with the desired transgene could be obtained successfully.

Infection by rAd is dependent on the viral titer and the incubation time. These conditions vary with the experimental design (e.g., transient expression in early embryos or transgenics) and need to be independently determined for each viral stock (**Table 1**).

Vector design and construction affect the expression level during early development. As in the case of a foreign DNA injected into the eggs, rAd-mediated transgene expression is also dependent on the nature of the transgene. Ad-mediated gene transfer method, however, is almost 100% efficient for the delivery of a foreign DNA into the preimplantation embryos (*also see* **Subheading 4.1.**).

### 3.3.1. Adenoviral Infection of the Zona-Free Fertilized Eggs

1. For each rAd determine the viral dosage independently. Virus titer ranging between $1 \times 10^7$ and $10^8$ pfu/mL (up to 10% of the medium) should be tested.
2. In addition to virus titer, standardize the incubation time for infection. Typically, 2–6 h of incubation is sufficient. In the case of incubation time longer than 2 h, aggregates of the zona-free eggs should be dispersed by pipeting every 2 h.

# Table 1
## Effect of Viral Dose on Expression of *LacZ* Gene and Integration Efficiency[a]

| AdexSR4LacZL (viral dosage [pfu/mL]) | Zona pellucida | Survival (%) Blastocysts/eggs | X-gal-stained positive embryos (%) Total eggs |
|---|---|---|---|
| 0 | – | 103/124 (83) | 0/80 (0) |
| $1 \times 10^1$ | – | 122/143 (85) | 2/108 (2) |
| $1 \times 10^3$ | – | 147/167 (89) | 6/103 (6) |
| $1 \times 10^5$ | – | 140/167 (84) | 64/95 (67) |
| $1 \times 10^7$ | – | 93/87 (84) | 64/68 (94) |
| $1 \times 10^7$ | + | 114/136 (96) | 0/136 (0) |

| AxCANLacZ (viral dosage [pfu/mL]) | Zona pellucida | Survival (%) Blastocysts/eggs | Survival (%) Fetuses/blastocysts | Integration efficiency (%) |
|---|---|---|---|---|
| 0 | – | 102/126 (81) | 41/81 (51) | ND |
| $1 \times 10^7$ | – | 286/419 (68) | 55/273 (20) | 2/55 (4) |
| $5 \times 10^7$ | – | 208/358 (58) | 32/194 (17) | 4/32 (13) |
| $1 \times 10^8$ | – | 232/587 (39) | 36/227 (16) | 4/36 (10) |

[a]Adex4SRLacZL contains the *E. coli LacZ* gene under the control of SRa promoter (SV40 early promoter fused with HTLV-I LTR). AxCANLacZ contains a nuclear localization signal (nls) fused to the *LacZ* gene under the control of CAG promoter (cytomegalovirus IE enhancer, chicken β-actin promoter, and rabbit β-globin poly [A] signal); ND, not determined (+) with zona pellucida; (–) without zona pellucida.

3. After incubation, remove the excess virus by four washes with WM-EDTA.
4. Select the pronuclear stage eggs under a stereomicroscope. Infected-eggs reach the blastocyst stage within 96 h of culture in microdrops.
5. If the embryos attach to the culture dish, wait until the compaction stage to free them from the surface by gentle pipeting. We use a nontreated dish for embryo culture (6-cm embryo culture dish; Corning).

### 3.3.2. β-Galactosidase Assay (X-Gal Staining)

Efficiency of gene transfer can be checked in the embryos after 96 h of culture. We examined the β-galactosidase activity in the embryos by histochemical staining according to the protocol of Takeda and Toyoda *(11)* with the following modifications.

1. Rinse the embryos twice with phosphate buffer (pH 7.5) containing 4 mg/mL of polyvinylpyrrolidone (PVP-40; Sigma).
2. Fix the embryos with 0.25% glutaraldehyde in phosphate buffer containing 4 mg/mL of PVP for 10 min on ice.
3. After washing twice in phosphate buffer, transfer the embryos into X-gal staining solution (*see* **Subheading 2.**, **item 22**) and incubate overnight at 37°C.

## 3.4. Transfer of Zona-Free Embryo Into the Uterus

1. Weigh the 2.5-d pseudopregnant foster mother and then anesthetize by ip injection of avertin (0.015 mL of 2.5% avertin/g of body wt).
2. Load a transfer pipet with 5–10 expanded blastocysts.
3. Expose and hold the top of the uterus with blunt forceps and use a 26-gauge, 1.3-in. needle to make a hole in the uterine wall.
4. Insert the transfer pipet into the uterus through the hole and gently transfer the blastocysts.

## 3.5. Analysis of Adenoviral DNA and Transgene DNA

To examine the efficiency of adenoviral integration into mouse chromosome, the embryos derived from blastocysts transferred into the uterus of pseudopregnant foster mother are analyzed.

1. Collect embryos with attached placenta at 11.5 and 13.5 d of gestation and wash with PBS.
2. Test the *LacZ* expression of the embryos by X-Gal staining (*see* **Subheading 2.**, **item 23**).
3. Prepare the genomic DNA from the placentas for polymerase chain reaction and Southern blot analysis (*see* Chapter 15).
4. Use *LacZ* expression and integration efficiency to determine the optimal conditions for generating a transgenic mouse line.
5. Confirm the germline transmission by Southern blot analysis of F1 tail DNA (*see* Chapter 15).

### 3.6. Other Applications

Further applications of the approaches described here relate to tissue- and stage-specific conditional transgenesis. When analyzing the molecular mechanisms of a gene function in vivo, one of the most powerful strategies is the use of mutations that confer gain or loss of function. The Cre/*loxP* site-specific recombination system provides one experimental strategy for this purpose. This system can mediate the "on/off" switching of transgene expression in vivo, where overexpression of a given gene can, for example, cause developmental defects. Recently, our colleagues have established a Cre expressing rAd *(13)*, which is able to mediate the switching of gene expression at specific embryo stages and in specific tissues.

Alternatively, conditional transgenesis may be mediated through the use of transgene bearing rAd by direct injection of rAd into the organ of choice in vivo *(14)*. This approach has also been used to deliver tissue-specific expression of gain of function mutants, which are particularly useful tools for the investigation of the molecular basis of vertebrate development *(15,16)*.

## 4. Note

1. It is important to check on viral toxicity before starting experiments and to consider the effects of the inserted gene products during early embryo development. Suitable control rAd's such as *LacZ (1)* or *GFP (12)* should be used. The most important parameter of toxicity appears to relate to the fundamental characteristics of viruses generated at high titre. We have noted some toxicity for developing embryos using particularly high viral titres, although other groups have not reported toxic effects even where high titers of virus were used in gene therapy protocols.

## Acknowledgments

We wish to thank Dr. Izumu Saito, Dr. Yumi Kanegae, and Dr. Sanae Miyake for generating the recombinant adenoviruses and for critically reading the manuscripts.

## References

1. Tsukui, T., Kanegae, Y., Saito, I., and Toyoda, Y. (1996) Transgenesis by adenovirus-mediated gene transfer into mouse zona-free eggs. *Nat. Biotech.* **14,** 982–985.
2. Tsukui, T., Miyake, S., Azuma, S., Ichise, H., Saito, I., and Toyoda, Y. (1995) Gene transfer and expression in mouse preimplantation embryos by recombinant adenovirus vector. *Mol. Reprod. Dev.* **42,** 291–297.
3. Miyake, S., Makimura, M., Kanegae, Y., Harada, S., Sato, Y., Takamori, K., Tokuda, C., and Saito, I. (1996) Efficient generation of recombinant adenovirus DNA-terminal protein complex and a cosmid bearing the full-length virus genome. *Proc. Natl. Acad. Sci. USA* **93,** 1320–1324.

4. Kanegae, Y., Lee, G., Sato, Y., Tanaka, M., Nakai, M., Sakaki, T., Sugano, S., and Saito, I. (1995) Efficient gene activation in mammalian cells by using recombinant adenovirus expressing site-specific Cre recombinase. *Nucleic Acids Res.* **23,** 3816–3821.
5. McGrory, W. J., Bautista, D. S., and Graham, F. L. (1988) A simple technique for the rescue of early region I mutations into infectious human adenovirus type 5. *Virology* **163,** 614–617.
6. Stratford-Perricaudet, L. D., et al. (1990) Evaluation of the transfer and expression in mice of an enzyme-encoding gene using a human adenovirus vector. *Hum. Gene Ther.* **1,** 241–256.
7. Niwa, H., Yamamura, K., and Miyazaki, J. (1991) Efficient selection for high expression transfectants with a novel eukaryotic vector. *Gene* **108,** 193–200.
8. Toyoda, Y., Yokoyama, M., and Hosi, T. (1971) Studies on the fertilization of mouse eggs in vitro. *Jpn. J. Anim. Reprod.* **16,** 147–151.
9. Hogan, B., Beddington, R., Constantini, F., and Lancy, E. (1994) *Manipulating the Mouse Enbryo*, 2nd Ed., Cold Spring Laboratory Press, Cold Spring Harbor, NY.
10. Hoshi, M. and Toyoda, Y. (1985) Effect of EDTA on the preimplantation development of mouse embryos fertilized in vitro. *Jpn. J. Zootech. Sci.* **56,** 931–937.
11. Takeda, S. and Toyoda, Y. (1991) Expression of SV40-lacZ gene in mouse preimplantation embryos after pronuclear microinjection. *Mol. Reprod. Dev.* **30,** 90–94.
12. Tavares, A. T., Tsukui, T., and Izpisua-Belmonte, J. C. (2000) Evidence that members of the Cut/Cux/CDP family may be involved in AER positioning and polarizing activity during chick limb development. *Development* **127,** 5133–5144.
13. Sato, Y., Tanaka, K., et al. (1998) Enhanced and specific gene expression via tissue-specific production of Cre recombinase using adenovirus vector. *Biochem. Biophys. Res. Com.* **244,** 455–462.
14. Shibata, H., Noda, T., et al. (1997) Rapid Colorectal Adenoma Formation Initiated by Conditional Targeting of the *Apc* Gene. *Science* **278,** 120–123.
15. Capdevila, J. Tsukui, T., Izpisua-Belmonte, J. C., et al. (1999) Control of vertebrate limb outgrowth by the proximal factor Meis2 and distal antagonism of BMPs by Gremlin. *Mol. Cell* **4,** 839–849.
16. Rodriguez-Esteban, C., Tsukui, T., Izpisua-Belmonte, C., et al. (1999) The T-box genes Tbx4 and Tbx5 regulate limb outgrowth and identity. *Nature* **398,** 814–818.

# 5

# Retroviral Infection

## Richard A. Bowen

## 1. Introduction

Retrovirus vectors have been used for a variety of applications requiring gene transfer, including the production of transgenic animals. In all such cases, the transgene is delivered through an infectious particle and as part of a retrovirus genome. In almost all cases, the retrovirus genome is replication defective because its genes have been replaced with the transgene. Following infection, the viral genome containing the transgene is reverse transcribed, and the resulting double-stranded DNA copy integrates into the host cell genome as a provirus, which functions thereafter as a host cell gene.

Production of a recombinant retrovirus is conducted in two steps. First, a plasmid form of the recombinant retroviral genome is constructed by sub-cloning the transgene, usually including a promoter, into a retroviral vector. A replication-defective retroviral vector is devoid of protein-encoding sequences but does contain transcriptional control sequences in its long terminal repeats, and a sequence that, in the form of RNA, allows encapsidation into a viral particle. Retrovirus vectors usually also contain the gene for a selectable marker (e.g., neo$^R$) that typically is expressed from the retroviral promoter, and an internal promoter followed by a multiple cloning site for insertion of the gene to be expressed (**Fig. 1**).

Second, the recombinant retrovirus plasmid is transfected into "packaging cells," which are cells that have previously been transfected with and express all the retroviral proteins necessary for encapsidation of RNA and production of virions. The envelope glycoprotein expressed in a given packaging cell line dictates host range of the virus it produces. Following transfection with the recombinant retrovirus plasmid, packaging cells are selected for expression of the marker gene and clones of "producer cells" screened for those that produce the highest quantity of recombinant virus.

From: *Methods in Molecular Biology, vol. 180: Transgenesis Techniques, 2nd ed.: Principles and Protocols*
Edited by: A. R. Clarke © Humana Press Inc., Totowa, NJ

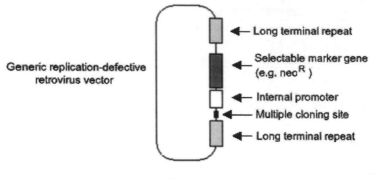

Fig. 1.

As a tool for stable gene transfer, retrovirus vectors are extremely effective, but they also have several distinct limitations. The amount of foreign DNA one can insert into a retroviral vector is limited to roughly 4 kb, and for practical purposes, it must not contain introns. More important, replication-defective retroviruses can undergo recombination with retroviral sequences in packaging cells and, at least conceptually in host cells, result in generation of replication-competent virus. With modern packaging cells and vectors, such productive recombination events are rare, but especially for certain types of transgenes (e.g., oncogenes), this potential must be fully appreciated.

## 2. Materials

1. A replication-defective retrovirus vector. Many different varieties are available, but most are derivatives of murine leukemia virus. The series of vectors constructed by Miller et al. *(1)* are widely used. These vectors incorporate within the retrovirus a marker such as the neomycin resistance gene for selection in mammalian cells and, within the plasmic backbone, an ampicillin resistance gene for selection in bacteria (*see* **Note 1**).
2. A cDNA of interest, usually ligated to a promoter appropriate to the experiment.
3. Packaging cells. The envelope glycoprotein expressed in the packaging cell dictates which cells can be infected with the resulting recombinant virus. For example, the PA317 cell line allows production of virus with amphotropic host range, which will efficiently infect cells from mice, rats, rabbits, dogs, cats, and humans *(1)*. By contrast, the PG13 cell line expresses the Gibbon ape leukemia virus envelope protein, and virus from those cells will not infect mouse cells but will infect cells from rats, hamsters, rabbits, dogs, cats, sheep, cattle, and humans *(2)*. These and other packaging cell lines are available from the American Type Culture Collection (Rockville, MD) and are grown in Dulbecco's modified Eagle's medium (DMEM) supplemented with 5–10% fetal or newborn bovine serum.

4. Reagents for subcloning and evaluating plasmid DNA, including restriction enzymes, DNA ligase, competent *Escherichia coli*, and agarose gel electrophoresis equipment.
5. Tissue culture plasticware (10- and 15-cm-diameter tissue culture dishes; 6-, 24-, and 96-well plates), $CO_2$ incubator.
6. Geneticin (G418, 100 mg of active drug/mL). Depending on the retrovirus vector used, alternatives include hygromycin B and histidinol.
7. Target cells for determination of virus titer. Many different cell types are acceptable, but they must be susceptible to infection with virus from the packaging cell used. D17 cells (a canine osteosarcoma cell line) are a good choice for virus from PA317 or PG13 cells. Growth medium for these cells is the same as for the packaging cells.
8. Polybrene (2 mg/mL in water).
9. Embryo recovery and culture supplies.
10. Micromanipulator and pipets capable of allowing injection of cells inside the zona pellucida.
11. Mitomycin C.

## 3. Methods
### 3.1. Construction of Retrovirus Vector and Producer Cells

For purposes of illustration, it is assumed here that the retrovirus will be constructed using the base vector pLNCX *(1)* and the PG13 packaging cell line *(2)*.

1. Subclone the cDNA of interest into pLNCX. In most cases, one will also replace the internal (cytomegalovirus) promoter in the retrovirus vector with a different promoter. Following confirmation of the construct, prepare purified DNA by standard methods such as Qiagen chromatography or equilibrium density centrifugation in cesium chloride.
2. Seed $10^6$ PG13 cells in a 10-cm-diameter tissue culture dish, grow overnight, and transfect with 10–20 μg of retrovirus plasmid. Transfection is typically conducted using a calcium phosphate precipitate, but other methods (lipofection, electroporation) will also work (*see* **Note 2**).
3. On the day following transfection, trypsinize the cells and seed 20 and 80% into each of two 15-cm dishes. If the transfection is good, the dish seeded with 80% of the cells will have too many colonies for easy cloning, but the other dish should be of appropriate density. Alternatively, split the transfected cells equally into 8–10 10-cm dishes.
4. Two days after transfection, initiate selection with G418 by supplementing the medium with 500 μg of active drug/mL.
5. Change the medium every 3 to 4 d, maintaining selection, until colonies reach 3 to 4 mm in diameter (10–14 d).
6. Collect cell clones. An easy means for cell cloning is to discard the medium, wash the cells once with phosphate-buffered saline (PBS), and harvest the clones

by gently "twisting" the colonies off the plate using a sterile cotton swab wet with trypsin solution. The swabs are immediately twirled into 1 mL of growth medium in wells of a 24-well plate. This method is much easier than using cloning cylinders and appears to be as efficient. A few cotton fibers are often seen in the wells, but do not appear to cause any problem.

7. When the clones approach confluence in the 24-well plates, they should be screened for virus titer, which can vary considerably among clones. Several screening tests are possible, but a standard technique is to determine the number of colony-forming units per volume of supernatant. On the day prior to screening, change the medium on the clones and seed target cells into 6-well plates (150,000 cells/well and three wells for each clone to be screened). Harvest 0.1 mL of supernatant from the clones and dilute in 0.9 mL of growth medium supplemented with polybrene (5 µg/mL final concentration), which facilitates virus adsorption. Centrifuge those samples to pellet any contaminating producer cells ($5000g$ for 5 min), and prepare the next dilution by taking from the top of the tube. From the centrifuged sample ($10^{-1}$ dilution), prepare serial dilutions of $10^{-2}$, $10^{-3}$, and $10^{-4}$, all in polybrene-supplemented medium, and inoculate 0.1 mL of these samples onto drained wells containing target cells. Allow adsorption to occur for 60 min with shaking every 15 min, and then add 1.9 mL of growth medium/well. The following day, begin selection by replacing the medium with medium containing 500 µg of G418/mL (or other appropriate drug). For each assay, include as a negative control at least one well that is not inoculated with virus. At 7–9 d after infection, drain the medium from the plates, wash once with PBS, and fix/stain cells with 0.5% crystal violet in 70% methanol for 30 min. Assess the number of colonies resulting, and expand and cryopreserve the clones that yield the most virus. A more accurate determination of titer can subsequently be acquired. The producer cells grow rapidly and will certainly need to be passed at least once or twice during the course of the screening assay. Alternatively, they can be frozen as the screening is begun.

8. Depending on the nature of the transgene and internal promoter, it is advisable to determine whether the transgene is expressed in target cells. This can be done by immunostaining, immunoblotting, or similar procedures.

## 3.2. Infection of Embryos by Coculture

The zona pellucida is impermeable to retroviruses and must be breached in order to allow infection of the embryo. One method of infection involves coculture of zona pellucida–free embryos with retrovirus producer cells. This technique has been useful for making transgenic mice (*3–5*), but is not broadly applicable, because the blastomeres from embryos of most species irreversibly dissociate when the zona pellucida is removed. Coculture also has been used to infect trophoblast of bovine blastocysts with recombinant retroviruses (*6*).

1. Seed retrovirus producer cells in multiple wells of 96-well tissue culture plates such that they will be 50–70% confluent the following day. For use in making transgenic mice, it may be best to have adapted the cells to growth in F10 medium containing 5% fetal bovine serum rather than the usual high-glucose DMEM. The medium should be supplemented with polybrene (5 µg/mL).
2. Collect two- to eight-cell embryos from superovulated mice. Remove the zonae pellucidae by brief incubation in acid Tyrode's solution, following by a rinse in medium.
3. Pipet the embryos individually into wells of the 96-well plate containing producer cells; placing more than one embryo per well often leads to their aggregating into chimeras. Incubate overnight, and then transfer the embryos to fresh mouse embryo culture medium for development to morulae.
4. Transfer the embryos to the uterine horns of pseudopregnant recipient females for development to term.

### 3.3. Infection of Embryos by Microinjection of Producer Cells

An alternative to coculture that is probably applicable to embryos from any species is to microinject producer cells inside the zona pellucida, where they continue to produce retrovirus in the immediate vicinity of the embryo. Prior to microinjection, the producer cells must be treated to prevent their continued replication; otherwise, they rapidly fill the inside of the zona pellucida, killing the embryo. The following protocol has been applied to embryos from cattle *(7)* and sheep *(see* **Note 3**):

1. Seed virus producer cells in small dishes or tissue culture flasks so that they will be 50% confluent the day before microinjection. On the afternoon before microinjection, supplement the medium of the producer cells with mitomycin C (final concentration of 10 µg/mL) and culture for 2 h. Discard the medium, and wash thoroughly to remove residual mitomycin C; drain and replace the medium at least three times at 20-min intervals, and then culture overnight and replace the medium again.
2. Collect embryos at the one- to four-cell stage.
3. Set up for microinjection using tools similar to those described for the injection of embryonic stem cells.
4. Scrape some producer cells off the dish or flask with a rubber policeman or plastic pipet tip, and place in a drop of medium under paraffin oil in a dish or slide to be used for microinjection. Pipet the embryos into the same drop.
5. Pick up an embryo with the holding pipet and aspirate a group of producer cells into the injection pipet. Inject two to five producer cells into the perivitelline space of the embryo. Repeat this procedure for all embryos.
6. Transfer the embryos into fresh medium supplemented with polybrene (5 µg/mL) and culture for 2–4 d (*see* **Note 4**).
7. Transfer the embryos to the uteri of recipient females.

### *3.4. Evaluation of Offspring*

Retroviruses integrate as single proviruses, but cells can harbor multiple proviruses and different cells may contain different numbers of proviruses at different locations. Initial characterization of the number of proviruses is conducted by standard Southern analysis, at times with tissues from different germ layers (*see* **Note 5**).

## 4. Notes

1. At a minimum, recombinant retroviruses and producer cells should be handled under Biosafety Level 2 conditions. Medium, unused cells, and contaminated plasticware should be autoclaved prior to disposal. Several other issues related to biosafety should be considered carefully. For example, if animals transgenic for a replication-defective retrovirus become infected with a wild-type retrovirus, the possibility exists that a replication-competent recombinant virus could be generated. This is probably quite unlikely if the provirus is of murine origin and the transgenic animal is not a rodent but would clearly be a concern for some types of transgenes. Several methods have been developed to determine whether producer cells are generating replication-competent or "helper" virus *(1)*.
2. An alternative method for generating producer cells, and one that seems to result in cell lines that yield higher titers of virus, is to transfect an ecotropic packaging cell line (e.g., PE501 or Psi2) with the retroviral plasmid, and then use the supernatant collected 2 to 3 d later to infect another packaging cell line (PA317 or PG13).
3. An alternative method for producing transgenic animals using retroviruses is to microinject into the perivitelline space virus that has been pseudotyped with the vesicular stomatitis virus envelope glycoprotein *(8)*. The host range of this type of virus is essentially universal among mammals. This technique also has been applied successfully to oocytes prior to fertilization, which should minimize mosaicism.
4. An unresolved problem with this technique is matching the culture medium for producer cells and embryos. The standard medium used to culture producer cells is DMEM containing 4.5 g of glucose/L and 10% serum, which is not a good medium for many types of embryos (i.e., early bovine embryos). A compromise that may be useful is F10 medium with 5% serum.
5. Transgenic animals produced by infection with retroviruses are usually mosaic, and the later the infection takes place in development, the more "severe" the mosaicism is likely to be. Animals can be mosaic for both integration site and number of proviruses.

## References

1. Miller, A. D., Miller, D. G., Garcia, J. V., and Lynch, C. M. (1993) Use of retroviral vectors for gene transfer and expression. *Methods Enzymol.* **217,** 581–599.

2. Miller, A. D., Garcia, J. V., von Suhr, N., Lynch, C. M., Wilson, C., and Eiden, M. V. (1991) Construction and properties of retrovirus packaging cells based on gibbon ape leukemia virus. *J. Virol.* **65,** 2220–2224.

3. Van der Putten, H., Botteri, F. M., Miller, A. D., Rosenfeld, M. G., Fan, H., Evans, R. M., and Verma, I. M. (1985) *Proc. Natl. Acad. Sci. USA* **85,** 6148–6152.

4. Soriano, P., Conde, R. D., Mulligan, R. C., and Jaenisch, R. (1986) Tissue-specific and ectopic expression of genes introduced into transgenic mice by retroviruses. *Science* **234,** 1409–1413.

5. Stewart, C. L., Schuetze, S., Vanek, M., and Wagner, E. F. (1987) Expression of retroviral vectors in transgenic mice. *EMBO J.* **6,** 383–388.

6. Kim, T., Leibfried-Rutledge, M. L., and First, N. L. (1993) Gene transfer in bovine blastocysts using replication-defective retroviral vectors packaged with Gibbon ape leukemia virus envelopes. *Mol. Reprod. Dev.* **35,** 105–113.

7. Haskell, R. E. and Bowen, R. A. (1995) Efficient production of transgenic cattle by retroviral infection of early embryos. *Mol. Reprod. Dev.* **40,** 386–390.

8. Chan, A. W. S., Homan, J. E., Ballou, L. U., Burns, J. C., and Bremel, R. D. (1998) Transgenic cattle produced by reverse-transcribed gene transfer into oocyte. *Proc. Natl. Acad. Sci. USA* **95,** 14,028–14,033.

# V

**TRANSGENESIS IN THE MOUSE: *THE ES CELL ROUTE***

# 6

## In Vitro Isolation of Murine Embryonic Stem Cells

### David Wells

### 1. Introduction

Embryonic stem (ES) cells are derived directly from those undifferentiated progenitor cells of early mouse embryos that have the developmental potential to subsequently form all the tissues of the fetus itself. With appropriate culture conditions, these embryonic cells can be maintained continuously in an undifferentiated state in vitro and form permanent ES cell lines *(1,2)* (*see* **Note 1**). When introduced into mouse blastocysts or aggregated with morulae, however, the ES cells are capable of responding to in vivo developmental signals and participate in normal embryogenesis. The stem cells differentiate and contribute to all the tissues of the fetus *(3)*, and occasionally to the trophectodermal and primitive endodermal lineages of the extraembryonic membranes *(4)*, leading to the formation of chimeric offspring. The full demonstration of their pluripotent potential is seen in the capacity of ES cells also to colonize the germline in chimeras, and form fully functional gametes *(3)*. More remarkably, some ES cell lines are capable of supporting complete fetal development, following aggregation with tetraploid embryos, and generate fertile adult mice that are entirely ES cell derived *(5)*. One of the major applications of ES cells is in providing a powerful approach for the introduction of novel genetic change into the mouse genome. While the ES cells are in culture, specific genes of interest can be modified in a desired manner with homologous recombination (gene targeting) technology, and by utilizing their pluripotent potential, the ES cells can be used as carriers of this genetic change through the germline and into subsequent generations of mice *(6)* (*see* Chapters 7 and 8).

In the mouse, ES cells are not the only pluripotent cell type that can be maintained in culture. Another source of cells that possess a similar developmental potential are the primordial germ cells, which under specific culture

From: *Methods in Molecular Biology, vol. 180: Transgenesis Techniques, 2nd ed.: Principles and Protocols*
Edited by: A. R. Clarke © Humana Press Inc., Totowa, NJ

conditions can be established as mouse embryonic germ (EG) cells, so called to identify their origin *(7,8)* (*see* **Note 1**). EG cells demonstrate similarities to ES cells in their ability to form germline chimeric mice, but subtle distinctions in terms of methylation imprinting patterns *(9,10)*.

The purpose of this chapter is to describe the materials and methods required for the isolation of murine ES cells and, to a lesser extent, EG cells. For details on the methods and progress toward the isolation of ES cells from farm animal species, *see* Chapters 1 and 14. The subsequent sections describe the tissue culture facility required, the media and standard culture procedures used, the production of mouse embryos from different developmental stages, the procedures used for the isolation of ES cells from each of these different embryonic stages, and routine in vitro ES cell culture techniques.

## 2. Materials

### 2.1. Tissue Culture Facility

The isolation and maintenance of stem cells should ideally be conducted in a laboratory established solely for sterile ES cell culture. Standard sterile tissue culture procedures should be employed throughout. Personnel entry should be restricted, and no mice should be handled in the tissue culture laboratory. The majority of tissue culture is performed within the vertical-flow cabinet. However, in the course of ES cell isolation, it is necessary to operate in the dissecting microscope. With care, microbial contamination can be kept to a minimum. Testing for mycoplasma-infected cells should be periodically undertaken. The basic equipment in this laboratory should include the following:

1. Vertical-flow tissue culture cabinet.
2. Humidified, 5% $CO_2$ (in air) incubators set at 37°C.
3. Benchtop centrifuge.
4. Water bath.
5. Inverted, phase-contrast microscope (magnification: ×40, ×100, and ×200).
6. Binocular, dissecting microscope with transmitted illumination (magnification: ×12–100) within a small still-air or laminar flow cabinet.
7. One gas Bunsen burner, with a pilot flame, within each culture cabinet.
8. Pipets, either glass or plastic disposables (e.g., Sterilin, Staffordshire, UK, or Costar, Cambridge, MA; pipet sizes: 1, 5, 10, and 25 mL).
9. Handheld pipet-aid.
10. Vacuum pump, with two waste traps, to aspirate spent medium through a tube connected to disposable, heat-sterilized, long-form glass Pasteur pipets.
11. Glassware designated specifically for stem cell culture. This glassware should be washed without any detergents. It should be soaked for 1 to 2 d in a bucket of double-distilled water and rinsed several times in ultrapurified water (e.g., Milli-Q Water Purification System; Millipore, Bedford, MA). After draining, general

glassware is heat sterilized (at 180°C for 2 h) while bottles with screw-on tops are autoclaved (at 101 kPa for 20 min).

12. Cryopreservation tubes (1 mL) (Nunclon, Roskilde, Denmark).

13. Tissue culture plasticware (*see* **Subheading 3.** for details).

14. Mouth-controlled, hand-pulled, plugged Pasteur pipets for transfer of embryos between culture dishes. These are constructed by rotating the pipet over a Bunsen burner, first to soften the glass. After withdrawal from the flame, the pipet is quickly pulled to produce a fine capillary, approx 6 cm long and with an internal diameter of 150–200 μm. The end of the capillary is broken cleanly to produce a square tip by scoring the glass with another finely pulled capillary. (Similar principles are used to produce finer pipets to manipulate ES cell colonies during isolation procedures, as outlined in **Subheading 3.3.**). A modified Pasteur pipet bulb may be utilized as an adapter to connect the end of the pulled pipet to a length of rubber tubing with a fitted mouthpiece. Fine control over the movement of embryos within the pipet is aided by first aspirating several minute volumes of medium interspaced with air bubbles.

## *2.2. Tissue Culture Reagents and Solutions*

1. All tissue culture media described here are based on a formulation of Dulbecco's modified Eagle's medium (DMEM) without sodium pyruvate but high in glucose (4500 mg/mL). This can be purchased in either liquid form (1 or 10X; dilute appropriately) or powdered form from tissue culture suppliers (e.g., Flow Laboratories, McLean, VA; Gibco; Life Technologies, Grand Island, NY). An alternative medium formulation favored by some researchers is Glasgow's MEM *(11)*.

2. It is essential that the water used for preparing tissue culture media and stock solutions be of the highest quality. The Millipore Milli-Q Water Purification System is satisfactory.

3. Serum: Both fetal calf serum (FCS) and newborn calf serum (NCS) are required. ES cells are very sensitive to the type and quality of the serum used to supplement the base DMEM media, and this is especially true for the isolation of new lines. Therefore, it is essential to select the most suitable batch of serum by testing several samples provided by suppliers for their respective plating efficiencies, with an established ES or feeder-dependent embryonal carcinoma cell line. A known number of ES cells are seeded onto replica plates with feeder cell layers (*see* **Subheading 3.1.2.**) in medium containing either 10 or 30% serum, from each test batch. Typically, 1000 ES cells may be seeded onto a 6-cm culture dish. The plates are then incubated for 7 d and the stem cell colonies stained and counted. After washing once with phosphate-buffered saline (PBS) (*see* **item 13**), the plates are stained with a 2% (w/v) solution of methylene blue for 2–5 min. Next, the plates are rinsed with water and allowed to dry. The average plating efficiency can then be calculated for each serum. The batch giving the highest plating efficiency, with no toxicity at 30% serum concentration, is then purchased in bulk order and stored at –20°C for several years. A good-quality FCS batch should give a plating efficiency of between 20 and 30%. It is generally not neces-

sary to heat-inactivate bovine serum used for mouse ES cell culture; however, with some batches used at high concentrations, serum toxicity may be observed owing to high levels of complement. Heat treatment at 56°C for 30 min to inactivate complement may markedly decrease toxicity.

4. 200 m$M$ L-Glutamine: This is added to medium to 1% (v/v), since this amino acid is unstable.

5. ES cell culture medium: This is supplemented with 1% (v/v) of a 100X stock solution of MEM nonessential amino acids (NEAAs) (e.g., Flow Laboratories). The final concentrations of the added NEAA are as follows: 0.1 m$M$ each of glycine, L-alanine, L-asparagine, L-aspartic acid, L-glutamic acid; 0.2 m$M$ each of L-proline, L-serine.

6. 100X β-Mercaptoethanol stock (10 m$M$). Dissolve 7 μL of the standard 14 $M$ stock solution of β-mercaptoethanol (Sigma, St. Louis, MO) in 10 mL of PBS (*see* **item 13**) and add freshly made to ES cell culture medium to a final concentration of 0.1 m$M$ to enhance stem cell attachment and growth *(12)*.

7. Antibiotics: These are sometimes used to a final concentration of 50 IU/mL of penicillin and 50 μg/mL of streptomycin (*see* **Subheading 3.2.** and **Note 2**).

8. HEPES buffer: This is used at a final concentration of 20 m$M$ to maintain the pH of the medium during manipulations outside the incubator for extended periods.

9. Buffalo rat liver (BRL) conditioned medium (optional; *see* **Subheading 3.5.1.** and **Note 3**).

10. LIF (*see* **Subheading 3.5.1.** and **Note 3**): Murine or human recombinant forms of leukemia inhibitory factor (LIF) (both are equally effective) may be prepared by transfecting the appropriate expression plasmids into either COS-7 cells (transformed African green monkey fibroblasts; *13*), yeast cells, or *Escherichia coli (14)*. In the situation in which LIF is not purified from the supernatant, a titration assay is necessary to determine the optimal volume of crude LIF supernatant to add to ES$_{10}$ medium (*see* **Subheading 2.3.**) in order to prevent ES cell differentiation. Recombinant forms of LIF are also commercially available (e.g., ESGRO, Gibco, Life Technologies).

11. Mitomycin C (Sigma): Appropriate precautions must be taken when handling this potential carcinogen. The stock solution is prepared by dissolving a 2-mg vial of mitomycin C in 10 mL of PBS (i.e., 200 μg/mL [20X stock]), which is stored at 4°C for no longer than 1 mo. Aliquots may be frozen at –20°C for longer-term storage. Mitomycin C is used at a final concentration of 10 μg/mL.

12. Trypsin/EGTA solution (TEG): To 1 L of sterile water add 7.0 g of NaCl, 0.3 g of Na$_2$HPO$_4$·12H$_2$O, 0.24 g of KH$_2$PO$_4$, 0.37 g of KCl, 1.0 g of D-glucose, 3.0 g of Tris-(2-amino-2-[hydroxymethyl] propane-1,3-diol), and 1.0 mL of phenol red. Dissolve ingredients, discard 100 mL of the salts solution, and then add 100 mL of 2.5% (10X) trypsin in modified Hank's balanced salt solution (Flow Laboratories), 0.4 g of EGTA, and 0.1 g of polyvinyl alcohol. Adjust TEG to pH 7.6; filter sterilize (0.22 μm); aliquot into 20-mL sterile, plastic universals; and store at –20°C. TEG is utilized for the subculture of both fibroblast and ES cell lines and for the disaggregation of embryonic outgrowths and putative stem cell colonies

in the process of ES cell isolation. The powdered ingredients should be of tissue culture purity and kept solely for ES cell culture.

13. PBS: $Ca^{2+}$- and $Mg^{2+}$-free PBS may be prepared either by dissolving preformulated tablets (e.g., Flow Laboratories) or by adding the following ingredients to 1 L of sterile water: 10.0 g of NaCl, 0.25 g of KCl, 1.44 g of $Na_2HPO_4 \cdot 12H_2O$, and 0.25 g of $KH_2PO_4$. Adjust PBS to pH 7.2. Aliquot into 500-mL bottles, sterilize by autoclaving, and store PBS at room temperature.

14. Gelatin (from porcine skin; Sigma): This is made as a 0.1% (w/v) solution in sterile water and dissolved and sterilized by autoclaving.

15. Pancreatin-Trypsin solution: $Ca^{2+}$- and $Mg^{2+}$-free PBS containing 2.5% (w/v) pancreatin (porcine; Sigma), 0.5% (w/v) trypsin (porcine 1:250; Sigma), and 0.5% (w/v) polyvinylpyrrolidone (PVP) ($M_r = 10,000$) further supplemented with 0.1% (w/v) D-glucose, 1% (v/v) of a 100X stock solution of NEAAs (Flow Laboratories), and 20 m$M$ HEPES buffer (pH 7.4) *(15)*.

## 2.3. ES Cell Culture Media

Two media formulations are used for ES cell culture. For ES cell isolation, medium contains 20% (v/v) serum ($ES_{20}$), whereas established ES cell lines may be maintained in medium with only 10% (v/v) serum ($ES_{10}$) and supplemented with recombinant LIF (*see* **Subheading 3.5.1.** and **Note 3**). In both instances, serum comprises a 50:50 mixture of selected batches of FCS and NCS (*see* **Subheading 2.2.**, **item 3**).

1. Medium for murine ES cell isolation ($ES_{20}$ medium): 64 mL of DMEM (1X DMEM), 10 mL of FCS, 10 mL of NCS, 1 mL of 200 m$M$ L-glutamine, 1 mL of 100X NEAAs stock (*see* **Subheading 2.2.**, **step 5**), 1 mL of 10 m$M$ β-mercaptoethanol stock, 13 mL of sterile water; osmolarity: about 290 mosM/kg of $H_2O$ (*see* **Note 4**).

2. Medium for murine ES cell maintenance ($ES_{10}$ medium): 74 mL of DMEM (1X DMEM), 5 mL of FCS, 5 mL of NCS, 1 mL of 200 m$M$ L-glutamine, 1 mL of 100X NEAAs stock, 1 mL of 10 m$M$ β-mercaptoethanol stock, 13 mL of sterile water; osmolarity: about 290 mosM/kg of $H_2O$ (*see* **Note 4**). To $ES_{10}$ medium, BRL-conditioned medium or recombinant LIF may be added (*see* **Subheading 3.5.1.** and **Note 3**).

All media are filter sterilized after preparation. For economy, it is possible to utilize resterilizable, 47-mm filter holders (Millipore) housing a glass-fiber prefilter overlying a 0.22-μm nitrocellulose filter. The medium is pushed through the filter with a 60- or 100-mL syringe. Alternatively, disposable Vacucap® filter units (Gelman, Ann Arbor, MI) are convenient, simply requiring a small vacuum pump to draw media through the unit. Media are stored at 4°C and warmed to 37°C immediately before use. Medium that is older than 2 wk should be supplemented with 1% (v/v) of 200 m$M$ L-glutamine. Note that no antibiotics are used for ES cell isolation or culture (*see* **Note 2**).

## 2.4. STO Cell Culture Media

DMEM$_{10}$ medium is used for the routine culture of the STO fibroblast cell line (and also primary mouse fibroblast cells), used to prepare feeder cells for coculture with ES cells. The medium for STO fibroblast culture (DMEM$_{10}$ medium) contains the following: 89 mL of DMEM (1X DMEM), 5 mL of FCS, 5 mL of NCS, 1 mL of 200 m$M$ L-glutamine. Media are filter sterilized and stored as described in **Subheading 2.3.**

## 2.5. Animals

Although ES cells have been isolated from a variety of inbred, outbred, and F$_1$ crosses, there does appear to be an effect of mouse genotype on the ease with which ES cells can be isolated *(16)*, with embryos from the 129 mouse strain yielding stem cells at significantly higher frequencies than other mouse genotypes *(17)* (*see* **Note 5**).

To produce embryos, a supply of 6- to 8-wk-old virgin females is required for superovulation. Following hormone treatment, these are mated with individually caged stud males. Mice are usually housed in a light cycle of 14 h light (4:00 AM to 6:00 PM) and 10 h dark, or a 12-h light (6:00 AM to 6:00 PM) and 12-h dark cycle. Mice are fed and watered ad libitum.

### 2.5.1. Superovulation

The hormones used to superovulate mice are pregnant mare's serum gonadotropin (PMSG) (Folligon; Intervet, Boxmeer, The Netherlands) to increase follicle production and human chorionic gonadotropin (hCG) (Chorulon; Intervet) to induce ovulation. Both PMSG and hCG are prepared by dissolving the lyophilized powders in sterile 0.9% (v/v) NaCl, to give a final concentration of 50 IU/mL. The hormones are then aliquoted, stored at –20°C, and replaced after 2 mo.

### 2.5.2. Anesthesia

Anesthetic is prepared by mixing separately 1 mL of Hypnorm (Janssen, Beerse, Belgium; containing 0.315 mg of fentanyl citrate and 10 mg of fluanisone) and 1 mL of Hypnovel (Roche, Basel, Switzerland; containing 5 mg of midazolam hydrochloride) each with 2 mL of distilled water, before combining the two solutions. The anesthetic is stored at room temperature and replaced after 4 wk. An alternative anesthetic may be used.

### 2.5.3. Additional Materials and Reagents

1. Standard surgical equipment and dissecting tools.
2. Cotton suture (5/0 Mersilk; Ethicon, Somerville, NJ).
3. Michel clips.
4. Progesterone solution (10 mg/mL dissolved in absolute ethanol or corn oil).

## 2.6. Karyotype Analysis

1. 0.56% (w/v) KCl.
2. Fixative: 3 vol of absolute methanol to 1 vol of glacial acetic acid, made fresh each time.
3. 2X Standard saline sodium citrate (SSC): 0.3 $M$ NaCl, 0.03 $M$ trisodium citrate.
4. Gurr's phosphate buffer, pH 6.8 (e.g., BDH, Poole, UK).
5. 0.25% (w/v) Trypsin (porcine; Difco, Detroit, MI) dissolved in Gurr's phosphate buffer.
6. 5% (v/v) Giemsa Gurr's R-66 stain (e.g., BDH) in Gurr's phosphate buffer.

## 3. Methods
### 3.1. Preparation of Feeder Cells

Historically, ES cells have been isolated and maintained on layers of mitotically inactivated, embryonic fibroblast "feeder" cells. If cultured in the absence of feeders, and without the presence of cytokine LIF, the ES cells rapidly differentiate *(1)* (*see* **Note 6**). The most commonly utilized fibroblast feeder layers have been those prepared from the continuous STO cell line *(18)*. However, there are a number of sublines in use in various laboratories, which may account for variable performance, especially in the isolation of ES cells. Other embryonic fibroblast cell lines also capable of maintaining stem cells include C3H 10T1/2 cells *(19)* and BALB-3T3/A31 cells *(20)*. Alternatively, primary mouse embryonic fibroblasts (PMEFs) may be isolated from fetuses in the third trimester of pregnancy. Although such feeders have been favored by some investigators *(21,22)*, a disadvantage in their use is the limited in vitro life span of these primary cells *(23)*. However, this inconvenience is outweighed by evidence that chromosome stability in ES cell lines may be greater on PMEFs than on STOs *(22)*. Certainly, it is recommended to isolate ES cell lines on feeder cell layers, since those isolated on gelatin-coated dishes with only soluble LIF possessed substantial populations of near-tetraploid cells *(24)*, indicating significant stress during the establishment phase in the absence of feeders.

To prepare feeder cell layers for coculture with embryos or ES cells, the fibroblast cells must be mitotically inactivated to prevent them from overgrowing the culture. Typically, this is achieved by treatment with the drug mitomycin C *(18)* or by exposure to irradiation *(25)*.

### 3.1.1. Routine Culture of STO Fibroblast Cells

1. Typically, the STO fibroblast line *(18)* is cultured in DMEM$_{10}$ medium in tissue culture plasticware. Flasks or dishes may be used with a range of growth area dimensions depending on the requirement for cells. For routine maintenance, flasks with a 25-cm$^2$ growth area surface may be used (e.g., Costar, Nunclon, Falcon).

2. Once confluent, passage or subculture the fibroblasts into fresh flasks. Aspirate the old medium from the flask, and wash the cells once with 10 mL of PBS.
3. After the PBS is aspirated, add 2 mL of TEG, and incubate the flask at 37°C for 3 min. After swirling the flask to obtain a single-cell suspension (checked in the inverted microscope), neutralize the trypsin solution by adding 3 mL of $DMEM_{10}$.
4. Typically, for routine culture, add a 1/10 aliquot (i.e., 0.5 mL) of this suspension to 10 mL of fresh $DMEM_{10}$ in a new flask. Under such a subculture regime, flasks may reach confluency after 4 to 5 d. The subculture ratio and the number and growth area of the flasks seeded may be varied to suit the demand for STOs.
5. Transfer the flasks into a $CO_2$ incubator with either their caps loosened or tightened after being gassed while in the flow cabinet via a sterile, plugged Pasteur pipet connected to a cylinder containing 5% $CO_2$, 20% $O_2$, and 75% $N_2$.

### 3.1.2. Preparation of STO Fibroblast Feeder Cell Layers

Mitotically inactivated STO feeder cell layers are prepared using the following protocol.

1. Aspirate the old medium from a subconfluent flask of STOs, and for a 25-$cm^2$ flask, replace with 9.5 mL of $DMEM_{10}$ and 0.5 mL of stock mitomycin C (final concentration: 10 μg/mL). Then incubate the STO flasks for 2 to 3 h.
2. After treatment, aspirate the mitomycin C medium, and wash the STO cells three times with 10 mL of PBS, before trypsinizing with 2 mL of TEG. After 3 min of incubation at 37°C, dissociate the STOs into single cells, and neutralize the TEG with 8 mL of $DMEM_{10}$.
3. Transfer the total 10-mL suspension into a centrifuge tube. Draw a minute volume of the suspension into a Pasteur pipet and transfer to a hemocytometer, in order to count the number of cells.
4. Centrifuge the STO cell suspension at 100 *g* for 5 min, and aspirate the supernatant. Then disrupt the cell pellet by carefully flicking the tube, and resuspend the cells in $DMEM_{10}$ medium, to one of three alternative densities (see below).
5. For convenience during the isolation of an ES cell line, feeder cell layers may be prepared on different sized surface areas: e.g., either in microdrops under toxicity-tested, lightweight paraffin oil; in 1.75-$cm^2$ four-well plates (Nunclon); or in 25-$cm^2$ flasks. To enhance STO feeder cell attachment, pretreat all culture surfaces with 0.1% gelatin. Microdrops may be prepared by placing two rows of five 10-μL drops of gelatin on 6-cm-diameter tissue culture dishes, overlayered with 5 mL of paraffin oil. For wells and flasks, add just enough gelatin so as to cover the culture surface. Store the gelatin for about 1 h in the tissue culture cabinet, before aspiration and replacement with the freshly prepared, inactivated STO cell suspension.
6. Plate the STOs at a density to ensure a uniform, confluent monolayer of cells. For microdrops, introduce 20 μL of a suspension containing $3 \times 10^5$ STO cells/mL. For each (1.75-$cm^2$) well of a four-well plate introduce 1 mL of a suspension

1. Anesthetize mice with a 0.15- to 0.20-mL ip injection of a Hypnorm and Hypnovel mixture (*see* **Subheading 2.5.2.**).
2. Once unconscious, wipe the back of the mouse with 70% alcohol. Use a pair of blunt forceps to pick up the skin and dressing scissors to make a 10-mm lateral incision across the midline, posterior to the last rib. Use paper tissues moistened with 70% alcohol to open the wound and remove any cut hair by wiping in a head-to-tail direction.
3. Sliding the skin from side to side, locate the position of the right-hand ovary beneath the body wall. Grasp the region directly overlying the ovary with pointed watchmaker's forceps, and make a 5-mm incision in the body wall with sharp iris scissors. By exerting gentle pressure on both sides of the incision, exteriorize the ovarian fat pad and grasp with blunt forceps.
4. Very carefully, pull out the associated ovary, oviduct, and the top of the uterine horn. The weight of the fat pad is generally sufficient to keep the ovary from slipping back inside the abdominal cavity, provided a small incision is made.
5. Then transfer the mouse, on absorbent tissues, to the stage of a binocular dissecting microscope with a fiberoptic incident light source. Focusing on the ovary at low magnification (about ×10), utilize two watchmaker's forceps to tear open the ovarian bursa delicately, encapsulating the ovary and the oviduct.
6. With the ovary released, slip a small, pretied loop (about 5 mm diameter) of fine suture cotton (5/0 Mersilk; Ethicon) between the ovary and the oviduct, and pull tight with forceps to ligate the ovarian blood vessels. Then cut away all the ovarian tissue using iris scissors.
7. Re-place the fat pad and oviduct inside the abdominal cavity by lifting up one edge of the incision and pushing in the fat pad with blunt forceps. Then repeat the ovariectomy on the other side. Close the incisions in the body wall with single sutures (5/0 Mersilk; Ethicon), and close the skin with either sutures or one or two Michel clips.
8. While the female is still unconscious, inject subcutaneously 0.1 mL of 10 mg/mL of progesterone into the flank region. To aid postoperative recovery, place the mouse onto a heated blanket.
9. Recover the delayed blastocysts from the uterine horns as outlined in **Subheading 3.2.3.** by flushing between d 6.5 and 8.5 p.c. (i.e., 4–6 d after ovariectomy or 3–5 d after blastocyst formation).

### 3.2.5. Recovery of Egg Cylinder–Stage Embryos on d 5.5 p.c.

1. Sacrifice females by cervical dislocation, and swab the belly region with 70% alcohol. Then open the abdominal cavity to reveal the reproductive tract as in **Subheading 3.2.2.** In turn, cut each uterine horn just below the uterotubal junction.
2. Grasping the end firmly with forceps, pull the uterus taut, and use scissors to tear away the mesometrium. Next, open the uterine horn by sliding the tip of a pair of iris scissors down along the antimesometrial wall of the uterus, moving toward the cervix. Keeping the uterus pulled taut, use forceps to "shell" the decidua (containing the embryos) out of the uterus.

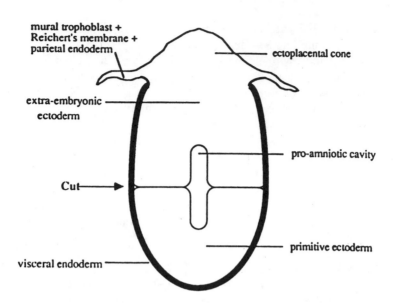

Fig. 1. Representative illustration of a d 5.5 p.c. egg cylinder–stage mouse embryo following dissection from the deciduum, showing the position of the lateral cut made in order to isolate the embryonic portion of the embryo.

3. Place the decidua into a dish containing $DMEM_{10}$ + antibiotics (50 IU/mL of penicillin and 50 µg/mL of streptomycin) and 20 m$M$ HEPES buffer, added to maintain the pH during subsequent manipulations outside the incubator.

4. In a dissecting microscope, delicately tease apart each deciduum with flame-sterilized watchmaker's forceps, to dissect out the early egg cylinder–stage embryo, easily identified as a dark speck from the trophoblast tissue. A representative illustration of the dissected embryo is shown in **Fig. 1**. The trophoblast and parietal endoderm tissues are torn away from the implantation site during the dissection procedure and, hence, appear ragged. A proamniotic cavity has often developed by this stage, and a clear division between the primitive ectoderm and the extraembryonic ectoderm (of trophectodermal origin) is often observed.

5. Isolate pure primitive ectoderm tissue by first cutting away the extraembryonic regions of the egg cylinder utilizing a fine-glass microneedle constructed from a hand-pulled Pasteur pipet, or a scalpel blade ("cut" in **Fig. 1**). Then, remove the overlying layer of visceral endoderm with an enzymatic pretreatment. After rinsing in PBS, incubate the tissue for 5 min in a 250-µL drop of a pancreatin-trypsin solution at 4°C.

6. Flood the dish with $DMEM_{10}$ to neutralize the crude proteases, and transfer the tissue to a fresh droplet of $ES_{20}$ medium under paraffin oil and incubate for 1 h at 37°C.

7. Following this recovery period, cleanly separate the endoderm layer from the primitive ectoderm by repeatedly aspirating the tissue gently up and down inside a fine, flame-polished, mouth-controlled Pasteur pipet.

8. This treatment generally yields relatively pure clumps of ectoderm, which may then be either explanted intact into tissue culture for ES cell isolation or first disaggregated into smaller pieces, comprising several cells each. To accomplish this, incubate the ectoderm in TEG for 3–5 min at 37°C followed by mechanical dissociation, using a fine pipet.

## 3.3. Isolation of Murine ES Cells

This section describes the typical procedures utilized to isolate ES cells from three different embryonic stages: from blastomeres of disaggregated morulae, from the inner cell mass (ICM) of delayed and nondelayed blastocyst-stage mouse embryos, and from d 5.5 p.c. primitive ectoderm. Emphasis is placed on the isolation of ES cells from blastocysts because this remains the most successful means of establishing new ES cell lines. However, apart from the timing of the disaggregation of the embryo outgrowth, the isolation procedures used are essentially identical for each of these developmental stages. ES cells are morphologically identified as small, rounded cells with a large nuclear-to-cytoplasmic ratio, containing one or more prominent nucleoli and no overtly specialized cellular structures, i.e., characteristics of a typically undifferentiated cell phenotype. For comments on the effect of mouse strain on ES cell isolation, *see* **Note 5**.

After embryo recovery, the blastomeres, blastocysts, or primitive ectoderm tissue are cocultured on STO feeder cell layers prepared in microdrops (*see* **Subheading 3.1.2.**). $ES_{20}$ medium is routinely used for ES cell isolation (*see* **Note 7**). The medium is changed on these microdrops 2–3 h before the embryos are introduced. Using a sterile, hand-pulled Pasteur pipet, embryos or embryonic cells are transferred to each microdrop. Microdrops allow the progress of each embryo to be monitored individually. The additional advantages of using microdrops are that any ES cell colonies arising from a single embryo can be pooled together, without the necessity of subcloning, as may be required for group cultures, and that the spread of any infection is minimized.

### 3.3.1. Isolation of ES Cells from Blastocyst-Stage Embryos

Two general strategies are used to divert the ICMs of blastocysts from their normal fate of differentiation and, instead, to encourage their continued proliferation in vitro in order to isolate ES cells. The method described here is based on that originally reported by Evans and Kaufman *(1)*, and subsequently detailed by Robertson *(23)*. Briefly, the technique involves the culture of intact blastocysts for a limited time until they grow to resemble egg cylinder–like structures. The embryonic portion of this outgrowth is then enzymatically and mechanically manipulated, and subsequently propagated in appropriate culture conditions. The second strategy, described by Martin *(2)*, involves the culture of ICMs following the immunosurgical destruction of the trophecto-derm from d 3.5 p.c. blastocysts *(27)*. A proportion of such ICMs was reported to give rise directly to colonies of stem cells, apparently without the necessity of any further specific manipulation *(2)*. Although it may be more efficient to isolate ES cells following immunosurgery compared to simply explanting blas-tocysts intact into culture (12 vs 6%) *(28)*, the latter procedure is more com-monly adopted because of its greater ease and general convenience.

### 3.3.1.1. CULTURE OF INTACT BLASTOCYSTS

Intact d 3.5 p.c. blastocysts explanted into microdrops hatch from the zona pellucida and attach to the STO feeder cells after 24–36 h by the outgrowth of the trophectodermal cells (**Fig. 2A**). These cells spread out and differentiate to form a monolayer of large, flat trophoblast cells. As a consequence of this disorganization, the ICM becomes exposed to the culture environment and appears initially as a small nest of cells in the center of the trophoblast out-growth after 2 to 3 d (**Fig. 2B**).

After embryo attachment, the medium on the microdrops should be replaced daily. The ICM cells continue to proliferate, and the outgrowths are allowed to grow to a suitably large size (determined by experience; *see* **Fig. 2C**), but with-out signs of endoderm differentiation, before they are disaggregated. Such out-growths are generally obtained after 4 to 5 d of culture with d 3.5 p.c. blastocysts. Delayed blastocysts, however, generally require a 6-d culture interval before they reach a suitable size because of their slower rates of in vitro growth initially. Variability between embryos exists in the rate of ICM growth, and, thus, each embryo must be monitored and assessed daily in the inverted microscope. If cultured without intervention, cells of the embryonic outgrowth differentiate (**Fig. 2D**) and have a much reduced capacity to yield ES cells. In nondelayed blastocysts, this typically occurs beyond d 5.

### 3.3.1.2. DISAGGREGATION OF ICM OUTGROWTHS

1. Two to 3 h before ICM outgrowths are to be disaggregated, the cultures should be refed with fresh medium. Perform the disaggregation procedure by transfer-ring the culture dish to the dissecting microscope, and with a blunt glass probe, prepared from a flame-polished, hand-pulled Pasteur pipet, "pick off" the ICM from the surrounding layer of trophoblast.
2. With a mouth-controlled, drawn-out Pasteur pipet, then transfer the ICM to a 20-µL wash-drop of TEG on a 6-cm dish (without oil) before placing into a fresh drop of TEG and incubate at 37°C for 5 min.
3. After returning the dish to the dissecting microscope, use a wide-bore pipet to transfer the loose ICM clump into a fresh STO feeder cell microdrop, containing $ES_{20}$ medium. Then use a finely pulled Pasteur pipet with a tip diameter approxi-

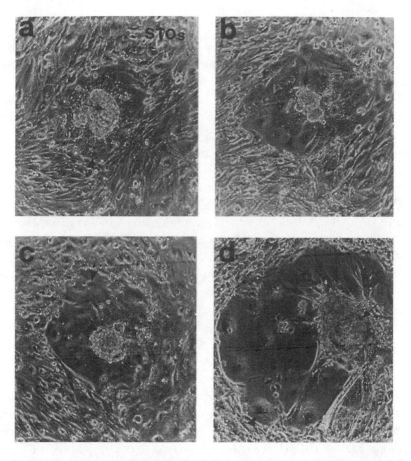

Fig. 2. Attachment and outgrowth of d 3.5 p.c. blastocyst-stage mouse embryos on feeder cell layers. (**a,b**) ICM is indicated by arrows and the trophoblast by arrowheads. Embryos hatch from the zona pellucida and attach to the STO feeder cells via the trophectoderm after 24–36 h of culture (A). In the process of attachment the blastocysts collapse, sequestering the ICM (A). Subsequently, the trophectoderm differentiates into giant trophoblast cells (A), which grow out radially to expose eventually the ICM after 2 or 3 d (B). The ICM outgrowth proliferates as a three-dimensional mass of cells and is disaggregated once it reaches a size similar to that in (**c**) after a total of 4 to 5 d in vitro. (**d**) ICM outgrowths should be disaggregated before they begin to differentiate into endoderm. Photographs were taken using phase-contrast microscopy.

mately one-fifth the size of the ICM clump to break up mechanically the partially digested outgrowth into several small pieces, each comprising about 15 cells, plus some single cells. This disaggregation is termed *passage one*.

4. Approximately 3–6 h after disaggregation, change the medium on the culture to remove any dead cells, as well as any residual traces of the trypsin solution.

### 3.3.1.3. IDENTIFICATION OF ES CELLS

The majority of colonies arising from the disaggregated outgrowths of blastomere, ICM, or primitive ectoderm origin will be of a differentiated morphology. Trophoblast and endodermal colonies are generally the most common (**Fig. 3A**); however, patches of fibroblast-like, neuronal (and with extended culture), beating muscle may also be observed. Undifferentiated cells possessing an ES-like morphology (so-called at this stage because many of the colonies initially displaying this cellular morphology may differentiate by the second or third passage) are identified as small, rounded cells with a large nuclear-to-cytoplasmic ratio and containing one or more prominent nucleoli (**Fig. 3B**). On the STO feeders, the ES cells tend to grow in tightly packed, three-dimensional colonies, in which clear intercellular boundaries are difficult to distinguish (**Fig. 3C**). At this stage, ES cells have to be identified on the basis of their cell morphology. This skill only comes with experience. A number of stem cell–specific (or pluripotent cell–specific) markers can be used to verify the ES cell phenotype and distinguish these cells from their differentiated derivatives. These cell markers include both immunohistologic and biochemical types. For example, high-alkaline phosphatase enzyme activities are observed in ES cells *(25)*. ES cells express the SSEA-1 antigen *(29)* and bind the monoclonal antibody (MAb) ECMA-7 *(30)* on the cell surface. Pluripotent stem cells fail to express cytoskeletal markers indicative of differentiated cell types, such as identified by the MAb TROMA-1 *(30)*. Extensive immunohistologic studies have gone some way toward characterizing the carbohydrate antigens present specifically on the surface of ES cells, or their differentiated cell derivatives *(31)*. ES cells also express the transcription factor Oct4 *(32)*.

The first passage in culture is used to screen out all the differentiated colonies and select only those displaying a stable ES-like cell morphology. This is achieved by employing a relatively long culture interval of 5–8 d and feeding the cultures initially every other day, but more frequently whenever the medium becomes acidic.

### 3.3.1.4. EXPANSION OF ES CELLS INTO PERMANENT CELL LINES

The first-passage cultures are monitored daily and disaggregated once ES-like cell colonies attain a "suitably large size," without having commenced differentiation (e.g., **Fig. 3C**). Because not all the colonies derived from each embryo will be of a similar size, the approach to disaggregation at the second passage depends on the number of stably growing colonies in each culture. If several ES-like colonies are present in the first-passage microdrop, then the entire drop may be trypsinized.

1. Aspirate the old medium, wash the cells *in situ* with PBS, and introduce 20 µL of TEG.

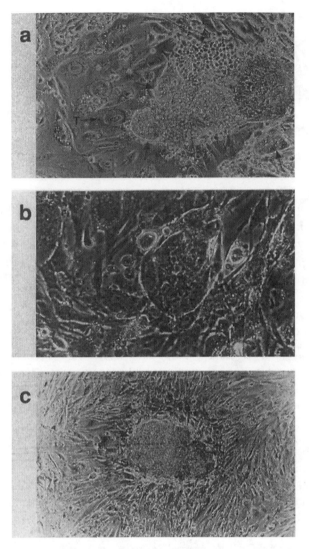

Fig. 3. Morphology of cell colonies following disaggregation of ICM-derived out-growths. **(a)** Giant trophoblast cells (T) and endodermal-like cells (E) are the most common cell colonies present in first-passage cultures of disaggregated ICMs. How-ever, some small ES-like colonies are present among the assortment of differentiated cells (A) (arrows indicate ES-like colonies in all panels). Higher magnification of a small ES-like colony 2 d after the first passage **(b)** shows the typical undifferentiated cell morphology. ES cells grow in close opposition to one another, with each cell having a high nuclear-to-cytoplasmic ratio and one or more prominent nucleoli (B). **(c)** After 7 d of culture on the STO feeders, some cells of the disaggregated ICM may grow into tightly packed, three-dimensional ES-like colonies, in which clear intercel-lular boundaries are difficult to distinguish. The colony shown in (C) was subsequently expanded and maintained as a permanent ES cell line. Photographs were taken using phase-contrast microscopy.

2. Following 3–5 min at 37°C, transfer the dish to the dissecting microscope and use a fine, mouth-controlled pipet to dissociate further the colonies into single cells. This is important, so as to prevent subsequent differentiation.

3. Transfer the cell suspension into an STO feeder well of a four-well plate containing 1 mL of $ES_{20}$ medium. After 1 or 2 d, small colonies of ES-like cells may be visualized in some of these cultures.

If only one or two ES-like colonies in the first-passage culture are ready for disaggregation, it is considered best to pick these out of the microdrop, whereas any remaining ES-like colonies, which are considered too small to survive readily the trypsinization procedure, are left to proliferate further. The larger ES-like colonies are picked off the STO feeder layer with a blunt glass probe, washed briefly in TEG, and then incubated at 37°C in a fresh drop of TEG for 3–5 min. After partial digestion, the colonies are disaggregated mechanically into single cells with a fine pipet and introduced into a fresh STO microdrop, before being eventually expanded into four-well plates.

From the four-well plates, the ES-like cells are subsequently expanded into 25-cm$^2$ flasks containing a layer of STO feeder cells:

1. Culture the ES-like cells within the wells for 4 to 5 d, feeding as required, by which time the colonies may have merged to form a monolayer of cells.

2. Working in the flow cabinet, aspirate the medium within the well, and wash the cells with 1 mL of PBS. Following a 3- to 5-min incubation in 0.25 mL of TEG, use a hand-pulled Pasteur pipet (tip diameter of about 1 mm) and bulb first to introduce about 0.5 mL of $ES_{20}$ medium to neutralize the trypsin, and then to vigorously pipet the suspension to disaggregate the cells (but avoiding the introduction of air bubbles into the well).

3. Transfer the single-cell suspension into a flask containing 10 mL of $ES_{20}$ medium. Then culture the flask in a $CO_2$ incubator with either the cap loosened or tightened after being gassed with a 5% $CO_2$, 20% $O_2$, and 75% $N_2$ mixture.

### 3.3.2. Isolation of ES Cells from Blastomeres

The following protocol is based on the method of Eistetter (33) for the establishment of ES cells from single blastomeres. (The author has attempted to repeat this work, but all initial ES-like colonies eventually differentiated into trophectoderm.)

1. To obtain single blastomeres from compacted 16- to 20-cell morulae, first remove the zonae pellucidae. This may be accomplished by treatment of embryos in either acid Tyrode's solution (pH 2.1) or with an enzymatic solution comprising 0.5% (w/v) pronase (grade B; Calbiochem) and 0.5% (w/v) PVP ($M_r = 10,000$) in HEPES-buffered M2 medium (15).

2. Incubate the zona-free morulae at 37°C in $Ca^{2+}$- and $Mg^{2+}$-free PBS containing 0.3% (w/v) EDTA for 3 min.

3. Complete dissociation of the blastomeres mechanically by utilizing a finely pulled Pasteur pipet with a flame-polished tip. All the blastomeres from each embryo may be explanted together into one feeder microdrop after washing extensively in $ES_{20}$ medium. Eistetter (33) reported that the blastomeres rapidly attach to the feeder cells; however, in the author's hands, the majority of blastomeres remain free-floating and continue to divide, forming miniblastocysts or trophectodermal vesicles.

4. After 3 or 4 d of culture, colonies with an ES-like morphology may be visualized. Depending on their size, pick them out and lightly disaggregate in TEG before plating them onto feeder layers away from the influence of differentiated colonies. Propagation of the blastomere-derived ES cells is the same as described previously for blastocysts.

### 3.3.3. Isolation of ES Cells from d 5.5 p.c. Primitive Ectoderm

Apart from the timing of the first passage, methods for the establishment of primitive ectoderm-derived ES cells are identical to those described previously for blastocyst-derived ES cell lines.

The isolation history of an ectoderm-derived ES cell line is illustrated in **Fig. 4**. The d 5.5 p.c. egg cylinder–stage embryo in **Fig. 4A** shows the position of the cut made to remove the extraembryonic region and the location of the visceral endoderm tissue overlying the primitive ectoderm. The primitive ectoderm is isolated as described in **Subheading 3.2.5.** and generally results in relatively pure clumps of tissue (**Fig. 4B**), which are then explanted onto STO feeder layers prepared in microdrops. After 2 or 3 d of culture, undifferentiated outgrowths are commonly observed (**Fig. 4C**), which are then lightly disaggregated with TEG, before any obvious signs of differentiation, into several small pieces each comprising about 10–15 cells. Despite this early first passage, extensive differentiation often results in the cultures, with many endodermal and flattened epithelial-like colonies developing. However, trophoblast colonies are rarely observed, except in instances of contamination with cells of extraembryonic origin, since d 5.5 p.c. primitive ectoderm no longer has the potential to develop trophectoderm (34). After a 6- to 7-d culture interval, first-passage colonies that remain undifferentiated should be trypsinized into single cells and passaged. From these cells, second-passage colonies of typical ES cell morphology may grow in some cultures, as in **Fig. 4D**. Although d 5.5 p.c. primitive ectoderms may often give rise to outgrowths that appear remarkably like established ES cells from the outset (as has been reported for immunosurgically isolated ICM outgrowths; [2]), in the author's experience, most of these differentiate extensively in subsequent passages. Because of the relatively large number of cells in the d 5.5 p.c. primitive ectoderm, differentiation may be minimized by lightly disaggregating the isolated ectoderms prior to explant culture.

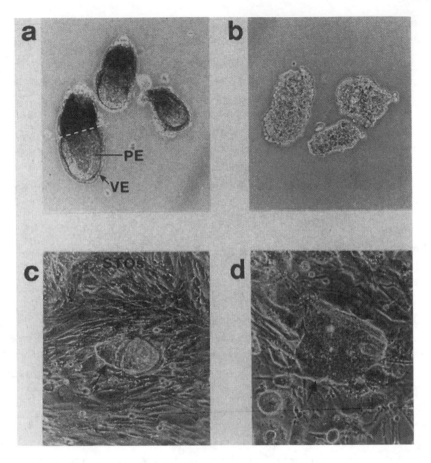

Fig. 4. Establishment of an ES cell line from d 5.5 p.c. primitive ectoderm. **(a)** Egg cylinder–stage embryos dissected from the decidua on d 5.5 p.c. Primitive ectoderm (PE) tissue **(b)** is isolated from the embryos by first cutting away the extraembryonic region along the dotted line in (A) and then utilizing an enzymatic procedure to remove the overlying visceral endoderm (VE). **(c)** Intact ectoderm explanted onto the STO feeder cells. From the undifferentiated outgrowth (see arrow) in (C), colonies of ES cell morphology were observed at the second passage **(d)** and subsequently expanded into a permanent stem cell line. Photographs were taken using phase-contrast microscopy.

### 3.4. Isolation of Murine Embryonic Germ Cells

When mouse primordial germ cells (PGCs) are cultured on feeder cell layers providing an adequate source of membrane-associated Steel factor (SF), in addition to a particular combination of soluble polypeptide signaling factors, including SF, LIF, and basic fibroblast growth factor (bFGF), they continue to proliferate in long-term culture and give rise to cell lines resembling ES cells *(7,8)* *(see* **Note 8**). These cells are termed *embryonic germ (EG) cells* to specify their origins and have been isolated from the PGCs obtained from d 8.5 and 12.5 p.c. embryos *(10)*. EG cells are morphologically similar to ES cells and share common cell-specific markers. They differentiate extensively in vitro; form teratocarcinomas when injected into immunocompromised mice; and participate in embryogenesis when injected into blastocysts, forming germline chimeras *(9,10)*. Thus, mouse EG cells derived from PGCs are also pluripotent.

A fundamental difference exists, however, between ES and EG cells regarding differences in methylation imprinting pattern *(10)* *(see* **Note 8**), perhaps illustrating their different embryonic origins. Also, by comparing the chimeras produced from either pregonadal or gonadal PGC-derived cell lines, it appears that EG cells derived from PGCs at different embryonic stages may be qualitatively different in terms of their developmental capacity *(10)* *(see* **Note 8**).

The following outlines the methods developed by Hogan and colleagues *(7,10)* to isolate EG cell lines from d 8.5 p.c. mouse PGCs:

1. Set up appropriate matings and sacrifice females on d 8.5 p.c.
2. Dissect embryos free of extraembryonic membranes.
3. Dissect the posterior third of the embryo, from the base of the allantois to the first somite. This region contains approx 150–350 PGCs at this stage. Either pool tissue fragments together or keep them separate, depending on the experiment.
4. Rinse the tissue fragments in $Ca^{2+}$- and $Mg^{2+}$-free PBS, and disaggregate in 0.25% trypsin and 1 m$M$ EDTA, aided with pipeting to achieve a single-cell suspension.
5. Plate the cell suspension onto a 1.75-cm$^2$ well containing a feeder cell layer comprising $Sl/Sl^4$ m220 cells, which express membrane-bound mouse SF *(10)* *(see* **Note 8**). Seed the cell suspension at a concentration of approx 0.5 embryo equivalents per well.
6. Culture the cells in DMEM with 4.5 g/L of glucose, 0.01 m$M$ NEAAs, 2 m$M$ glutamine, 50 µg/mL of gentamycin, 15% FCS, and 0.1 m$M$ β-mercaptoethanol. In these explant cultures, supplement the medium with soluble recombinant rat SF (60 ng/mL), 20 ng/mL of LIF, and 20 ng/mL of bFGF. Change the medium daily.
7. After 6 d, stain a few cultures with alkaline phosphatase to assess PGC survival and growth.
8. After 10 d, trypsinize the remaining cell cultures to a single-cell suspension, and plate onto PMEF feeder cell layers in the DMEM base medium (**step 6**) but supplemented only with soluble LIF (10 ng/mL) *(see* **Note 8**).
9. After this first passage, monitor the cultures for the appearance of EG cell colonies, which are tightly packed and morphologically resemble ES cells. Once at a suitable size, isolate individual EG cell colonies by picking and establish cell lines and propagate.

10. Maintain EG cell cultures in a similar fashion as for ES cells, with PMEF feeder cells and LIF added to the medium.

The isolation of EG cells from later developmental stages is similar. However, because of the greater number of cells, finely minced fragments of genital ridges are trypsinized and plated at 0.1 embryo equivalents per well *(10)*. The frequency of EG cell isolation is lower with cultures derived from d 12.5 p.c. gonadal PGCs. No EG cell lines were isolated from genital ridges obtained from d 15.5 p.c. fetuses, nor from 8-d-old newborn mice *(10)*. These are stages by which PGCs have ceased proliferating in vivo, and they do not appear to respond to the combination of polypeptide factors that have been successfully used with PGCs from d 8.5 p.c.

## 3.5. Culture of Established Murine ES Cells

This section describes the procedures used for the routine maintenance of ES cells, the freezing and thawing of cell lines, the induction of in vitro differentiation, and the karyotypic analysis of stem cells.

### 3.5.1. Maintenance of ES Cell Lines

Cell lines maintaining a stable ES cell morphology may be routinely cultured in $25\text{-cm}^2$ flasks, or 6-cm dishes. After the fifth passage, ES cells can be "weaned" from the STO feeder layer and the serum concentration in the medium reduced by half. ES cells may be subsequently maintained on gelatin-coated flasks in medium ($ES_{10}$) supplemented with LIF to prevent ES cell differentiation (*see* **Notes 3** and **6**). This factor may be supplied from either BRL-conditioned medium or a recombinant source. The supplementation of LIF to medium provides a complete substitute for feeder cells preventing the differentiation of isolated stem cells and, hence, greatly simplifies the routine culture of ES cell lines. For BRL-conditioned medium, the optimal concentration minimizing spontaneous ES cell differentiation is a 60% (v/v) strength, diluted with fresh ES medium *(35)*. ES cells are generally grown in medium containing 10 ng/mL of LIF (equivalent to 1000 U/mL) to maintain the undifferentiated stem cell morphology *(13,14)*. In the absence of feeders, the ES cells grow as monolayers, compared to the "tight nests" of cells observed on feeders. Despite this morphologic change in colony appearance, established ES cells retain their full developmental potential and capacity to colonize the germline in chimeras when maintained in the "crude" BRL-conditioned medium *(36)* or in the presence of purified recombinant LIF *(37)*.

### 3.5.2. Subculture of ES Cells

Established ES cell lines generally are passaged every 3 to 4 d, and the medium is changed every other day or whenever it becomes acidic.

1. When confluent, refeed the cells 2 to 3 h before the passage to maximize subsequent cell survival. Then aspirate the medium and wash the cells once in PBS.
2. After removing the PBS, add 2 mL of TEG, and incubate the flask at 37°C for 3–5 min. Periodically monitor the extent of trypsinization in the inverted microscope. To minimize spontaneous differentiation of ES cells, the aim is to produce a single-cell suspension. To achieve this, it may sometimes be necessary to knock the flask gently against the bench, or the palm of the hand.
3. Once the cells have been dissociated, neutralize the trypsin with 3 mL of $ES_{10}$ + LIF medium, and mix the suspension thoroughly. Routinely, the cells can be subcultured by transferring a 1/10 aliquot of this suspension (i.e., 0.5 mL) into a pregelatinized flask (0.1% [w/v] gelatin in sterile water for about 1 h) containing 10 mL of fresh $ES_{10}$ + LIF medium. The volume of the aliquot dispensed can be varied depending on the requirement for ES cells; however, it should not be reduced below a 1:10 ratio.

## 3.5.3. Freezing and Thawing of ES Cells

Generally, it is wise to freeze several ampoules of cells as soon as possible after isolating a new ES cell line. For long-term storage, the ampoules are kept under liquid nitrogen. The cryoprotectant used in the medium is dimethylsulfoxide (DMSO) at a final concentration of 10%. Although the procedures described here are for ES cells, similar methods can be used to freeze and thaw other tissue culture cell lines, such as STO and BRL cells.

### 3.5.3.1. FREEZING CELLS

The ES cells should be harvested during log phase growth, to maximize subsequent survival. Each 25-cm$^2$ flask should yield about $1 \times 10^7$ cells. Cells can be frozen in 1-mL aliquots, at any desired concentration between $5 \times 10^5$ to $1 \times 10^7$ cells/mL.

1. Collect cells by trypsinization (**Subheading 3.5.2.**) and take a sample to determine the number of cells, using a hematocytometer.
2. Pellet the cells in a centrifuge tube by spinning at 100 $g$ for 5 min.
3. Aspirate the supernatant, and resuspend the cells in $ES_{10}$ + LIF medium, to half of the desired final volume (i.e., the cell suspension is at twice the final concentration of cells at this stage).
4. Make up the final volume with the freezing medium (20% [v/v] DMSO in FCS), which is added slowly, while gently flicking the tube, to ensure constant mixing of the two solutions.
5. Dispense 1 mL of the final cell suspension into each 1-mL cryotube (Nunclon). Then wrap the cyrotubes in a few layers of paper tissue in a polystyrene freezing box, before placing them in a –70°C freezer overnight.
6. For long-term storage, store the cryotubes in a suitable liquid nitrogen container.

### 3.5.3.2. Thawing Cells

1. Once retrieved from the liquid nitrogen, thaw the cryotube quickly in a 37°C water bath, until the ice crystals have all melted.
2. Sterilize the cryotube by wiping with paper tissues moistened with 70% alcohol.
3. Working at room temperature within the flow cabinet, use a sterile Pasteur pipet and bulb to transfer the 1-mL cell suspension in the cryotube into 9 mL of $ES_{10}$ + LIF medium in a centrifuge tube, while constantly flicking the tube to mix the two solutions. Then pellet the cells at 100 $g$ for 5 min.
4. Aspirate the supernatant and resuspend the cells in an appropriate volume of $ES_{10}$ + LIF medium in order to seed a culture plate of a suitable size depending on the number of cells originally frozen. The ES cells should preferentially be seeded onto a feeder cell layer to minimize cell stress, following the thawing process.
5. Culture the cells in the $CO_2$ incubator, and change the medium after approx 6 h to remove any cellular debris.
6. Seed the thawed ES cells at a relatively high density, to ensure that the culture dish attains confluence within 1 to 2 d.

### 3.5.4. Induction of In Vitro Differentiation in ES Cells

With the suspension culture of cellular aggregates, ES cells can be induced to differentiate along pathways thought to be analogous to those of early embryonic development, leading to the formation of embryoid bodies. These may be simple structures comprising an outer layer of endodermal cells or may progress into fluid-filled, cystic embryoid bodies. These comprise an inner layer of ectodermal-like cells, with a Reichert's membrane separating a presumed outer layer of parietal endodermal cells. These cystic embryoid bodies contain α-fetoprotein and transferrin and are thus analogous to the visceral yolk sac of the postimplantation-stage mouse embryo *(21)*. For more detailed information on the in vitro differentiation of ES cells, the reader is referred to **ref. 38**.

1. Suspension culture is conducted in agarose-coated dishes. Prepare these by applying a base layer consisting of 2% (w/v) agarose (Type 1; Sigma) in PBS, which is dissolved and sterilized by autoclaving. Add approx 1.5 mL per 6-cm dish to give an even layer and leave to set at room temperature. Then apply a second, thin layer using 1% (w/v) agarose in PBS. Once this has set, add 5 mL of $DMEM_{10}$ + 0.1 m$M$ β-mercaptoethanol to each dish and incubate to allow equilibration. Before use, replace the medium.
2. Lightly trypsinize ES cells with TEG for 1 to 2 min. By gently rocking the flask, large clumps of cells will detach. Then immediately neutralize the trypsin with $DMEM_{10}$ medium + 0.1 m$M$ β-mercaptoethanol.
3. Dispense an approx 1/20 aliquot of this aggregate suspension into a 6-cm agarose-coated dish. At higher seeding densities, the individual aggregates adhere to each other.

Cultures are fed regularly, and the old medium should be aspirated either by transferring the suspension into a conical universal to allow the embryoid bodies to settle, or by simply tilting the dish, before adding fresh medium. Utilizing these procedures, simple embryoid bodies form within 2–4 d and become cystic after 7–10 d. If simple embryoid bodies are allowed to attach to a tissue culture surface, the resulting differentiation is chaotic and a wide range of different cell types form, which may be identified utilizing standard histology or specific cell markers.

### 3.5.5. Karyotypic Analysis of ES Cells

Karyotypic analysis is used to determine the sex and chromosome complement of a cell line. To obtain germline transmission of the ES cell genotype in chimeras (*see* Chapter 7), it is vital that a high proportion of cells within the cell line have a euploid chromosome complement and a modal number of 40 chromosomes. It is necessary to examine the karyotype of newly established ES cell lines, since some of them may in fact be abnormal (*39*). This may be influenced by suboptimal in vitro culture conditions during isolation, as observed with the karyotypic abnormalities resulting from the isolation of stem cells in LIF, without the support of feeder cells (*24*). Additionally, it is important to check the karyotype of ES cell lines routinely with extended time in culture, because there is the risk of selecting aneuploid cells, which exist within most ES cell lines. To regenerate a karyotypically normal cell line, the ES cells may be single-cell cloned, and euploid cultures identified and reexpanded.

This section describes the methods for preparing cells in metaphase in order to perform chromosome counts and G-banding analysis, which involves the denaturation of the chromosomes with hot saline and trypsin in order to identify individual chromosomes on the basis of their unique banding pattern, and to determine whether any abnormalities are present.

#### 3.5.5.1. PREPARATION OF METAPHASE SPREADS

Metaphase spreads of ES cells may be prepared either from cultures exposed to colcemid or, preferably, by utilizing cultures in an exponential phase of cellular growth, in order to maximize the number of cells in mitosis. The quality of the mitotic spreads is generally superior when colcemid is not used, since the chromosomes tend to be longer, facilitating their identification following banding. However, for simple chromosome counts, it may be advantageous to use colcemid because this treatment often results in more contracted chromosomes, resulting in fewer overlapping chromosomes in the spreads and making counting easier.

1. Trypsinize ES cells in log phase growth from a 25-cm$^2$ flask and pellet in a centrifuge tube as described previously (*see* **Subheading 3.5.2.**). Alternatively,

colecmid may be added to the medium at a final concentration of 0.02 µg/mL and left for 1 h in the incubator (or longer, to arrest more cells in metaphase if they are only required for chromosome counting).

2. Aspirate the medium from the tube and disrupt the cell pellet before using a Pasteur pipet and bulb to introduce dropwise about 0.5 mL of a 0.56% (w/v) KCl solution. Once mixed, add excess hypotonic KCl solution to make 5 mL and leave for 15 min at 37°C, to allow the cells time to swell.

3. Add an equal volume (i.e., 5 mL) of freshly prepared, ice-cold fixative (3 vol of absolute methanol to 1 vol of glacial acetic acid; taking care not to inhale the vapor), mixing constantly, and then centrifuge at 40 *g* for 5 min.

4. Aspirate the supernatant, disrupt the pellet of cells, add 10 mL of acetic methanol fixative, and leave on ice for 10 min.

5. Centrifuge the tube at 40 *g* for 5 min and repeat **steps 4** and **5**.

6. Resuspend the cells in about 1.0 mL of fixative. They are now ready for producing mitotic spreads.

7. Metaphase spreads of the fixed, swollen cells are made on glass microscope slides that have been chilled on ice. For the cells to spread adequately, it is important that the slides be thoroughly cleaned by wiping them previously with soft paper tissues moistened with the fixative solution. Draw a small quantity of the fixed cell suspension into a Pasteur pipet with a bulb. Release two drops of the suspension onto different locations on the slide, from a height of between 1 and 10 cm.

8. Evaporation of the fixative is aided by warming the slide over a beaker of boiling water. Examine the first slide to determine the density of cells present, and dilute as necessary if too thick. The rate of drying, cleanliness of the slides, and height from which the cells fall are some of the important variables in maximizing the rupture of the swollen cells and, hence, the spreading of the chromosomes.

9. To determine the modal chromosome number, stain the slides immediately in a 3% (v/v) solution of Gurr's Giemsa stain in PBS for 15 min. Then rinse the slides in two changes of distilled water and allow to air-dry before counting the chromosomes. For the purposes of G-banding, the best staining results are obtained by first storing the slides in a dust-free location for 10–14 d after preparing the mitotic spreads.

### 3.5.5.2. G-BANDING ANALYSIS

1. Incubate slides in 2X SSC at 60°C for 1 h. Then rinse the slides four to five times in distilled water and store temporarily in a rack under water.

2. Individually immerse each slide in a 0.25% (w/v) trypsin solution in Gurr's phosphate buffer (pH 6.8) for between 7 and 15 s at room temperature.

3. Neutralize the trypsin remaining on the slide after this digestion in Gurr's buffer containing 5% (v/v) NCS.

4. Rinse the slides further in two changes of buffer before staining in freshly prepared 5% (v/v) Giemsa Gurr's R-66 stain in Gurr's buffer (pH 6.8) for 8–10 min.

5. Finally rinse the slides in two changes of buffer, followed by two changes of distilled water, and allow to air-dry. The incubation times of the slides in trypsin

and stain should be determined empirically in order to optimize the clarity of the banding patterns.
6. Examine G-banded metaphase spreads with a standard format, bright-field microscope utilizing oil immersion, objective lenses (maximum magnification of about ×1000). Several suitable mitotic spreads with no or minimal overlapping of chromosomes should be photographed. Identify G-banded chromosomes according to the Standardized Genetic Nomenclature for mice *(40)*, and construct karyograms to determine the sex of the ES cell line and whether populations exist within the cell line that possess chromosomal abnormalities.

## 4. Notes

1. In the mouse, pluripotent ES cells have been isolated from the blastomeres of 16- to 20-cell morulae *(33)* from the ICM of blastocyst-stage embryos *(1,2)*, and from the primitive ectoderm of the d 5.5 p.c. egg cylinder–stage embryo *(17)*. ES cell lines are most commonly derived from the ICM and have been obtained from fertilized d 3.5 p.c. or implantationally delayed blastocysts, and from parthenogenetically *(41,42)* and androgenetically produced blastocyst embryos *(43)*. Furthermore, EG cells have been isolated in the mouse from d 8.5 and 12.5 p.c. PGCs *(7–10)* (*see also* **Note 8**).

   Historically, the majority of investigators have isolated ES cell lines from the readily available d 3.5 p.c. blastocyst-stage embryo. ES cell cultures have been obtained from both intact blastocysts and immunosurgically isolated ICMs with equal success (about 10%; *[28]*). From experience, blastocysts that have undergone a period of implantational delay are expected to generate, on average, a threefold increase in the efficiency of ES cell isolation compared to nondelayed d 3.5 p.c. embryos *(17,39)*. The increase in the potential of delayed embryos to yield ES cells either may arise as a consequence of the small yet significant increase that occurs in the number of cells in the ICM *(44)*, or may be a result of some epigenetic change in gene expression. Although primitive endoderm has formed in the implantationally delayed embryo, no further differentiation of the ICM occurs *(45)*. Therefore, the normal pattern of expression of genes responsible for the differentiation of the ICM is halted, and this may be an essential feature in the establishment of ES cell lines in vitro.

   There appears to be a period of embryonic development, from at least d 2.5 to 12.5 p.c., during which pluripotent cells present within the mouse embryo can be successfully isolated and maintained in culture either as ES or EG cells. Since detailed chimeric analyses have not been reported, it is not known whether the blastomere- and primitive ectoderm–derived cell lines are of the same lineage as ICM-derived ES cells, or if their respective in vivo developmental potentials are "frozen" at the embryonic stage from which they were each isolated. From the studies with EG cells, however, it does appear that subtle differences exist in the characteristics and developmental potential of cell lines isolated from difference sources within the mouse embryo *(10)* (*see* **Note 8**).

2. For both ES cell isolation and routine culture, no antibiotics are used. If a culture becomes infected (especially if by fungus or yeast), it is generally wise to discard it from the laboratory. If, however, a particularly valuable culture becomes infected with bacteria, it may be extensively washed in $Ca^{2+}$- and $Mg^{2+}$-free PBS before medium containing 50 IU/mL of penicillin (sodium salt) and 50 μg/mL of streptomycin is added.

3. Smith and Hooper *(35)* found that medium conditioned by BRL cells contained a factor that was a potent inhibitor of stem cell differentiation. This factor was termed *stem cell differentiation inhibitory factor* (DIA). Structural and functional comparisons have shown that DIA is identical to the murine myeloid LIF *(13,14)*. Furthermore, a third growth factor, human interleukin for DA cells (a murine leukemic cell line), has been shown to be essentially identical to the human LIF protein *(46,47)*. In this chapter, this factor is simply referred to as LIF.

4. The osmolarity of DMEM, as supplied by Gibco, is approx 345 mosM/kg of $H_2O$, and is still considered very high after all the media supplements and sera have been added ($ES_{20}$: about 335 mosM/kg of $H_2O$; $ES_{10}$: about 340 mosM/kg of $H_2O$; *[48]*). Although an initial study found no significant differences in the efficiency of ES cell isolation with media of varying osmolalities (unpublished data), the plating efficiency of an established ES cell line was observed to have been optimal with medium of 290 mosM/kg of $H_2O$ (J. McWhir, Institute of Animal Physiology and Genetics Research, Roslin, Scotland, personal communication). Hence, it is recommended that the osmolality of both $ES_{10}$ and $ES_{20}$ media be reduced to about 290 mosM/kg of $H_2O$ by the addition of 13% (v/v) sterile water (*see* **Subheading 2.3.**).

5. Some mouse strains are considered permissive for ES cell isolation and readily yield lines (e.g., 129 and C57BL/6 strains; *[22]*). By contrast, other mouse strains are considered more difficult (e.g., CBA and C57BL/6 x CBA; *[48,49]*). It appears that the genetic background determines the level of suppression of ES cell proliferation. Despite the manipulations used to disaggregate ICM outgrowths, in some mouse strains the ES cell precursors appear unable to escape the regulatory control of the differentiated cell phenotypes of the embryo.

   A novel approach has been used to establish ES cell lines from mouse strains that are refractory to stem cell isolation. McWhir et al. *(49)* have developed a strategy to selectively ablate differentiated cells in primary cultures, allowing the ES cells to more readily proliferate in their absence. They first generated transgenic mouse embryos (genotypically: 87.5% CBA, 12.5% C57BL/6) that carried the G418 (neo) resistance gene under the control of the Oct3/4 promoter (the transcription factor Oct3/4 is only expressed in undifferentiated mouse cells and is strongly downregulated following cellular differentiation). ES cell lines were then isolated by culturing these transgenic embryos and cells, in the continual presence of G418. On cell differentiation, the expression of Oct3/4 ceases, and hence G418 resistance was lost, thus killing the differentiated cells and removing their influence from the culture. By using this approach, ES cell lines were isolated from a mouse strain considered nonpermissive for ES cell isolation *(49)*.

6. The function of the feeder cells, in addition to providing a more suitable attachment surface for direct coculture *(18)*, is in the active suppression of stem cell differentiation. It has been shown that a factor known as LIF (*see* **Note 3**), which inhibits the differentiation of stem cells, is produced by these feeder cells as both a diffusible protein and in an immobilized form, associated with the extracellular matrix *(19)*. Furthermore, in the coculture system, ES cells secrete a heparin-binding growth factor responsible for the stimulation of LIF expression in the feeder cells *(50)*. Recombinant LIF, added in the culture medium, can replace feeder cells in the isolation of ES cell lines, which may retain their capacity for germline transmission *(24,51)*. However, it is strongly recommended that feeder cell layers be used for the isolation of new ES cells lines, because in the absence of feeder cells the chromosome complement of most cell lines was abnormal *(22)*. The use of the coculture feeder system may aid in isolating cell lines with a higher probability of being euploid.

   In the author's experience, following the isolation of ES cells on STO feeders, it has generally been satisfactory to subsequently maintain ES cell lines without feeder cell layers but on gelatin-coated dishes, in the presence of soluble LIF. However, it is acknowledged that cells do adapt to their in vitro culture environment *(25)* and may not readily tolerate changes in culture conditions without altering their karyotype *(52)*. This tolerance appears to be greatly dependent on the individual cell line *(52)*.

7. $ES_{20}$ medium is used routinely for ES cell isolation. However, others have demonstrated that the use of medium conditioned on a variety of other cell lines (including PSA-1 embryonal carcinoma cells *[2]*, BRL cells *[11]*, and the 5637 human bladder carcinoma cell line *[52]*), in addition to the presence of feeder cells, facilitates the isolation of ES cells. However, Robertson and Bradley *(39)* have observed that although conditioned medium may enhance the growth of primary colonies, the majority tend to be of a "pretrophoblast" lineage and ultimately differentiate. Furthermore, a comparative study has shown no significant effect of cell-conditioned medium on the efficiency of ES cell isolation *(28)*.

8. PGCs can be isolated at various stages of fetal development, during their migration to the gonad. On d 7 p.c., PGCs are identified as a group of about 100 alkaline phosphatase–positive cells present in the extraembryonic mesoderm, just posterior to the definitive primitive streak *(53)*. From the base of the allantois, they migrate along the hindgut and reach the genital ridges by d 11.5. In the genital ridge, PGCs stop dividing by d 13.5, by which time they have multiplied to 25,000 cells that will establish the meiotic population *(54)*.

   It appears important for the initiation of PGC cultures that they be grown on feeder cell layers that provide membrane-associated SF, such as with the $Sl/Sl^4$ m220 cells *(7,10)*. PMEFs do not appear to make significant quantities of SF *(55)*. In addition to membrane-bound forms, soluble SF is also needed to optimize PGC survival and, in combination with soluble LIF and bFGF, promotes proliferation. However, this initial feeder requirement is in marked contrast to that following the first subculture. At this stage, PMEF feeder cell layers are in fact superior in supporting the growth of the PGCs *(10)*, which by now have altered their characteristics and become more ES-like in cell morphology and can be routinely maintained in a manner similar to ES cells.

Some EG cell lines derived from d 8.5 p.c. PGCs have colonized the germline of chimeric mice *(9,10)*, demonstrating their pluripotent potential. However, EG cell lines from older d 12.5 p.c. gonadal PGCs have only formed somatic cell chimeras at low frequency and with low contributions of the EG cells in the animal *(10)*. Furthermore, some chimeras have had skeletal abnormalities *(10)*, reminiscent of the defects associated with androgenetic ES cell chimeras *(43)*.

A significant difference has been observed between ES cells and EG cells in terms of the methylation imprint in region 2 of the insulin-like growth factor-2 receptor (*Igf2r*) gene *(10)*. ES cells have the typical somatic cell methylation pattern, where only the maternal allele is methylated. However, half the d 8 and 8.5 p.c. EG cell lines, and all the 12.5 p.c. EG cell lines, have both alleles unmethylated. This particular methylation pattern is stable with extended culture. There appears to be no correlation between methylation pattern and the ability of the cells to contribute to the germline of mice, since EG cell lines with both alleles unmethylated or one that is methylated have both yielded germline chimeras *(10)*.

The differences in methylation pattern may illustrate a fundamental difference between ES and EG cells, owing to their different embryonic origins. Also, it seems likely that EG cell lines derived from different developmental stages may be qualitatively different, as seen in the chimera data between pregonadal and gonadal PGC-derived lines *(10)*.

## Acknowledgments

Many of methods described in this chapter were learned and developed during my PhD studies with Dr. Ian Wilmut (Institute of Animal Physiology and Genetics Research, Roslin, Scotland). I am grateful to Dr. Martin Hooper (Department of Pathology, University of Edinburgh) for the opportunity to train in his laboratory. I am also indebted to Drs. Jim McWhir and Alan Clarke, who taught me the "key" to ES cell isolation, and to Dr. Gerry O'Neill for advice on karyotyping.

## References

1. Evans, M. J. and Kaufman, M. H. (1981) Establishment in culture of pluripotential cells from mouse embryos. *Nature* **292,** 154–156.
2. Martin, G. R. (1981) Isolation of a pluripotent cell line from early mouse embryos cultured in medium conditioned by teratocarcinoma stem cells. *Proc. Natl. Acad. Sci. USA* **78,** 7634–7638.
3. Bradley, A., Evans, M., Kaufman, M. H., and Robertson, E. (1984) Formation of germ-line chimaeras from embryo-derived teratocarcinoma cell lines. *Nature* **309,** 255, 256.

4. Beddington, R. S. P. and Robertson, E. J. (1989) An assessment of the developmental potential of embryonic stem cells in the midgestation mouse embryo. *Development* **105,** 733–737.
5. Nagy, A., Rossant, J., Nagy, R., Abramow-Newerly, W., and Roder, J. C. (1993) Derivation of completely cell culture-derived mice from early-passage embryonic stem cells. *Proc. Natl. Acad. Sci. USA* **90,** 8424–8428.
6. Hooper, M. L. (1992) in *Embryonal Stem Cells: Introducing Planned Changes into the Germline* (Evans, H. J., ed.), Harwood Academic, Switzerland.
7. Matsui, Y., Zsebo, K., and Hogan, B. L. M. (1992) Derivation of pluripotential embryonic stem cells from murine primordial germ cells. *Cell* **70,** 841–847.
8. Resnick, J. L., Bixler, L. S., Cheng, L., and Donovan, P. J. (1992) Long-term proliferation of mouse primordial germ cells in culture. *Nature* **359,** 550, 551.
9. Stewart, C. L., Gadi, I., and Bhatt, H. (1994) Stem cells from primordial germ cells can reenter the germ line. *Dev. Biol.* **161,** 626–628.
10. Labosky, P. A., Barlow, D. P., and Hogan, B. L. M. (1994) Mouse embryonic germ (EG) cell lines: transmission through the germline and differences in the methylation imprint of insulin-like growth factor 2 receptor (*Igf2r*) gene compared with embryonic stem (ES) cell lines. *Development* **120,** 3197–3204.
11. Handyside, A. H., O'Neill, G. T., Jones, M., and Hooper, M. L. (1989) Use of BRL-conditioned medium in combination with feeder layers to isolate a diploid embryonal stem cell line. *Roux's Arch. Dev. Biol.* **198,** 48–55.
12. Oshima, R. (1978) Stimulation of the clonal growth and differentiation of feeder layer dependent mouse embryonal carcinoma cells by β-mercaptoethanol. *Differentiation* **11,** 149–155.
13. Smith, A. G., Heath, J. K., Donaldson, D. D., Wong, G. G., Moreau, J., Stahl, M., and Rogers, D. (1988) Inhibition of pluripotential embryonic stem cell differentiation by purified polypeptides. *Nature* **336,** 688–690.
14. Williams, R. L., Hilton, D. J., Pease, S., Willson, T. A., Stewart, C. L., Gearing, D. P., Wagner, E. F., Metcalf, D., Nicola, N. A., and Gough, N. M. (1988) Myeloid leukaemia inhibitory factor maintains the developmental potential of embryonic stem cells. *Nature* **336,** 684–687.
15. Hogan, B., Constantini, F., and Lacy, E., eds. (1986) Chemicals, supplies and solutions, *Manipulating the Mouse Embryo.* Cold Spring Harbor Laboratory Press, Cold Spring Harbor, NY, pp. 269–277.
16. Martin, G. R., Jakobovits, A., and Joyner, A. (1984) Factors that affect the growth of teratocarcinoma and embryonic stem cells. *J. Embryol. Exp. Morphol.* **82(Suppl.),** 147.
17. Wells, D. N., McWhir, J., Hooper, M. L., and Wilmut, I. (1991) Factors influencing the isolation of murine embryonic stem cells. *Theriogenology* **35,** 293.
18. Martin, G. R. and Evans, M. J. (1975) Differentiation of clonal lines of teratocarcinoma cells: formation of embryoid bodies *in vitro. Proc. Natl. Acad. Sci. USA* **72,** 1441–1445.
19. Rathjen, P. D., Toth, S., Willis A., Heath, J. K., and Smith, A. G. (1990) Differentiation inhibiting activity is produced in matrix-associated and diffusible forms that are generated by alternate promoter usage. *Cell* **62,** 1105–1114.

20. Ogiso, Y., Kume, A., Nishimune, Y., and Matsushiro, A. (1982) Reversible and irreversible stages in the transition of cell surface marker during the differentiation of pluripotent teratocarcinoma cell induced with retinoic acid. *Exp. Cell Res.* **137,** 365–372.

21. Doetschman, T. C., Eistetter, H., Katz, M., Schmidt, W., and Kemler, R. (1985) The *in vitro* development of blastocyst-derived embryonic stem cell lines: formation of visceral yolk sac, blood islands and myocardium. *J. Embryol. Exp. Morphol.* **87,** 27–45.

22. Suemori, H. and Nakatsuji, N. (1987) Establishment of the embryo-derived stem (ES) cell lines from mouse blastocysts: effects of the feeder cell layer. *Dev. Growth Diff.* **29,** 133–139.

23. Robertson, E. J. (1987) Embryo-derived stem cell lines, in *Teratocarcinomas and Embryonic Stem Cells: A Practical Approach* (Robertson, E. J., ed.), IRL, Oxford, pp. 71–112.

24. Nichols, J., Evans, E. P., and Smith, A. G. (1990) Establishment of germ-line-competent embryonic stem (ES) cells using differentiation inhibiting activity. *Development* **110,** 1341–1348.

25. Wobus, A. M., Holzhausen, H., Jakel, P., and Schoneich, J. (1984) Characterization of a pluripotent stem cell line derived from a mouse embryo. *Exp. Cell Res.* **152,** 212–219.

26. Bergstrom, S. (1978) Experimentally delayed implantation, in *Methods in Mammalian Reproduction* (Daniel, J. C., ed.), Academic, NY, pp. 419–435.

27. Solter, D. and Knowles, B. B. (1975) Immunosurgery of mouse blastocyst. *Proc. Natl. Acad. Sci. USA* **72,** 5099–5102.

28. Axelrod, H. R. and Lader, E. (1983) A simplified method for obtaining embryonic stem cell lines from blastocysts, in *Cold Spring Harbor Conferences on Cell Proliferation. Volume 10. Teratocarcinoma Stem Cells* (Silver, L. M., Martin, G. R., and Strickland, S., eds.), Cold Spring Harbor Laboratory Press, Cold Spring Harbor, NY, pp. 665–670.

29. Martin, G. R. and Lock, L. F. (1983) Pluripotent cell lines derived from early mouse embryos cultured in medium conditioned by teratocarcinoma stem cells, in *Cold Spring Harbor Conferences on Cell Proliferation. Volume 10. Teratocarcinoma Stem Cells* (Silver, L. M., Martin, G. R., and Strickland, S., eds.), Cold Spring Harbor Laboratory Press, Cold Spring Harbor, NY, pp. 635–646.

30. Mummery, C. L., Feyen, A., Freund, E., and Shen, S. (1990) Characteristics of embryonic stem cell differentiation: a comparison with two embryonal carcinoma cell lines. *Cell Diff. Dev.* **30,** 195–206.

31. Brown, D. G., Warren, V. N., Pahlson, P., and Kimber, S. J. (1991) Carbohydrate antigens expressed in embryonic stem cells. I. Lacto and neo-lacto determinants. *Histochem. J.* **25,** 452–463.

32. Pesce, M. and Schöler, H. R. (2000) Oct-4: Control of totipotency and germline determination. *Mol. Reprod. Dev.* **55,** 452–457.

33. Eistetter, H. R. (1989) Pluripotent embryonal stem cell lines can be established from disaggregated mouse morulae. *Dev. Growth Diff.* **31,** 275–282.
34. Gardner, R. L. (1985) Clonal analysis of early mammalian development. *Philos. Trans. Royal Soc. London* **312,** 163–178.
35. Smith, A. G. and Hooper, M. L. (1987) Buffalo rat liver cells produce a diffusible activity which inhibits the differentiation of murine embryonal carcinoma and embryonic stem cells. *Dev. Biol.* **121,** 1–9.
36. Hooper, M., Hardy, K., Handyside A., Hunter, S., and Monk, M. (1987) HPRT-deficient (Lesch-Nyhan) mouse embryos derived from germline colonization by cultured cells. *Nature* **326,** 292–295.
37. Gough, N. M., Williams R. L., Hilton, D. J., Pease, S., Willson, T. A., Stahl, J., Gearing, D. P., Nicola, N. A., and Metcalf, D. (1989) LIF: a molecule with divergent actions on myeloid leukaemic cells and embryonic stem cells. *Reprod. Fertil. Dev.* **1,** 281–288.
38. Wobus, A. M., Guan, K., and Pich, U. (2001) In vitro differentiation of embryonic stem cells and analysis of cellular phenotypes, in: Gene Knockout Protocols (Tymms, M. J. and Kola, I., eds.), Humana Press, NJ, *Meth. Mol. Biol.* **158,** 263–286.
39. Robertson, E. J. and Bradley, A. (1986) Production of permanent cell lines from early embryos and their use in studying developmental problems, in: *Experimental Approaches to Mammalian Embryonic Development* (Rossant, J. and Pederson, R. A., eds.), Cambridge University Press, Cambridge, UK, pp. 475–49.
40. Nesbitt, M. N. and Franke, U. (1973) A system of nomenclature for band patterns of mouse chromosomes. *Chromosoma* **41,** 145–158.
41. Kaufman, M. H., Robertson, E. J., Handyside, A. H., and Evans, M. J. (1983) Establishment of pluripotent cell lines from haploid mouse embryos. *J. Embryol. Exp. Morphol.* **73,** 249–261.
42. Evans, M., Bradley, A., and Robertson, E. (1985) EK cell contribution to chimeric mice: from tissue culture to sperm, in *Genetic Manipulation of the Early Mammalian Embryo*, Banbury report 20 (Costantini, F. and Jaenisch, R., eds.), Cold Spring Harbor Laboratory, Cold Spring Harbor, NY, pp. 93–102.
43. Mann, J. R., Gadi, I., Harbison, M. L., Abbondanzo, S. J., and Stewart, C. L. (1990) Androgenetic mouse embryonic stem cells are pluripotent and cause skeletal defects in chimeras: implications for genetic imprinting. *Cell* **62,** 251–260.
44. Copp, A. J. (1982) Effect of implantational delay on cellular proliferation in the mouse blastocyst. *J. Reprod. Fertil.* **66,** 681–685.
45. Gardner, R. L., Davies, T. J., and Carey, M. S. (1988) Effect of delayed implantation on differentiation of the extra-embryonic endoderm in the mouse blastocyst. *Placenta* **9,** 343–359.
46. Moreau, J.-F., Donaldson, D. D., Bennet, F., Witek-Giannotti, J., Clark, S. C., and Wong, G. G. (1988) Leukaemia inhibitory factor is identical to the myeloid growth factor human interleukin for DA cells. *Nature* **336,** 690–692.
47. Gough, N. M., Gearing, D. P., King, J. A., Willson, T. A., Hilton, D. J., Nicola, N. A., and Metcalf, D. (1988) Molecular cloning and expression of the human homo-

logue of the murine gene encoding myeloid leukemia-inhibitory factor. *Proc. Natl. Acad. Sci. USA* **85,** 2623–2627.

48. Wells, D. N. (1991) Studies on the isolation of murine and ovine embryonic stem cells. PhD thesis, University of Edinburgh, Edinburgh, Scotland.

49. McWhir, J., Schnieke, A. E., Ansell, R., Wallace, H., Colman, A., Scott, A. R., and Kind, A. J. (1996) Selective ablation of differentiated cells permits isolation of embryonic stem cell lines from murine embryos with a non-permissive genetic background. *Nat. Genet.* **14,** 223–226.

50. Rathjen, P. D., Nichols, J., Toth, S., Edwards, D. R., Heath, J. K., and Smith, A. G. (1990) Developmentally programmed induction of differentiation inhibiting activity and control of stem cell populations. *Genes Dev.* **4,** 2308–2318.

51. Pease, S., Braghetta P., Gearing, D., Grail, D., and Williams, R. L. (1990) Isolation of embryonic stem (ES) cells in media supplemented with recombinant leukemia inhibitory factor (LIF). *Dev. Biol.* **141,** 344–352.

52. Lederman, B. and Burki, K. (1991) Establishment of a germ-line competent C57BL/6 embryonic stem cell line. *Exp. Cell. Res.* **197,** 254–258.

53. Ginsberg, M., Snow, M. H. L., and McLaren, A. (1990) Primordial germ cells in the mouse embryo during gastrulation. *Development* **110,** 521–528.

54. Tam, P. P. L. and Snow, M. H. L. (1981) Proliferation and migration of primordial germ cells during compensatory growth in mouse embryos. *J. Embryol. Exp. Morphol.* **64,** 133–147.

55. Flanagan, J. G. and Leder, P. (1990) The kit ligand: a cell surface molecule altered in steel mutant fibroblasts. *Cell* **63,** 185–194.

# 7

## Production of Chimeras Derived from Murine Embryonic Stem Cells

### David Wells

## 1. Introduction

Embryonic stem (ES) cells are undifferentiated cells derived from early mouse embryos, which under appropriate culture conditions proliferate continuously in vitro. ES cells have been demonstrated to be pluripotent in vivo from their capacity to form teratocarcinomas and germline chimeric mice (*1–3*; *see* **Fig. 1**), depending on the environment into which the stem cells are introduced. When ES cells are introduced under the kidney capsule, in vivo differentiation is chaotic with the teratocarcinoma composed of a wide variety of different cell types. If, however, the stem cells are returned into a preimplantation mouse embryo, in vivo differentiation proceeds in a normal and organized manner, and the ES cells colonize the three primary cell lineages of the developing embryo: the primitive ectoderm, endoderm, and mesoderm. This leads to the formation of chimeric offspring composed of cells of two different genetic constitutions: the host embryonic cells and those derived from the ES cells.

ES cells are capable of contributing to every tissue in the fetus, including the germ cells (*3,4*). Occasionally, the trophectodermal and primitive endodermal derivatives in the extraembryonic tissues of the conceptus may also be colonized by ES cells (*5,6*). Most significantly, some mouse ES cell lines possess the potential to support complete fetal development, with adult fertile mice being generated following manipulations that combine tetraploid embryos and a small number of stem cells (*7–9*).

Although ES cell–derived chimeras may be generated relatively easily (an average of 35% of pups born were chimeric in a study of 17 independently derived ES cell lines *[4]*), obtaining germline transmission is much more difficult and

From: *Methods in Molecular Biology, vol. 180: Transgenesis Techniques, 2nd ed.: Principles and Protocols*
Edited by: A. R. Clarke © Humana Press Inc., Totowa, NJ

Fig. 1. A male germline chimera transmitting the dominant black agouti coat color genotype derived from the 129/*Sv* ES cells to all offspring, after mating with an albino MF1 female. The chimera was produced following the injection of XY stem cells into a presumed XX host blastocyst, from the albino MF1 mouse strain. The male stem cells resulted in the "sex conversion" of the mouse.

requires the use of euploid ES cells (*see* **Notes 1** and **2**). It is important to monitor regularly the karyotype of cell lines to check that they do not drift significantly from the norm. However, the loss of potential for germline transmission may also be a consequence of more minor genetic or epigenetic changes that may accumulate with extended culture and vary with individual cell lines *(10)*. It is generally found that the frequency of germline chimerism is greatest with early passage cells, typically up to 10–15 passages, and normally declines thereafter. Under normal circumstances, ES cell–derived chimeras are not associated with high incidences of midgestational fetal loss or tumor formation in mice *(4)*, as often occurs with chimeras derived from embryonal carcinoma cells (the undifferentiated cells isolated from teratocarcinomas; reviewed in **ref.** *11*). However, these abnormalities may still occur if one or more of the originally injected stem cells "escapes" the regulative control of the host embryo *(12)*. Therefore, it is important that excessive numbers of ES cells not be incorporated with early embryos during the manipulations used to produce chimeras; otherwise, abnormal fetal development may result *(7,12–14)*.

ES cells are ideally suited to genetic modification and subsequent selection in vitro. In combination with their capacity to colonize the germline of mice, ES cells have been extensively utilized to introduce precise mutations into the mouse genome via the application of homologous recombination (gene targeting) technology, in order to study the developmental function of specific genes *(15)*. To achieve the objectives of these experiments, efficient production of germline chimeras is an important component.

This chapter describes two general methods for generating chimeric mice: the physical injection of ES cells into host embryos (morula and blastocyst-stages), and the aggregation of stem cells with cleavage-stage embryos. The following sections consider the general requirements for mice, culture media, embryo injection and aggregation procedures, embryo transfer to recipient females, and analysis of chimeric offspring.

## 2. Materials

### 2.1. Mouse Requirements

Mice are required to act as donors to supply host embryos and pseudo-pregnant recipients to foster the manipulated embryos.

### 2.1.1. Donor Mice

The choice of mouse strain to use as donors for host embryos depends on the genotype of the stem cells and the combination of cell markers necessary to distinguish the host embryonic cells from those cells in the chimera of stem cell origin (*see* **Note 3**). Furthermore, consideration should be given to the compatibility between the two mouse strains used, to allow the stem cells a competitive advantage over the cells of the host embryo, during subsequent development (*see* **Note 2**). The procedure for superovulation is described in Chapter 6, Subheading 3.2.1. (*see* **Note 4**). To produce embryos of the correct developmental stage at a convenient time of day for the experimenter, the light cycle and the timing of the gonadotropin injections are important parameters. Mice are normally housed in a cycle of 12 h of light and 12 h of dark. Mice normally mate in the midpoint of the dark cycle. With superovulated females, it is generally assumed that ovulation occurs 12 h after the administration of human chorionic gonadotropin, which should be timed to occur around the midpoint of the dark cycle to optimize fertilization. The approximate time of initiation of embryonic development is determined by checking females for copulation plugs the following morning. This is designated d 0.5 postcoitum (p.c.).

Eight-cell-stage embryos are found in the oviduct on d 2.5 p.c. By d 3.5 p.c. in the mouse, the embryos have entered the uterus and developed to the expanded blastocyst stage. The methods for recovering these two embryo stages from the reproductive tract are described in Chapter 6. Following recovery, embryos are group cultured in microdrops (25 µL) of the appropriate medium (*see* **Subheading 2.2.**) under oil, until they are required for manipulation.

## 2.1.2. Recipient Mice

Pseudo-pregnant recipient females are used as surrogate mothers for the manipulated embryos. Recipients can be of any genetic strain that is available; however, to optimize embryonic survival, it is best to use F1 hybrid strains (e.g., C57BL/6 x CBA), because a higher proportion of F1 recipients tend to maintain a pregnancy, and they also tend to be better mothers.

Pseudo-pregnant recipients are obtained from natural matings between females and vasectomized males. The approximate time of ovulation is determined by checking for copulation plugs the following morning, which is designated d 0.5 p.c. These matings must be set up so that the recipient females are 1 d less advanced developmentally than the donor embryos that they will receive.

## 2.2. Embryo Culture

Cleavage- and blastocyst-stage mouse embryos are cultured at 37°C in a humidified 5% $CO_2$ (in air) incubator. However, these two embryonic stages must be cultured in separate media because of differences in their energy substrate requirements.

### 2.2.1. Media for Cleavage-Stage Embryos

Cleavage-stage embryos do not have a glucose-based metabolism and are cultured in Medium 16 (M16) with pyruvate as an energy source, supplemented with 4 mg/mL of bovine serum albumin (BSA) immediately before use. For M16 *(16)*, to 100 mL of Analar water, add 0.5534 g of NaCl, 0.0356 g of KCl, 0.0162 g of $KH_2PO_4$, 0.0294 g of $MgSO_4 \cdot 7H_2O$, 0.0252 g of $CaCl_2 \cdot 2H_2O$, 0.32 mL of sodium lactate (60% syrup), 0.0036 g of sodium pyruvate, 0.1000 g of D-glucose, 0.0010 g of phenol red, and 0.2106 g of $NaHCO_3$. Filter sterilize (0.22 µm), and store at 4°C for not more than 3 wk.

For manipulations conducted outside the $CO_2$ incubator, it is recommended that Medium 2 (M2) with 4 mg/mL of BSA, supplemented with HEPES buffer, be used to maintain the pH. For M2 *(17)*, to 100 mL of Analar water, add 0.5534 g of NaCl, 0.0356 g of KCl, 0.0162 g of $KH_2PO_4$, 0.0294 g of $MgSO_4 \cdot 7H_2O$, 0.0252 g of $CaCl_2 \cdot 2H_2O$, 0.32 mL of sodium lactate (60% syrup), 0.0036 g of sodium pyruvate, 0.1000 g of D-glucose, 0.0010 g of phenol red, 0.0337 g of $NaHCO_3$, and 0.5000 g of HEPES (free acid). Filter sterilize (0.22 µm), and store at 4°C for not more than 3 wk.

### 2.2.2. Media for Blastocysts

Blastocysts have a glucose-based metabolism and may be cultured in the medium used for ES cells ($ES_{10}$ medium; *see* Chapter 6). For manipulations conducted outside the incubator, it is recommended that medium supplemented with 20 m$M$ HEPES buffer be used to maintain the pH.

## 2.2.3. Media for ES Cell Culture

These are described in detail in Chapter 6.

## 2.2.4. Media for Removing Zonae Pellucidae

1. Acidified Tyrode's solution: To 100 mL of Analar water, add 0.800 g of NaCl, 0.020 g of KCl, 0.024 g of $CaCl_2 \cdot 2H_2O$, 0.010 g of $MgCl_2 \cdot 6H_2O$, 0.100 g of D-glucose, 0.400 g of polyvinylpyrrolidone (PVP 40). Adjust to pH 2.1 with Analar HCl. Filter sterilize (0.22 μm), and store aliquots at −20°C.
2. Pronase solution:
   a. Make 0.5% (w/v) pronase solution (Calbiochem) in M2 (*see* **Subheading 2.2.1.**) with 4 mg/mL of BSA (M2 + BSA).
   b. Sprinkle 1.0% (w/v) PVP 40 on top of the solution and dissolve slowly.
   c. Incubate at 37°C for 2 h to digest contaminating nucleases.
   d. Centrifuge the solution to remove the insoluble material.
   e. Filter sterilize (0.22 μm) and store aliquots at −20°C.

## 2.2.5. Decompaction and Aggregation Media

Decompaction of morulae is assisted by incubation of embryos in $Ca^{2+}$- and $Mg^{2+}$-free phosphate-buffered saline (PBS) containing 0.3% (w/v) EDTA, until such time that the blastomeres become distinguishable. PBS may be prepared either by dissolving preformulated tablets (e.g., Flow Laboratories) or by adding the following ingredients to 1 L of sterile water: 10.0 g of NaCl, 0.25 g of KCl, 1.44 g of $Na_2HPO_4 \cdot 12H_2O$, 0.25 g of $KH_2PO_4$. PBS is adjusted to pH 7.2. Aliquot into bottles, sterilize by autoclaving, and store PBS at room temperature.

Aggregation of cleavage-stage embryos may be assisted by the inclusion of phytohemagglutinin (0.2 μg/mL) in media.

## 2.3. ES Cell Injection into Mouse Embryos

The injection chamber used here is prepared simply by introducing a 250-μL drop of medium into the center of a 90-mm plastic bacteriologic dish and overlayering this with lightweight paraffin oil. The dish is then transferred to the microscope stage. The medium comprises $ES_{10}$ + LIF (*see* Chapter 6) with 20 m*M* HEPES buffer to maintain the pH with the manipulations conducted outside the incubator. For blastocyst injection, some researchers recommend the use of equipment to cool the injection chamber to 8–10°C (e.g., *see* **ref. 15**). The principal advantages are that the embryo becomes more rigid and thus easier to penetrate with the injection pipet; and cell lysis is reduced, thereby minimizing the resulting "stickiness." However, the simple procedure described here works quite satisfactorily at room temperature.

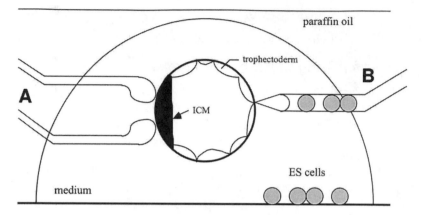

Fig. 2. Orientation of micropipets within the manipulation chamber. A holding pipet (a) is used to immobilize the blastocyst by gentle suction, while an injection pipet (b) introduces the ES cells into the blastocoel cavity of the embryo (figure not to scale).

## 2.3.1. Preparation of Micropipets

Two micropipets, constructed from glass capillary tubing, are utilized for the embryo manipulations. The embryo is immobilized by gentle suction with a holding pipet, and an injection pipet is used to introduce the ES cells either into the blastocoel cavity of blastocysts (**Figs. 2** and **3**; *see* **Subheading 3.2.1.**) or under the zona pellucida of morula-stage embryos (*see* **Subheading 3.2.2.**).

### 2.3.1.1. INJECTION PIPETS

Injection pipets are prepared by pulling thin-walled, 1.0-mm-diameter glass capillary tubing (e.g., GC100T-15; Clark Electromedical) on a horizontal or vertical pipet puller (e.g., Campden or Kopf). The settings are adjusted to produce a capillary with a gradual taper over a 15- to 20-mm length of glass, from the shoulder to the needle point. The pulled capillary tube is then broken at right angles at an external diameter of 18–20 μm utilizing a microforge (e.g., Defonbrune, Narishige, or Research Instruments). This diameter is generally just large enough to accommodate the stem cells. The capillary is positioned horizontally on the microforge with the small glass bead (0.2–0.5 mm in diameter) on the platinum wire heating filament, directly below the capillary at the required diameter (determined by a graticular eyepiece). With the glass bead heated slightly (the temperature has to be determined empirically), the equatorial plane of the capillary is slowly brought into contact with the glass bead, so that the two just barely fuse together. The filament is then switched off, and as the glass bead retracts as it cools, it causes the capillary to break cleanly at the point of contact.

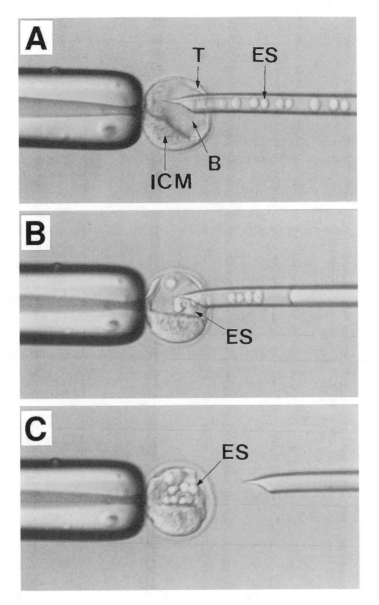

Fig. 3. Procedure for injecting ES cells into a blastocyst. (**A**) The blastocyst is immobilized by microsuction on the tip of the holding pipet. The injection pipet is pushed through the zona and trophectoderm (T) and into the blastocoel cavity (B), which is expanded. (**B**) Ten to 15 ES cells, previously loaded into the injection pipet, are slowly expelled into the blastocoel cavity. (**C**) The injection pipet is withdrawn from the embryo, once a slight negative pressure has been applied, aiding the collapse of the trophectoderm. Once inside the embryo, the ES cells become incorporated into the inner cell mass (ICM) and participate in subsequent embryogenesis.

To allow the injection pipet to enter the embryo cleanly, a sharp tip is required. With a good deal of patience and skill, some operators are able to snap the glass capillary by hand in the dissecting microscope using a sharp scalpel blade and a spongy silicon rubber mat *(15)*. Although it may be a little more time-consuming, using a microgrinder (Narashige) to bevel the pipets is certainly a more reproducible method to construct the desired injection tip. The tip of the pipet is beveled to an angle of 45°. This is accomplished by grinding the pipets for approx 5–10 s on the grinding wheel lubricated with distilled water and with a stream of air flowing through the inside of the pipet, to keep the internal bore free of debris. The outside of the beveled tip is then cleaned by immersing the tip in 15% (v/v) hydrofluoric acid for 15 s with a constant stream of air flowing through the capillary using a syringe and plastic tubing adapter, to prevent the acid from entering the bore of the pipet. The pipet tip is then rinsed in two changes of distilled water. With the aid of a microforge, a sharp point is then put onto the leading edge of the beveled pipet. With the pipet positioned horizontally, the filament is heated to a low tempera-ture, at which the pipet tip just barely fuses to the glass bead. Once fused, the pipet is quickly moved away from the filament to draw out a short glass spike. It is important that the spike not be too long, since during the injection proce-dure, the blastocyst might collapse before the pipet has completely entered the blastocoel cavity.

To provide a "siliconizing" effect, the tips of the injection pipets are washed in a 1.25% (v/v) detergent solution of Tween-80 in distilled water using a syringe and a plastic tubing adapter, to draw the solution inside the fine capillary repeatedly and then blown out with a stream of air. The pipets are then allowed to dry at room temperature before use.

### 2.3.1.2. Holding Pipets

The holding pipet is constructed from a 10-cm-long capillary tube (external diameter of 1.0–1.5 mm; e.g., GC100-10, Clark Electromedical). The taper may be fashioned either by utilizing a pipet puller as already described, or by hand. A small section of the capillary tubing may be uniformly heated over a small flame of a propane gas burner. Once softened, the capillary is withdrawn from the flame and pulled quickly by hand, to produce a gradual taper. It is desirable that the tip of the holding pipet have an external diameter of 90–100 µm, which is about 2–5 cm from the shoulder of the capillary. This pulled capillary is then broken at right angles by positioning the capillary vertically on the microforge, at the required diameter, adjacent to the glass bead on the heating filament. The capillary is brought into contact with the glass bead in an equato-rial position and delicately scored at this point with a fine diamond pencil. The bottom of the capillary (which is waste) is gently tapped, and the glass usually breaks cleanly at the desired diameter. Because of the relatively large diameter of the holding pipet, it is generally easier to break the pipet squarely using the aforementioned procedure than trying to use the microforge as described for the injection pipet, especially if the glass bead is small, as is necessary for constructing the finer injection pipets. The tip of the holding pipet is then heat polished by placing the tip above, but just clear of, the hot glass bead on the microforge, causing the glass to melt and produce a concentric orifice of 20–25 µm in diameter.

### 2.3.1.3. SECONDARY STRUCTURE OF PIPETS

The appropriate bends need to be created in the pipets with the microforge, so that they can enter the injection chamber and allow the tips of the pipets to remain at the same focal level when they are being moved in a horizontal plane. The exact angles depend on the type of chamber being used. To allow entry into the injection chamber constructed from a bacteriologic dish, both the injection and holding pipets are bent at an angle of about 30°, approx 4 mm from the respective tips (**Fig. 1**). It is important to orientate the injection pipet correctly on the microforge, so that once the bend has been created, the bevel of the pipet will be in the desired orientation when assembled on the micromanipulation arm.

## 2.3.2. Micromanipulator Assembly

This section describes the micromanipulatory, optical, and suction systems, and the orientation of the pipets in the injection chamber.

### 2.3.2.1. MANIPULATORS AND OPTICS

Embryo injections are performed using hand-operated micromanipulators to control the three-dimensional movement of the two pipets in the injection chamber precisely. Two brands commonly used by researchers are Leitz and Narishige. One micromanipulation arm is mounted on each side of the microscope. The injection and holding pipets can be mounted on either of these micromanipulators depending on comfort for the operator. For embryo injections, ideally an inverted, phase-contrast microscope with a magnification range of ×40 to ×200 and with a fixed stage should be used (e.g., Nikon Diaphot or Leitz Diavert). Such microscopes provide a large working distance between the stage and the condenser to accommodate the injection chamber, and to facilitate simple entry of the pipets into the chamber. Phase-contrast optics are desirable in order to better distinguish between live and dead cells. If, however, differential interference contrast objectives are used, the injection chamber must be constructed of glass and not plastic. This may be the typical "hanging drop" injection chamber, constructed from a normal glass microscope slide with two lengths of glass, 22 mm long and 2 mm square, adhered to the longitudinal edges of the slide with Vaseline to support a glass cover slip, also held in place with Vaseline. A small volume of medium is introduced into the middle of the cavity, and each side is backfilled with mineral oil to prevent evaporation.

### 2.3.2.2. MICROSUCTION

The two instrument holders on the manipulation arms are each connected to micrometer spring-loaded syringes using thick-walled (polyethylene) tubing (internal diameter of about 1.0 mm). The tubing is filled with lightweight inert mineral oil (Fluorinert 77 [Sigma], paraffin, or silicon oils). For ease of operation, each microsyringe should be placed on the side opposite to the manipulator that controls the movement of the pipet. In this way, the movement and suction of one pipet can be controlled simultaneously with both hands. For the holding pipet, a 0.5- to 1.0-mL ground glass syringe may be used. To allow finer control of the suction with the injection pipet, a 50- to 250-μL Hamilton syringe is ideal. Before fixing the pipets into the respective instrument holders, the tubing must be purged of air bubbles using a 20-mL reserve syringe filled with inert oil, connected into the hydraulic line via a three-way tap. The pipets are locked into the instrument holders, and using the reservoir syringes, each pipet is filled with inert oil.

### 2.3.2.3. ORIENTATION WITHIN INJECTION CHAMBER

The pipets are aligned in the injection chamber with the tips parallel to the bottom surface, so that they remain in the same focal level when moving in a horizontal plane (**Fig. 1**). The type of injection chamber used dictates the secondary structure of the pipets and the arrangement of the micromanipulation arms to allow entry into the chamber. The injection chamber recommended here is prepared by simply introducing a 250-μL drop of medium into the center of a 90-mm plastic bacteriologic dish, overlayered with lightweight paraffin oil. Thus, a 30° bend created at the tip of the pipets (*see* **Subheading 2.3.1.3.**) means that with the instrument holders angled at 60°, the tips of the pipets are parallel to the bottom of the chamber. With the hanging-drop injection chamber, the pipets must enter the chamber horizontally, and, thus, the secondary bends must allow for this different orientation.

Once the pipets are positioned within the chamber, medium from the chamber is drawn a short distance into each pipet using the micrometer-controlled syringes. Improved microsuction control with the injection pipet is often achieved by drawing up several short columns of medium interspaced with oil. Several hundred freshly trypsinized ES cells are then introduced into the chamber and allowed to settle to the bottom before commencing injections.

# 3. Methods

## 3.1. Preparation of ES Cells

For routine culture and maintenance of ES cells, refer to Chapter 6. Briefly, the cells are cultured on a 3-d passaging regime, with ES cells being used for chimera production on d 1 or 2 of the growth cycle. To maximize stem cell viability, the medium should be changed every day, and 2 to 3 h before use (Important: *see* **Note 5**). If the ES cells have been cocultured with feeder cells, the majority of feeders may be removed before manipulation by seeding thinly a gelatin-coated dish with the cell suspension and incubating the dish for 30 min at 37°C, by which time many of the feeders have preferentially attached to the culture surface.

### 3.1.1. Preparation of ES Cells for Injection into Embryos

For injection into host embryos, the ES cells are trypsinized into a single-cell suspension and a small volume, containing several hundred cells, is introduced into the injection chamber on the microscope stage. After 1 to 2 h of micromanipulation, these stem cells should be discarded and a new injection chamber prepared with freshly trypsinized ES cells.

### 3.1.2. Preparation of ES Cells for Aggregation with Morulae

For aggregation with morulae (*see* **Subheading 3.3.3.1.**), the ES cells are prepared by lightly trypsinizing them for 1 to 2 min to give a suspension consisting of small clumps of cells. A mouth-controlled pipet is then used to pick out healthy clumps individually comprising 5–10 cells each, which are transferred to microdrops of ES cell medium and incubated for 1 h. This allows the cells to recover and form rounded masses. The microdrops should be prepared in bacteriologic dishes to prevent the cells from adhering to the plastic.

## 3.2. Injection of ES Cells into Mouse Embryos

The procedure described here is a simplified version of the techniques detailed by Bradley *(18)*.

### 3.2.1. Blastocyst Injection Procedure

1. Transfer embryos to be injected from the incubator into the injection chamber only in small groups to prevent them from being exposed to lower temperatures for prolonged periods.
2. Lower the injection pipet to the base of the chamber, and individually select and gently drawn 10–15 healthy ES cells for one injection into the pipet by microsuction, so that they are positioned one directly behind the other. With experience, up to 100 ES cells may be aspirated into the pipet at one time, to allow for the serial injection of 10 blastocysts.

3. Hold the expanded blastocysts by suction onto the holding pipet. By patiently blowing and sucking medium through the orifice of the holding pipet, the embryo can be rolled along the base of the chamber until it is eventually orientated in the desired position. It is preferable either that the ICM region of the blastocyst be positioned over the orifice of the pipet (as in **Fig. 1**) or that the ICM be at either the "north or south pole" positions (as in **Fig. 2A**).

4. Focus, at high magnification, on a "junction" between where two trophectodermal cells join together around the equatorial plane of the embryo. Then raise the tip of the injection pipet to this same focal level and bring the injection pipet to the point just adjacent to this cellular junction (**Fig. 1**).

5. With a controlled forward movement on the manipulation joystick, push the pipet into the blastocoel cavity (**Fig. 2A**). This movement is critical to the success of the operation. If the pipet is moved too slowly, the blastocyst may collapse before the injection pipet has fully entered the blastocoel cavity. If, on the other hand, the injection pipet is moved too rapidly in an uncontrolled manner, it may either damage the embryo extensively or break the tip of the pipet if it makes contact with the holding pipet. It is generally easier to inject blastocysts when they are very expanded, with the trophectoderm "stretching" the zona. If necessary, blastocysts may be cultured until they reach this stage (alternatively, the use of a cooling stage increases the rigidity of the trophectoderm and improves ease of injection).

6. With the injection pipet inside the blastocoel cavity, gently blow the ES cells out of the pipet and into the embryo (**Fig. 2B**). It is important to have a controlled microsuction system. Air or leaks in the hydraulic line often cause the stem cells to rush in and out of the pipet (*see* **Note 2** for comments on the number of stem cells introduced).

7. Once the ES cells have been introduced, apply a small amount of suction with the injection pipet, in order to remove the small buildup of pressure created from the media introduced along with the cells. This minimizes any loss of injected stem cells through the hole in the trophectoderm, following the withdrawal of the pipet from the blastocyst. Shortly thereafter, the blastocyst will collapse owing to the punctured trophectoderm (**Fig. 2C**). Then release the injected embryo from the tip of the holding pipet, in a remote part of the injection chamber, and repeat the process with a new blastocyst.

8. During the manipulations, cellular debris and cells often unavoidably adhere to the tip of the injection pipet. If this occurs, these may be removed by simply passing the tip repeatedly through the media/oil meniscus. Alternatively, tools may be cleaned periodically by using PBS containing 0.025% trypsin, 20 µg/mL of proteinase K, and 300 IU/mL of DNase I, and then rinsing in medium before using again.

9. Once the embryos in the chamber have all been used, transfer the successfully injected blastocysts to the incubator in a droplet of $ES_{10}$ medium under oil, where they will slowly begin to reexpand. After 30–60 min of culture, the ES cells may be seen adhering to the ICM of the host blastocyst. Injected embryos should not

be cultured for more than 6 h before being transferred to recipients; otherwise, many of them may begin to hatch from the zona and are therefore more difficult to handle. Injected blastocysts should be transferred regardless of whether they have fully reexpanded.

### 3.2.2. Morula Injection Procedure

The injection of ES cells into earlier cleavage-stage embryos may result in the stem cells contributing more extensively to the tissues of the chimera than from blastocyst injections *(19,20)* (*see* **Note 2**). The procedure uses the same micromanipulatory setup as described for blastocyst injections. It is technically straightforward, simply requiring the injection pipet to enter through the zona pellucida and introduce about 5–10 ES cells close to, or even between, individual blastomeres.

It has been preferred to inject the ES cells before the onset of compaction. To decompact eight-cell embryos that are in the process of compacting, *see* **Subheading 2.2.5.**

To minimize physical damage to the blastomeres, it is beneficial to add 1 µg/mL of cytochalasin D to the injection medium to increase the plasticity of the blastomere cell membranes, making them less prone to lysis following contact with the injection pipet. It is important to remove the residual cytochalasin following the injection of a batch of embryos.

Following injection, the morulae may be cultured overnight in M16 medium and transferred at the blastocyst stage.

### 3.3. Aggregation of ES Cells with Morulae

This procedure does not require the sophisticated and expensive micromanipulatory equipment necessary for embryo injections, nor the personnel with the necessary expertise to perform the manipulations. All that is required are the elementary skills necessary for the routine handling of preimplantation-stage embryos. Essentially, there are two methods of aggregation: sandwiching a clump of ES cells between two eight-cell embryos *(7)*, or coculture of eight-cell embryos on a "lawn" of ES cells *(14)*. The use of these aggregation methods can result in similar proportions of germline-transmitting chimeras compared to blastocyst injection; however, more of the males tend to transmit the ES cell genotype to all of their offspring *(13)* (*see* **Note 2**).

Since the zona is removed from the eight-cell embryos to enable aggregation, the ES cell–embryo aggregates must be cultured to the blastocyst stage before they are transferred to recipients. As a result, many original protocols suggested the use of a compromise medium catering for the different nutrient requirements of both ES cells and cleavage-stage embryos *(18)*. The medium therefore contained pyruvate, to support the cleavage-stage embryos, and serum, rather than BSA, for the maintenance of the stem cells. However, if the embryo culture conditions excessively favor the ES cells, they tend to pre-dominate the tissues of the chimeric fetus and may result in severely abnormal development *(13)*. Therefore, following the attachment of stem cells to the embryo by either aggregation method, the aggregates should be cultured in a medium such as M16 + BSA to favor embryo development. Also, the number of ES cells aggregated to the embryo should be limited to avoid developmental problems in the fetus (*see* **Note 2**).

### 3.3.1. Removal of Zonae Pellucidae

The zona may be digested from eight-cell embryos by treatment with either acidified Tyrode's solution or with a pronase enzymatic solution.

1. Using a finely pulled, mouth-controlled Pasteur pipet, transfer a few embryos at a time into a 25-μL drop of either Tyrode's or pronase solution (these drops do not have to be under oil).
2. While monitoring the progress in the dissecting microscope, quickly remove the embryos from the drop just before the last remnants of the zona are about to dissolve. This usually takes between 20 and 30 s for the Tyrode's solution and 2 to 3 min for the pronase solution, with digestions performed at room temperature.
3. Wash the zona-free embryos through several microdrops of M2 + BSA to remove traces of the digesting agents. Then culture the zona-free eight-cell embryos in pairs at 37°C in microdrops of M16 + BSA for 1 h to increase the "stickiness" of the cell surfaces. It is important that the microdrops be set up in bacteriologic dishes and not tissue culture dishes to prevent the embryos from adhering to the plastic surface.

### 3.3.2. Extent of Embryo Compaction

Chimeric mice have been produced following aggregation of ES cells with both precompacted eight-cell embryos and those having just completed the compaction process *(7,13)*. There are two advantages to using eight-cell embryos undergoing compaction: the ES cells adhere more strongly to these embryos; and they are not as fragile to handle without a zona, compared to uncompacted stages *(13)*. If necessary, however, morulae may be decompacted by briefly incubating embryos at 37°C in $Ca^{2+}$- and $Mg^{2+}$-free PBS containing 0.3% (w/v) EDTA.

### 3.3.3. Aggregation Procedures

#### 3.3.3.1. MORULA–ES CELL CLUMP AGGREGATION

1. Transfer a small clump of 5–10 ES cells (prepared as described in **Subheading 3.1.2.**) and two zona-free eight-cell embryos to each 5-μL drop of M16 with 10% (v/v) fetal calf serum (FCS) under oil, in a bacteriologic dish.

2. Sandwich the ES cells between the two embryos by blowing streams of medium through a fine, mouth-controlled Pasteur pipet. Since the cells are very sticky, it is important not to manipulate them with the pipet itself; otherwise, they might adhere to the glass.

3. To ensure that the cell surfaces remain sticky, it is necessary that the manipulations be performed at or near 37°C and that the cells do not cool extensively. Cooling can be minimized by having a limited number of microdrops per dish, and having a heated microscope stage. The aggregates should be checked after 30 min to ensure that they are beginning to adhere together.

4. If there are problems in getting the aggregates to form, some of the medium in the microdrop can be aspirated to lower the oil/medium meniscus down onto the embryos, forcing them together. Alternatively, small depressions may be made in the plastic bottom of the bacteriologic dish with a needle to allow the clump of ES cells and embryos to nestle together *(10)*. Aggregation also can be enhanced by chemical agents such as phytohemagglutinin (0.2 μg/mL) *(21)*.

5. Culture the aggregates overnight to the blastocyst stage in M16, and transfer to the uteri of d 2.5 p.c. recipients (*see* **Subheading 3.4.**). Although the resultant blastocysts are composed of more cells than normal (originating from the aggregation of two eight-cell embryos and a group of ES cells), the fetus is not larger than usual, since size regulation occurs in the primitive ectoderm shortly after implantation *(22)*.

### 3.3.3.2. COCULTURE OF EIGHT-CELL EMBRYOS AND ES CELLS

The following is an outline of the technically simple procedure developed by Wood et al. *(14)*, which enables eight-cell embryos to be handled en masse and minimizes the manipulations and time necessary to generate chimeras.

1. Prepare a single-cell suspension of ES cells at a concentration of $1 \times 10^6$ cells/mL in Dulbecco's modified Eagle's medium containing 5% (v/v) FCS and 23 m$M$ lactate (which also serves as the coculture medium).

2. Place 15-μL droplets of the ES cell suspension on a bacteriologic dish and overlayer with oil.

3. Digest the zonae from eight-cell embryos either undergoing or having just completed compaction, and place 5–10 embryos into each drop onto the "lawn" of ES cells.

4. Incubate the dish at 37°C for 2 to 3 h.

5. Following coculture, culture those embryos with approx 5–10 ES cells attached in M16 medium overnight to the blastocyst stage and then transfer to recipient mice.

### 3.3.3.3. AGGREGATION OF TETRAPLOID EMBRYOS AND ES CELLS

The full developmental potential of mouse ES cells has been investigated by aggregating ES cells with developmentally compromised tetraploid four-cell-stage embryos. In this way, fertile adult mice have been generated that are entirely derived from the cultured ES cells *(8–10)*. While ES cells have the potential to contribute to all tissues of the fetus proper, they develop poorly into the cell lineages of the extraembryonic membranes *(5)*, in contrast to tetraploid embryonic cells *(23)*. Thus, the two cell types provide a complementary function when combined. The stem cells are able to differentiate and form the entire fetus, while the tetraploid cells, because of their ploidy, are excluded and establish the placental connection instead.

Although these studies have demonstrated the remarkable developmental potential of ES cells, the methodology is not at the stage of providing a practical alternative in generating genetically modified animals compared to the chimeric route. The reason is that the tetraploid approach is inefficient, with between 0 and 6% of manipulated embryos transferred yielding offspring that may survive to adulthood *(8–10)* although recent results with F1 hybrid ES cell lines suggest higher efficiencies *(24)*. The technique is associated with high incidences of early postimplantation resorptions, midgestational fetal deaths, the requirement of Caesarean section to deliver pups and subsequent fostering, and high postnatal death rates both immediately after birth and during subsequent growth. Typically, about 60% of newborns that are delivered may fail to sustain respiration or have other developmental abnormalities, resulting in their death after Caesarean section.

The success of the procedure appears very dependent on the cell line used, and most of the 100% stem cell–derived pups have originated from early passage cultures. It appears that the proportion of ES cells that maintain full developmental potential diminishes with extended passage, and some cell lines may lose this potential quicker than others. Subcloning, however, may be necessary to isolate developmentally competent cells, and in this way, viable mice have been produced from the equivalent of ES cells at passage 24 *(10)*.

A variety of methods have been used to produce these ES cell–derived mice. The tetraploid mouse embryos may be created either by the electrofusion of two-cell embryos *(7,9)* or following the cytochalasin B treatment of fertilized oocytes *(8)*.

Both aggregation and injection methods have been utilized to combine stem cells with embryos. ES cells have been aggregated with compacting four-cell tetraploid embryos either with small clumps of cells *(7,10)* or following coculture methods using a lawn of ES cells *(8)*. Alternatively, ES cells have been injected into tetraploid embryos cultured to the blastocyst stage *(9)*. All methods have had similar levels of success.

### 3.4. Embryo Transfer

Manipulated embryos are surgically transferred into the reproductive tracts of pseudopregnant female recipients to allow the embryos to develop in vivo.

Pseudopregnant females are produced from matings with vasectomized male mice (Vasectomy is described in Chapter 3). Blastocyst-stage embryos are transferred to the uterus, whereas morulae are transferred to the oviducts. In the case of manipulated embryos, it is best to transfer the embryos to pseudopregnant recipients that are developmentally 1 d less advanced than the embryos themselves. This allows the transferred embryos time to resynchronize with the reproductive tract, following the trauma of the micro-surgery and in vitro culture. Thus, blastocysts are transferred to recipients that are d 2.5 p.c., whereas injected morulae may be transferred to recipients that are d 0.5–1.5 p.c.

The procedure for uterine embryo transfer is outlined in this section. (Oviduct transfer is described in Chapter 3). Surgical operations may be performed on the laboratory bench. However, it is important to minimize animal stress and, hence, losses owing to embryonic resorptions.

1. Anesthetize the mouse (recipient females should be between 6 and 12 wk of age).
2. Swab the back of the mouse with 70% (v/v) alcohol.
3. Use a pair of blunt forceps to pick up the skin, and use dressing scissors to make a 10-mm lateral incision across the midline, posterior to the last rib. Use paper tissues, moistened with 70% (v/v) alcohol, to open the wound and remove any cut hair, by wiping in a head-to-tail direction.
4. By sliding the skin from side to side, visualize the position of the ovaries beneath the body wall. Grasp the region of the body wall directly overlying the ovary with pointed watchmaker's forceps, and make a 5-mm incision in the body wall with sharp iris scissors. By exerting gentle pressure on both sides of the incision, exteriorize the ovarian fat pad and grasp with blunt forceps.
5. Very carefully, pull out the associated ovary, oviduct, and the top of the uterine horn. The weight of the fat pad is generally sufficient to keep the ovary from slipping back inside the abdominal cavity.
6. Transfer the mouse, on absorbent tissues, to the stage of a binocular dissecting microscope, with an associated fiberoptic incident light source.
7. In another dissecting microscope, with transmitted illumination, aspirate the embryos to be transferred into a hand-pulled Pasteur pipet, with an internal diameter just larger than the embryos themselves and with a square tip. The pipet should be preloaded with alternate media and air bubbles, which allows fine control over the movement of the embryos. Draw up the embryos into the pipet, one directly behind the other, so as to introduce a minimum of medium into the reproductive tract when the embryos are transferred. Typically, nine manipulated embryos are transferred unilaterally. If more embryos are to be transferred, it is best to transfer them bilaterally. However, no more than 12 embryos should be transferred per recipient, and no fewer than five embryos should be transferred to each side of the reproductive tract, since they should be crowded to their normal density to prevent oversized fetuses from developing, which may result in pre- or early postnatal death.

8. While focusing on the region of the uterotubal junction (magnification about ×10), use a pair of watchmaker's forceps to grasp the top of the uterine horn, gently lift it up slightly, and use a 25-gage hypodermic needle to puncture the uterine wall and enter the lumen.

9. While still holding the top of the uterus and observing down the microscope, insert the tip of the transfer pipet containing the embryos 5 mm inside the uterine lumen, through the hole created with the needle.

10. Gently blow the embryos into the uterine lumen, using the air bubbles along the pipet as markers. After transfer, the pipet should be checked to ensure that all the embryos were expelled into the reproductive tract.

11. Return the reproductive organs to the abdominal cavity by lifting up one edge of the incision and gently pushing in the ovarian fat pad with blunt forceps. The incision in the body wall can be closed with a single suture (5/0 Mersilk; Ethicon), whereas sutures or Michel clips may be used to close the skin. To aid postoperative recovery, mice should be placed on a heated blanket before they are returned to a clean cage. If pregnant, the females will litter 17 d later.

## 4. Notes

1. In achieving germline chimerism, male ES cell lines have a greater potential than female lines, because the XY chromosome constitution is more stable in vitro *(11)*, sex conversion produces some chimeric males that transmit only the ES cell genotype *(3,25)*, and breeding from males is more rapid. There are only a few reports of germline female chimeras from either XX ES cells *(26)*, embryonic germ cells (derived from primordial germ cells; *[27]*), or in vitro isolated XX embryonal carcinoma cells *(28,29)*. Female ES cell lines often suffer either a complete or partial loss of one X chromosome, thought to be a compensatory mechanism for X-inactivation *(30)*. A normal, stable chromosome complement is vital for transmission through the gametes. This generally implies the use of ES cells with a short culture history, since the risk of selecting aneuploid cells increases the longer the cells remain in vitro. This has practical implications for obtaining germline transmission following gene-targeting experiments. However, germline transmission from ES cells following 260 cell generations in vitro has been demonstrated, but not without periodic subcloning to identify and reexpand euploid cells *(31)*.

   Because the Y chromosome is principally responsible for sex determination in mammals, the integration of XY stem cells with an XX host embryo results in sex conversion in a proportion of formerly female embryos. These sex-converted animals transmit only the XY stem cell–derived genotype in their sperm *(3)*. However, up to two-thirds of sexually converted phenotypic males may in fact be sterile hermaphrodites from a low contribution of ES-derived XY somatic cells in the female germinal ridge, which exert only a mild masculinizing influence *(25)*. In the situation in which XY ES cells are combined with an XY embryo, the phenotypic male chimeras produced may transmit the ES genotype in only a small fraction of their sperm (typically 0.3–3%; *[25]*).

The vast majority of investigators who have demonstrated germline transmission from ES chimeras have utilized ES cells derived from embryos of the 129 mouse strain (including a variety of substrains; *see* **ref. 32**). However, there are reports describing ES cells derived from other mouse strains and producing germline chimeras, including CD-1 *(31)*; C57BL/6 x CBA *(20)*; C57BL/6 *(33)*; and Balb/c *(24)*. There is no biologic reason to assume that euploid ES cells derived from the embryos of any mouse strain would not be capable of colonizing the germline. However, some mouse strains are less permissive toward the isolation of ES cell lines than others.

2. Factors that may increase the contribution of ES cells into the germline of chimeras include the following:

   a. Increasing the number of stem cells injected into blastocysts (from 3–5 to 10–15 cells; *[11]*): However, it may be important not to introduce more than 20 ES cells into the blastocyst, since there is the potential for abnormal fetal development and tumor formation if the stem cells either are karyotypically abnormal, possess minor genetic or epigenetic changes, or escape the regulatory control of the ICM *(12)*. Similar considerations apply for morula stages, be they introduced via injection or aggregation methods. It is recommended that 5–10 cells be introduced in these earlier embryo stages, because this is compatible with normal development and highly chimeric mice *(13)*.

   b. The choice of mouse strain from which to obtain host embryos: ES cells isolated from the 129 mouse strain have entered the germline at a greater frequency when injected into inbred C57BL/6 host embryos, compared to outbred MF1 or CD-1 strain embryos *(15,34,35)*. This emphasizes that the stem cells from certain mouse strains have competitive advantages over the embryonic cells from other mouse strains in their subsequent development.

   c. Utilization of strains of mice carrying specific fertility mutations to produce host embryos (e.g., the *W* mutation ["dominant white spotting"]; *[11]*).

   d. The developmental stage of the host embryo: Both the injection and aggregation of ES cells with morula-stage embryos have shown more extensive chimerism, compared with those produced from the injection of cells at the blastocyst stage *(13,19)*. The proportion of total germline transmitters tends to be the same, when comparing the two embryo stages of chimera production. However, the proportions of sex-converted animals, and those absolute germline transmitting chimeras, may be greater at the morula stage, emphasizing the stem cells to have dominated *(13)*.

The introduced stem cells tend to migrate into the center of the morula mass following injection or aggregation, owing to their inherent cell-surface properties. It is because the ES cells assume a more internal position that they contribute more to the resultant ICM and, hence, more extensively to all the tissues of the subsequent fetus. Because of this phenomenon, it is important to limit the number of stem cells introduced to the morula, in order to avoid developmental problems in the resulting fetus that may arise, dependent on the normality of the ES cell line. Apart from major karyotypic abnormalities, more minor genetic or epigenetic changes may possibly restrict the developmental function of some stem cells *(10)*. Under circumstances in which this impaired function is adequately compensated for by the host embryonic cells, normal chimeric development proceeds. However, in the situation in which the stem cell contribution is excessive, this may be incompatible with fetal development and midgestation death occurs. At the extreme, this effect is observed in the developmental abnormalities associated with the 100% stem cell–derived fetuses and mice, produced using the tetraploid method *(10)* (*see* **Subheading 3.3.3.3.**).

3. Depending on the nature of the study, pigmentation, biochemical, or histochemical markers (introduced into the ES cells) may be used:

   a. Pigmentation markers: An appropriate combination of coat color markers is the simplest way to detect overt chimeras, and to assay for germline contributions. For instance, if the ES cells are derived from a strain of mouse with homozygous dominant pigmentation alleles, and these stem cells are injected into an albino mouse (homozygous recessive), then by backcrossing the chimera to an albino mouse, any transmission of the stem cell genotype through the germline will be detected by the presence of pigmented progeny (which will be heterozygous at the pigmentation locus). If suitable coat color combinations are not available, the glucose phosphate isomerase (GPI) isoenzymes may be used instead. Pigmentation markers are only useful postnatally; however, pigment can be detected in the eye of the fetus from d 10 p.c. onward.

   b. Biochemical markers: Biochemical markers are required for midgestation analyses and for analyzing the extent of chimerism in internal body organs. The most commonly utilized marker is GPI, because these isoenzymes (present in one of three forms in mice) are ubiquitously expressed and can be easily separated electrophoretically and stained. For methods of electrophoresis and staining of the GPI isoenzymes, *see* **refs.** *16*, *36*, and *37*.

   c. Histochemical markers: The fate of ES cells introduced into the preimplantation embryo can be followed as they participate in embryogenesis and the formation of chimeric offspring, if the stem cells are tagged by a genetic tracer. ES cells stably expressing β-galactosidase, e.g., under the control of an endogenous promoter following electroporation, have been used to generate chimeras, and the pattern of colonization by the stem cells followed by *in situ* histochemical staining of whole fetuses *(6,19)*. An extension of this approach is to use β-galactosidase in so-called enhancer trap and promoter trap vectors to identify novel developmentally regulated genes, on the basis of temporally and spatially restricted patterns of enzyme staining in the chimeric conceptus *(38,39)*.

4. In producing embryos from donor mice, some investigators have preferred to use natural matings, because superovulation tends to yield some embryos that are morphologically abnormal or developmentally retarded. Although this may be true, these disadvantages are outweighed by the advantages superovulation gives in terms of synchronizing females and maximizing the utilization of mice when mouse stocks are limited.

5. To obtain germline chimerism, a high proportion of the stem cells in a cell line must have a normal chromosomal constitution. Thus, the karyotype should be checked before beginning any attempts to generate chimeras (*see* Chapter 6). The longer cells remain in tissue culture, the greater the chance that the proportion of aneuploid cells in the cell line will increase. By single-cell cloning, however, euploid cell cultures can be identified and reexpanded. ES cell cultures should also be checked regularly for mycoplasma contamination. Such infection causes fetal death and drastically reduces the success in obtaining chimeras *(18)*. Infected cell lines should be discarded and improved aseptic tissue-culture techniques adopted.

## Acknowledgments

I am grateful to Dr. Ian Wilmut (Institute of Animal Physiology and Genetics Research, Edinburgh) for allowing me the opportunity to learn many aspects of the techniques described here, during my time under his PhD supervision. I am also indebted to Dr. Jim McWhir for useful discussions on micromanipulative procedures.

## References

1. Evans, M. J. and Kaufman, M. H. (1981) Establishment in culture of pluripotential cells from mouse embryos. *Nature* **292,** 154–156.
2. Martin, G. R. (1981) Isolation of a pluripotent cell line from early mouse embryos cultured in medium conditioned by teratocarcinoma stem cells. *Proc. Natl. Acad. Sci. USA* **78,** 7634–7638.
3. Bradley, A., Evans, M., Kaufman, M. H., and Robertson, E. (1984) Formation of germ-line chimaeras from embryo-derived teratocarcinoma cell lines. *Nature* **309,** 255, 256.
4. Evans, M., Bradley, A., and Robertson, E. (1985) EK cell contribution to chimeric mice: from tissue culture to sperm, in *Genetic Manipulation of the Early Mammalian Embryo*, Banbury report 20 (Costantini, F. and Jaenisch, R., eds.), Cold Spring Harbor Laboratory, Cold Spring Harbor, NY, pp. 93–102.
5. Beddington, R. S. P. and Robertson, E. J. (1989) An assessment of the developmental potential of embryonic stem cells in the midgestation mouse embryo. *Development* **105,** 733–737.
6. Suemori, H., Kadodawa, Y., Goto, K., Araki, I., Kondoh, H., and Nakatsuji, N. (1990) A mouse embryonic stem cell line showing pluripotency of differentiation in early embryos and ubiquitous β-galactosidase expression. *Cell Diff. Dev.* **29,** 181–186.
7. Nagy, A., Gocza, E., Merentes Diaz, E., Prideaux, V. R., Ivanyi, E., Markkula, M., and Rossant, J. (1990) Embryonic stem cells alone are able to support fetal development in the mouse. *Development* **110,** 815–821.
8. Ueda, O., Jishage, K., Kamada, N., Uchida, S., and Suzuki, H. (1995) Production of mice entirely derived from embryonic stem (ES) cell with many passages by

coculture of ES cells with cytochalasin B induced tetraploid embryos. *Exp. Anim.* **44,** 205–210.

9. Wang, Z.-Q., Kiefer, F., Urbánek, P., and Wagner, E. F. (1997) Generation of completely embryonic stem cell-derived mutant mice using tetraploid blastocyst injection. *Mech. Dev.* **62,** 137–145.

10. Nagy, A., Rossant, J., Nagy, R., Abramow-Newerly, W., and Roder, J. C. (1993) Derivation of completely cell culture-derived mice from early-passage embryonic stem cells. *Proc. Natl. Acad. Sci. USA* **90,** 8424–8428.

11. Robertson, E. J. and Bradley, A. (1986) Production of permanent cell lines from early embryos and their use in studying developmental problems, in *Experimental Approaches to Mammalian Embryonic Development* (Rossant, J. and Pederson, R. A., eds.), Cambridge University Press, Cambridge, UK, pp. 475–491.

12. Hardy, K., Carthew, P., Handyside, A. H., and Hooper, M. L. (1990) Extragonadal teratocarcinoma derived from embryonal stem cells in chimaeric mice. *J. Pathol.* **160,** 71–76.

13. Wood, S. A., Allen, N. D., Rossant, J., Auerbach, A., and Nagy, A. (1993) Non-injection methods for the production of embryonic stem cell-embryo chimaeras. *Nature* **365,** 87–89.

14. Wood, S. A., Pascoe, W. S., Schmidt, C., Kemler, R., Evans, M. J., and Allen, N. D. (1993) Simple and efficient production of embryonic stem cell-embryo chimeras by coculture. *Proc. Natl. Acad. Sci. USA* **90,** 4582–4585.

15. Hooper, M. L. (1992) in *Embryonal Stem Cells: Introducing Planned Changes into the Germline* (Evans, H. J., ed.), Harwood Academic, Switzerland, p. 147.

16. Whittingham, D. G. (1971) Culture of mouse ova. *J. Reprod. Fertil.* **Suppl. 14,** 7–21.

17. Fulton, B. P. and Whittingham, D. G. (1978) Activation of mammalian oocytes by intracellular injection of calcium. *Nature* **273,** 149–151.

18. Bradley, A. (1987) Production and analysis of chimaeric mice, in *Teratocarcinomas and Embryonic Stem Cells: A Practical Approach* (Robertson, E. J., ed.), IRL, Oxford, UK, pp. 113–151.

19. Lallemand, Y. and Brulet, P. (1990). An *in situ* assessment of the routes and extents of colonisation of the mouse embryo by embryonic stem cells and their descendents. *Development* **110,** 1241–1248.

20. Yagi, T., Tokunaga, T., Furuta, Y., Nada, S., Yoshida, M., Tsukada, T., Saga, Y., Takeda, N., Ikawa, Y., and Aizawa, S. (1993) A novel ES cell line, TT2, with high germline-differentiating potency. *Anal. Biochem.* **214,** 70–76.

21. Fujii, J. T. and Martin, G. R. (1983) Developmental potential of teratocarcinoma stem cells *in utero* following aggregation with cleavage-stage mouse embryos. *J. Embryol. Exp. Morphol.* **74,** 79–96.

22. Lewis, N. E. and Rossant, J. (1982) Mechanism of size regulation in mouse embryo aggregates. *J. Embryol. Exp. Morph.* **72,** 169–181.

23. Tarkowski, A. K., Witkowska, A., and Opas, J. (1977) Development of cytochalasin B-induced tetraploid and diploid/tetraploid mosaic mouse embryos. *J. Embryol. Exp. Morphol.* **41,** 47–64.

24. Eggan, K., Akutsu, H., Loring, J., Jackson-Grusby, L. Klemm, M., Rideout 3rd, W. M., Yanagimachi, R., and Jaenisch, R. (2001) Hybrid vigor, fetal overgrowth, and viability of mice derived by nuclear cloning and tetraploid embryo complementation. *Proc. Natl. Acad. Sci. USA* **98,** 6209–6214.
25. Robertson, E. J. (1986) Pluripotential stem cell lines as a route into the mouse germ line. *Trends Genet.* **2,** 9–13.
26. Wells, D. N. (1991) Studies on the isolation of murine and ovine embryonic stem cells. PhD thesis, University of Edinburgh, Edinburgh, UK.
27. Stewart, C. L., Gadi, I., and Bhatt, H. (1994) Stem cells from primordial germ cells can reenter the germ line. *Dev. Biol.* **161,** 626–628.
28. Stewart, T. A. and Mintz, B. (1981) Successive generations of mice produced from an established culture line of euploid teratocarcinoma cells. *Proc. Natl. Acad. Sci. USA* **78,** 6314–6318.
29. Stewart, T. A. and Mintz, B. (1982) Recurrent germ-line transmission of the teratocarcinoma genome from the METT-1 culture line to progeny in vivo. *J. Exp. Zool.* **224,** 465–469.
30. Robertson, E. J., Evans, M. J., and Kaufman, M. H. (1983) X-chromosome instability in pluripotential stem cell lines derived from parthenogenetic embryos. *J. Embryol. Exp. Morphol.* **74,** 297–309.
31. Suda, Y., Suzuki, M., Ikawa, Y., and Aizawa, S. (1987) Mouse embryonic stem cells exhibit indefinite proliferative potential. *J. Cell. Physiol.* **133,** 197–201.
32. Baribault, H. and Kemler, R. (1989) Embryonic stem cell culture and gene targeting in transgenic mice. *Mol. Biol. Med.* **6,** 481–492.
33. Lederman, B. and Burki, K. (1991) Establishment of a germ-line competent C57BL/6 embryonic stem cell line. *Exp. Cell. Res.* **197,** 254–258.
34. Schwartzberg, P. L., Goff, S. P., and Robertson, E. J. (1989) Germ-line transmission of a *c-alb* mutation produced by targeted gene disruption in ES cells. *Science* **246,** 799–803.
35. DeChiara, T. M., Efstratiadis, A., and Robertson, E. J. (1990) A growth-deficiency phenotype in heterozygous mice carrying an insulin-like growth factor II gene disrupted by targeting. *Nature* **345,** 78–80.
36. Eicher, E. M. and Washburn, L. L. (1978) Assignment of genes to regions of mouse chromosomes. *Proc. Natl. Acad. Sci. USA* **75,** 946–950.
37. McLaren, A. and Buehr, M. (1981) GPI expression in female germ cells of the mouse. *Gen. Res. Cambridge* **37,** 303–309.
38. Gossler, A., Joyner, A. L., Rossant, J., and Skarnes, W. C. (1989) Mouse embryonic stem cells and reporter constructs to detect developmentally regulated genes. *Science* **244,** 463–465.
39. Rossant, J. and Joyner, A. L. (1989) Towards a molecular-genetic analysis of mammalian development. *Trends Genet.* **5,** 277–283.

# 8

## Gene-Targeting Strategies

### David W. Melton

### 1. Introduction

The vast majority of DNA integrations in mammalian cells are random nonhomologous events, probably occurring between the free ends of an incoming DNA molecule and a preexisting or transient break, or nick, in a host chromosome. This random process is exploited in the production of conventional transgenic animals by pronuclear DNA injection. Conventional transgenics have made an invaluable contribution to the study of the control of mammalian gene expression, to the dissection of complex processes such as development and oncogenesis, and have also demonstrated considerable commercial potential. However, powerful as this system is, it does suffer from two inherent disadvantages: there is no control over the copy number of the integrated transgene, and the random integration can render expression of the transgene particularly susceptible to its local chromosomal environment. This variability can be overcome by integrating the transgene into a particular chromosomal location. This is gene targeting, in which homologous recombination between an introduced vector and a chromosomal target locus is exploited to make a specific change to an endogenous gene. This change can, as with conventional transgenics, involve the expression of a transgene or, more important, can affect the expression of the endogenous gene itself. This is the key attraction of gene targeting, because it offers the ability to carry out the same sort of genetic analysis in the mouse as had previously only been possible in lower organisms—to make a particular genetic change and study the functional consequences. It is not surprising that gene targeting has become so widely used in the study of a wide range of complex mammalian processes and in the development of animal models for the study of human disease.

From: *Methods in Molecular Biology, vol. 180: Transgenesis Techniques, 2nd ed.: Principles and Protocols*
Edited by: A. R. Clarke © Humana Press Inc., Totowa, NJ

## 1.1. Early Landmarks

In the early 1980s, the first demonstrations of homologous recombination in cultured mammalian cells involved recombination between two introduced plasmid molecules (e.g., *see* **ref. *1***) and proceeded to situations in which one of the plasmid molecules was randomly integrated into the genome and recombined with a second introduced plasmid. The first endogenous locus, β-globin, was targeted by Oliver Smithies in 1985 *(2)*. The same low frequency of gene targeting ($\sim 10^{-3}$ of the frequency of random integrations) was observed in two human somatic cell lines, only one of which expressed the target gene. This experiment was particularly significant because it demonstrated that gene expression was not a prerequisite for successful targeting. Attention then shifted from somatic cells to gene targeting in cultured mouse embryonic stem (ES) cells, because of the powerful attraction of being able to generate new strains of mice with specific genetic alterations from cultured cells manipulated in vitro. A series of experiments, using the hypoxanthine phosphoribosyltransferase (HPRT) gene as target, evaluated the two basic types of targeting vector (insertion and replacement) that remain in current use and the requirements for successful gene targeting. Two features of the HPRT system made it particularly favorable for these early model experiments: convenient procedures exist to select either for, or against, HPRT expression in cultured cells; and the gene is X-chromosome linked so that, in male cells, only a single locus need be targeted to produce a change in phenotype that could then be selected for. In 1989, this work culminated with the production of the first strain of gene-targeted mice containing an HPRT locus that had been corrected by homologous recombination in ES cells *(3)*.

## 1.2. Gene-Targeting Frequency and Need for Selection

Contrary to the situation in prokaryotes and yeast, in which gene inactivation and replacement are routine, typically only $\sim 10^{-2}$ to $10^{-3}$ of DNA integrations in ES cells are homologous events. This, coupled with the relatively low frequency with which ES cells take up, integrate, and express added DNA, makes gene targeting a rare event. Typically, the frequency of targeting will be $\sim 10^{-5}$ to $10^{-6}$ of the ES cells treated with vector DNA. Far less gene targeting has been carried out in somatic cells, and there have been few systematic attempts to compare gene targeting frequencies between ES and somatic cells in which the same vector is used to target the same gene in two different, but isogenic, cell types. However, the indications are that, although many somatic cell types have a higher gene transfer frequency than ES cells, the ratio of homologous to random integrations may be lower, so that overall the frequency of targeting per cells treated with vector DNA is lower for somatic than ES

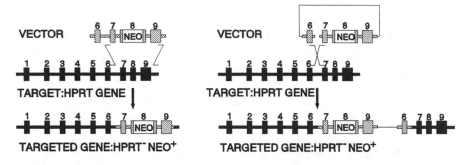

Fig. 1. HPRT gene inactivation with sequence replacement and insertion vectors. The structure of the mouse HPRT gene is shown schematically, with the nine exons numbered. For sequence replacement, the vector carries a selectable (neo) marker, flanked on either side by sequences homologous to the target (HPRT) gene. After homologous recombination between the vector and the target gene as shown, the incoming sequences replace the genomic sequence by a double crossover event, resulting in HPRT gene inactivation with no additional changes outside the gene. For sequence insertion, the vector is linearized in the region of homology. After homologous recombination at the double-strand break, the entire vector is introduced into the target gene, resulting in HPRT gene inactivation and partial gene duplication. Solid boxes, endogenous HPRT sequences; dotted boxes, vector-derived HPRT sequences; thin line, plasmid sequences.

cells. The low frequency of gene targeting necessitates the use of methods to select for those cells that have integrated vector DNA and, within this population, to select, or at least enrich, for clones with homologous integrations. Efficient screening procedures are also required to identify what are often rare targeted clones in a population.

### 1.3. Insertion and Replacement Vectors

The use of the two basic types of plasmid targeting vector, insertion and replacement, is illustrated in **Fig. 1** for the inactivation of the HPRT gene. Insertion vectors contain a region of homology to the target locus and are linearized by restriction within this region of homology before being introduced into cells. This double-strand break stimulates a classic reciprocal recombination reaction within the region of homology between vector and target locus, leading to the insertion of the vector into the locus, with a resulting partial gene duplication. Replacement vectors typically contain two separate arms of homology to the target locus, separated by a region of nonhomologous DNA. These vectors are cut at the end of one of the homologous arms before being

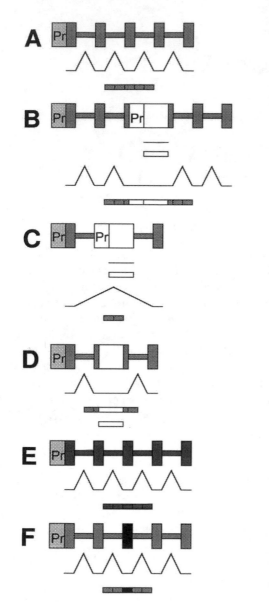

**A** Target gene
Transcript
Protein

**B** Insertional mutagenesis
Marker gene transcript
Marker protein
Target gene transcript
Non-functional target protein

**C** Deletion
Marker gene transcript
Marker protein
Target gene transcipt
Non-functional target protein

**D** Reporter gene insertion
Target gene transcript
Non-functional target protein
Reporter protein

**E** Knock-in
Target gene transcript
Knocked-in protein

**F** Subtle gene alteration
Target gene transcript
Altered protein

Fig. 2. Alterations that can be introduced by gene targeting. (**A**) Structure of a typical target locus with 5 exons. The encoded transcript is processed to remove introns as indicated and then translated to give the protein product. (**B**) Insertional mutagenesis. The target gene is knocked out by the insertion of a selectable marker into exon 3, which results in the production of a nonfunctional protein from the targeted gene. The marker gene is expressed from its own promoter. (**C**) Deletion. The target gene is knocked out by the removal of exons 2–4 and their replacement by a selectable marker. A nonfunctional protein is produced from the targeted gene. The marker gene is expressed from its own promoter. (**D**) Insertional mutagenesis with introduction of a

introduced into cells. A nonreciprocal homologous recombination event between each arm of the vector and the target locus results in the replacement of part of the target locus with vector sequences. With both types of vector, the use of DNA for vector construction from a mouse strain isogenic with the target cells results in a fivefold stimulation of targeting frequency *(4)*. Provided isogenic DNA is used and the vectors contain equivalent regions of homology, then insertion and replacement vectors give the same targeting frequency *(4)*. With both types of vector, increasing the length of homology improves the targeting frequency, although the precise mathematical description (linear or exponential increase) remains the subject of disagreement. Successful targeting is difficult to achieve with replacement vectors when the total length of homology drops below 2 kb, although 500 bp of homology on the short arm has proved sufficient. The partial target gene duplication associated with the use of insertion vectors and the resulting possibility that alternative patterns of RNA splicing could interfere with the intended consequences of the targeted change have led to replacement vectors being used much more commonly for most current gene-targeting applications.

## 2. Gene-Targeted Changes

The types of gene-targeted change that can be introduced are illustrated in **Fig. 2**. Simple gene knockout by insertional mutagenesis remains the most common form of gene targeting (*see* **Fig. 2B**). Here, a selectable marker is inserted into a coding exon of the target locus, ideally resulting in the production of a completely nonfunctional protein. There are many instances in which such a situation results instead in an incomplete knockout, with the protein

---

*(Figure 2 caption continued from previous page)* reporter gene. The target gene is knocked out as in (b), by the insertion of a reporter gene into exon 3. The reporter gene is expressed from the promoter of the target gene and contains signals for its efficient translation from the hybrid mRNA. (**E**) Knock-in. The target gene is replaced with the equivalent gene from a different species that is expressed from the promoter of the targeted gene. (**F**) Subtle gene alteration. A missense mutation is introduced into exon 3, resulting in the production of an altered protein from the targeted locus. Lightly shaded boxes, target locus (promoter, exons, introns, and encoded protein); open boxes, marker gene/reporter gene; heavily shaded boxes, knocked-in gene; solid boxes, exon with a missense mutation. Note that the indicated full-length targeted gene transcripts and encoded (nonfunctional) proteins may not be produced if the insertion of the marker gene affects the normal pattern of RNA processing, or introduces a polyadenylation signal, or introduces an in-frame translational stop signal.

encoded by the target locus retaining partial function. Unless one can be certain that the exon to be targeted by insertional mutagenesis is essential for normal function, it is safer to achieve complete gene knockout by deleting all, or part, of the coding sequence (*see* **Fig. 2C**).

By inserting a reporter gene, rather than a simple selectable marker, into the target locus, it is also possible to achieve a gene knockout as well as to monitor the expression of the targeted gene simultaneously (*see* **Fig. 2D**). In this example, the reporter has been inserted into an exon of the target locus and thus becomes part of the endogenous transcription unit. The reporter gene contains signals required for its translation, and the reporter gene product is thus translated from the hybrid mRNA.

With a typical knock-in alteration, the coding sequence of the target locus is replaced by that of a related gene, or the equivalent gene from another species (*see* **Fig. 2E**). This approach is particularly useful to investigate issues of gene redundancy and the precise role of individual genes in complex control circuits such as that controlling myogenesis.

Targeting also can be used to introduce more subtle changes, such as single nucleotide alterations (*see* **Fig. 2F**). There are many cases, particularly involving the modeling of some human inherited diseases, in which a simple knockout is not adequate. For instance, the most common cystic fibrosis mutation ($\Delta$F508) results from the loss of a single phenylalanine codon, rather than from the lack of the entire cystic fibrosis transmembrane conductance regulator protein, while some of the familial prion diseases result from single amino acid alterations to the prion protein.

## 3. Selectable Markers

Many of the most frequently used selectable markers for gene targeting are available commercially.

### *3.1. neo*

The bacterial neomycin phosphotransferase (neo) gene remains the most commonly used marker for simple gene knockouts, often expressed from the promoter of the phosphoglycerate kinase-1 (PGK-1) gene *(5)*, or from the pMC1 promoter, which consists of a mutant polyoma enhancer linked to the promoter from the herpes simplex virus thymidine kinase (HSVTK) gene *(6)*. Expression of the neo gene in cultured mammalian cells confers resistance to the synthetic antibiotic G418. The structure of MT-neo is shown in **Fig. 3A**. Here, the neo gene, with the translational initiation region modified to give a strong match to the Kozak consensus, is under the control of the mouse metallothionein-1 gene promoter *(7)*.

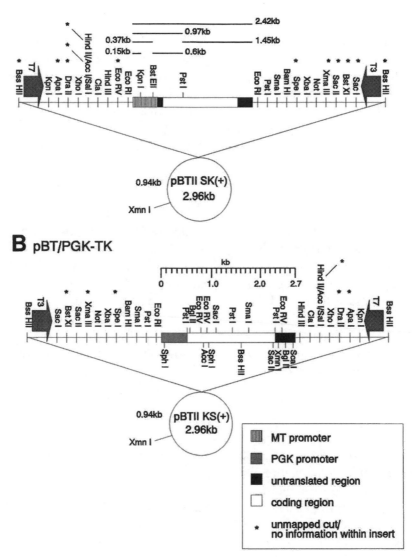

Fig. 3. Structure of selectable markers for gene targeting. (**A**) pBT/MT-neo(RI). The 2.4-kb marker is inserted into the *Eco*RI site of pBluescriptII SK(+). It is under the control of a 0.77-kb fragment from the mouse metallothionein-1 (MT) gene promoter. The neo coding region has a strong synthetic translational start motif, with a 0.65-kb fragment from the human growth hormone gene providing the 3'-untranslated region. (**B**) pBT/PGK-TK. The 2.7-kb marker is cloned as an *Eco*RI-*Hin*dIII fragment in pBluescript II KS(+). The HSVTK gene is under the control of the mouse PGK-1 promoter and 3'-untranslated region. An asterisk against a site within the pBluescript polylinker denotes either that there is an unmapped site (or sites) for this enzyme within the marker gene or that there is no information on sites for this enzyme within the marker gene.

### 3.2. β-geo

β-geo is valuable in situations in which there is a requirement to both knock out a gene and monitor its expression by the insertion of a reporter *(8)*. It consists of a fusion of the *Escherichia coli* lacZ and neo genes, so that the expressed fusion protein encodes both β-galactosidase and neomycin phosphotransferase activity. The marker is not expressed from its own promoter but relies instead on integration into the transcription unit at the target locus and is transcribed as part of a hybrid transcript. The marker contains an internal ribosome entry site just upstream of the β-geo reading frame, which is required for the efficient translation of the β-geo fusion protein from the hybrid mRNA.

### 3.3. HSVTK

The main application of the HSVTK marker is in the positive-negative selection procedure *(9)*, which has proved extremely valuable as a method to enrich for gene-targeted clones among the population that have taken up vector DNA. This procedure is described in **Subheading 4.** Typically, the HSVTK gene is expressed from the PGK (*see* **Fig. 3B**) or pMC1 promoter. The viral enzyme has a different substrate specificity from the mammalian form, such that cells expressing the viral enzyme will convert thymidine analogs, such as ganciclovir or 5-FIAU, into a toxic product that kills the cell. Cells expressing only the mammalian enzyme do not metabolize the analog and are resistant to its toxic effect. Thus, cells expressing HSVTK are killed in the presence of ganciclovir.

### 3.4. neo/TK

More sophisticated gene-targeting procedures, such as those used to achieve subtle gene alteration, require two sequential selection steps. While some vectors for this purpose contain separate neo and HSVTK markers, a fused neo/TK transcription unit has also been developed *(10)*. This expresses a bicistronic mRNA from the PGK-1 promoter, with an internal ribosome entry site upstream of the TK reading frame ensuring efficient translation of both neo and HSVTK coding regions. Selection for this marker is achieved in G418, while selection for loss of the marker is achieved with ganciclovir.

### 3.5. HPRT

Most of the HPRT markers in use are truncated derivatives (minigenes) of the mouse HPRT gene, under the control of the endogenous mouse HPRT promoter or the PGK promoter *(11)*. When used in conjunction with HPRT-deficient ES cells, it is possible to select both for the presence of the marker and for its loss. Selection for HPRT expression is achieved in HAT medium *(12)*. Mammalian cells have two ways of making purine nucleotides: they can synthesize them *de novo*, or build them up from existing purines via a salvage pathway where HPRT catalyzes the first step. If the *de novo* pathway is blocked with aminopterin (the A in HAT medium), cells become dependent on the salvage pathway (i.e., HPRT) for survival. Hypoxanthine and thymidine are also present in HAT medium to provide substrates for the purine salvage pathway and the pyrimidine salvage pathway (which is also blocked by aminopterin). Selection against HPRT expression is achieved by growth in the presence of the purine analog 6-thioguanine (6-TG), which is phosphoribosylated by HPRT to give a toxic product that kills HPRT-expressing cells. The key attraction of HPRT markers is their use in HPRT-deficient ES cells as part of an efficient two-step gene-targeting procedure (known as double replacement; *[13]*) for the introduction of subtle gene alterations. Two-step targeting procedures place greater demands on the stability of marker gene expression than simple gene knockout. HPRT markers have proved more effective in these procedures than combinations of neo and HSVTK, or the neo/TK fusion. Two HPRT minigenes are illustrated in **Fig. 4**, one under the control of the PGK promoter (**Fig. 4A**), and the second under endogenous HPRT promoter control (**Fig. 4B**). This latter minigene (pDWM110) has proved particularly valuable in situations in which long-term stability of marker gene expression is essential *(14)*.

## 4. Design of Gene-Targeting Strategies

### 4.1. Knockout

A more complete knockout is likely to result from the deletion of one or more exons from the target gene than the simple insertion of a marker into one exon. The most common method to achieve such a knockout is through the use of a replacement vector and the positive-negative selection procedure *(9)*. The example given is for neo (positive marker) and HSVTK (negative marker), although the same design considerations would apply equally to other markers. For optimum targeting frequencies, the targeting vector should be constructed using DNA isogenic with the ES cell line to be used. Most ES cell lines are derived from strain 129, but more than one substrain has been used and there are differences among them *(15)*. 129/Ola (equivalent to 129/ReJ) and 129/SvJ are the most common; be sure to use DNA from a library that matches your ES cells. It is strongly recommended that the design allow for the detection of targeted clones by screening with a robust PCR reaction for a product that is specific for the targeting event. For this reason, the targeting vector should have 1–1.5 kb of short-arm homology and 4–8 kb of long-arm homology to the target locus. The short arm is then ideal for short-range PCR reactions, while the total length of homology should give acceptable targeting frequencies. If the amount of homology is increased much beyond this, the plasmid vectors become very large and difficult to propagate and manipulate.

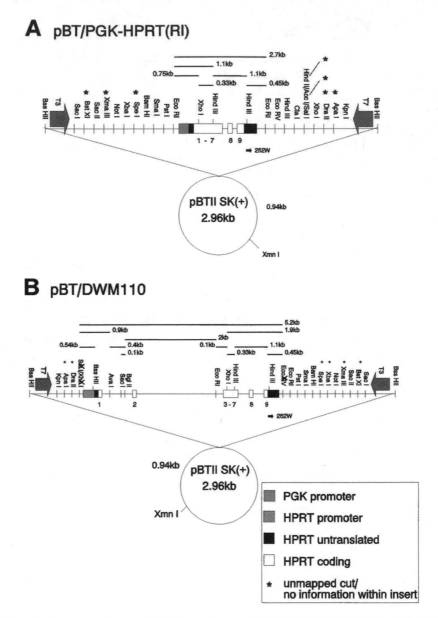

Fig. 4. Structure of mouse HPRT minigenes. **(A)** pBT/PGK-HPRT(RI). This 2.7-kb minigene, inserted into the *Eco*RI site of pBluescript II SK(+), is under the control of a 0.55-kb fragment from the mouse PGK-1 gene promoter. **(B)** pBT/DWM110. This 5.2-kb minigene is under the control of the mouse HPRT promoter itself. The HPRT sequences in the minigene are partly derived from genomic sequences and are, in part, cDNA derived. The numbers indicate the location in the minigenes of the nine exons from the HPRT gene itself. An asterisk against a site within the pBluescript polylinker denotes either that there is an unmapped site (or sites) for this enzyme within the minigene or that there is no information on sites for this enzyme within the minigene.

The principle of positive-negative selection is that it is not only possible to select for cells that have integrated and expressed the neo gene, but also to select against random integrants that have retained the HSVTK gene. In this way the population is enriched for gene-targeted clones. On homologous recombination, the vector replaces target sequences, leading to target gene inactivation by deletion, and integration and expression of the neo gene, but loss of the HSVTK marker because it is outside the region of homology to the target locus. Targeted clones are thus neo⁺ HSVTK⁻ and survive in G418 and ganciclovir. By contrast, most random integrations occur via the free ends of the targeting vector, so that random integrants are neo⁺ HSVTK⁺ and die in ganciclovir.

Ensure that the vector can be linearized as indicated in **Fig. 5**, so that the HSVTK sequence is protected by the plasmid DNA. By doing this and positioning the HSVTK marker at the end of the short arm of homology (as opposed to the long arm), the efficiency of negative selection will be increased. Typical enrichment values following positive-negative selection (G418$^R$ clones/ G418$^R$+GANC$^R$ clones) range from 5 to 20. Even values at the lower end of the scale are useful because they reduce the number of clones to be screened. Design a PCR strategy to screen G418$^R$ GANC$^R$ clones to identify homologous recombinants. Use a primer from the end of the neo marker in combination with a primer from the target locus, lying just outside the region of short-arm homology in the vector. Only homologous recombinants should give a PCR product of the predicted size. This is the most efficient way to screen a large number of clones while they are still at the initial colony-forming stage. Positive clones can then be expanded and the structure of the targeted locus confirmed by Southern blotting. Using this basic design strategy (albeit with HPRT-deficient HM-1 cells and HPRT markers), we have successfully targeted a number of different loci, with the frequency of targeted clones among the survivors of positive-negative selection varying from 1 in 4 to 1 in 50.

Most of the same design considerations apply to the use of replacement vectors to achieve a knockout with the simultaneous insertion of a reporter gene such as β-geo. However, in this case, since the reporter relies for expression on being part of the transcription unit of the target gene, targeting should not lead to the deletion of any parts of the target locus, but instead to the insertion of β-geo into one of the exons of the target locus to achieve insertional mutagenesis

---

*(Figure 4 caption continued from previous page)* A cross through a restriction site indicates loss of this site during the cloning process. The location of primer 262W, which is used in a polymerase chain reaction (PCR) assay to detect gene-targeted clones, is indicated by the arrow.

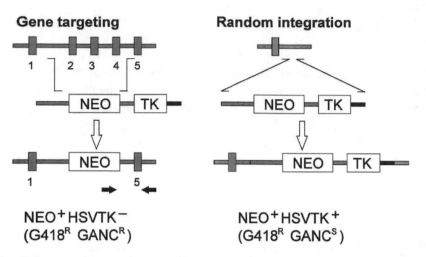

Fig. 5. Strategy for gene knockout by deletion utilizing positive-negative selection. A typical target gene containing five exons (numbered shaded boxes) is shown. The replacement vector contains sequences homologous to introns 1 and 4 of the target gene (note the positions of the long and short arms of homology) and carries a positive selectable marker, the neo gene (NEO, open boxes), replacing exons 2–4. An HSVTK gene (TK, open boxes) is located at the terminus of one region of homology. Note that the vector is cut so that the HSVTK gene is protected by plasmid sequence (solid line). For gene targeting, homologous recombination between the incoming vector and chromosomal target gene sequences results in replacement of the target gene by the incoming vector and loss of the flanking HSVTK sequence. Targeted cells are neo+ HSVTK− and are thus resistant to G418 (G418$^R$) and ganciclovir (GANC$^R$). Note the location of primers (solid arrows) that are used in a PCR assay to detect gene-targeted clones. For random integration, as a result of random integration into a different region of the genome (vertical shaded boxes), the entire targeting vector integrates into the genome. Such cells are neo+ HSVTK+, which renders them resistant to G418 but sensitive to ganciclovir.

and simultaneous reporter gene expression. Since the marker is reliant for expression on insertion into an active transcription unit, this will reduce the number of random integrations resulting in β-geo expression and G418$^R$. As a consequence, the frequency of targeted clones in the G418$^R$ population will be elevated, often to the level where they can be isolated without the need for a negative selection (HSVTK) component.

## 4.2. Knock-in

Although knock-in, in its most basic form, can be achieved in a single step, this is not recommended because it leaves a selectable marker deposited either in or adjacent to the target locus (*see* **Fig. 6**). The example in **Fig. 6** shows

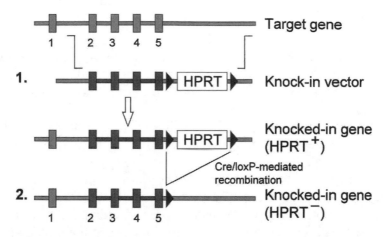

Fig. 6. Strategy for gene knock-in. Targeting is carried out in the HPRT-deficient ES cell line, HM-1. A target gene with five exons (numbered boxes, light shading) is shown schematically. Exons 2–5 of the target gene are to be replaced with the equivalent region from a related gene (heavy shading). The knock-in vector contains the region flanked by two arms of homology (intron 1 and 3'-flanking sequences) to the target locus. The knock-in vector also contains an HPRT marker (open boxes) in the 3'-flanking sequence with loxP sites (solid arrowheads) to each side. The first (gene-targeting) step involves selection for HPRT$^+$ cells. The knocked-in gene has a flanking HPRT marker. The marker can be removed in a second step, involving Cre-mediated recombination between the loxP sites, with selection for HPRT$^-$ cells.

knock-in with a replacement vector using an HPRT marker. Despite attempts to place the marker in expression-neutral locations, there are an increasing number of examples of such markers having unanticipated effects on the expression of both the target locus and adjacent genes *(16)*. If the marker has been flanked by loxP sites, then site-specific recombination with Cre recombinase can subsequently be used to eliminate it (*see* Chapter 9). In either case, the frequency of targeted clones, containing both the knocked-in sequence and the selectable marker (essential since there is no selection for the knocked-in sequence alone), will be increased if the distance between the two is minimized. This will favor homologous recombinations occurring within the homology arms flanking both the knock-in and marker sequences, rather than one occurring between the knock-in and marker sequence.

### 4.3. Subtle Alterations

Two main two-step gene-targeting methods have been developed for the introduction of subtle gene alterations. Both result in the ideal situation in which the only change to the target locus is the subtle alteration; neither selectable markers nor loxP sites are left behind (*see* **Subheading 4.2.**). The disadvantages of these procedures are that two sequential rounds of targeting are required (as opposed to one round of targeting, followed by site-specific recombination with a Cre/loxP-based system) and that the prolonged ES cell culture may reduce the potential for germline colonization (*see* **Subheading 9.**).

The first of these two-step procedures to be developed is termed *hit-and-run* *(17)*. In the first step, a targeting vector carrying the desired gene alteration inserts into the target locus, causing a partial gene duplication. This duplication is resolved in a second step, by an intrachromosomal recombination, which ideally restores the target locus to its original state, except for the introduction of the subtle alteration. The disadvantages of this method are that there is no control over the position of the intrachromosomal recombination within the gene duplication, so that the altered sequence, rather than the endogenous one, is often lost, leaving the locus unmodified; and that a completely new round of targeting must be carried out for each different subtle alteration required to a locus.

The second method for making subtle gene alterations does not suffer from these disadvantages and is illustrated in **Fig. 7** for use in the HPRT-deficient ES cell line, HM-1, with HPRT markers. The first step, in which a region of the target locus is deleted and replaced with an HPRT marker, using a positive-negative selection strategy to give a simple knockout, is identical to the situation described in **Subheading 4.1.**, and exactly the same design considerations apply. By making a deletion in the first step, this strategy permits control over the site of homologous recombination occurring during the second step, to ensure that the subtle alteration is introduced as intended.

In the second step, another replacement vector is used to replace the HPRT marker with the altered gene segment. Starting from the same initial targeting event, a series of different second-step vectors can be used to produce a whole series of mouse strains with different subtle gene alterations at the same target locus. The second-step (alteration) vectors should contain the same regions of homology as the first-step (knockout) vector. The inclusion of an HSVTK cassette for negative selection is not necessary, but stable expression of the HPRT marker is essential for the success of the second step. If this is not the case, then a high frequency of 6-TG$^R$ colonies, arising from loss of expression, rather than physical loss of the marker as a result of successful second-step targeting, is likely to occur. For this reason, use of the HPRT marker, pDWM110, is recommended over PGK-HPRT *(14)*. Design the vector so that a cassette containing the region of the target locus to be altered can be readily excised and reinserted. This permits the use of the same vector backbone for the introduction of different gene alterations. To detect the presence of the gene alteration,

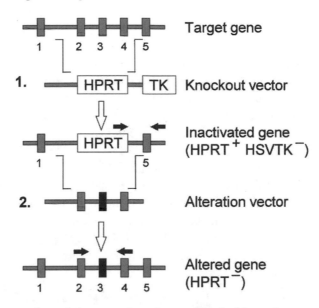

Fig. 7. Strategy for subtle gene alterations using double-replacement gene targeting. Targeting is carried out in the HPRT-deficient ES cell line, HM-1. A target gene with five exons (numbered shaded boxes) is shown schematically. The first targeting step is designed to inactivate the target gene by replacing exons 2–4 with an HPRT minigene (HPRT). In the knockout vector, the HPRT marker is flanked by regions of homology (shaded lines) to the target locus. The knockout vector also contains an HSVTK (TK) gene for positive-negative selection. Selection in the first step is for HPRT+ HSVTK− cells. Cells containing random rather than targeted integrations of the vector will contain the HSVTK gene and can be killed by the thymidine analog ganciclovir. Targeted clones, where the HSVTK gene has been lost, will survive in ganciclovir. Targeted clones are identified using PCR (solid arrows indicate the positions of primers that will only generate a product in targeted cells). In the second step, the HPRT marker is itself replaced by the altered gene segment, containing a subtle alteration in exon 3 (solid box). The alteration vector contains the same regions of homology to the target locus as the first vector. Selection in step two is for HPRT− cells. The presence of the alteration is monitored by PCR, the positions of appropriate primers are indicated by solid arrows. From a single initial knockout in step one, a parallel series of different second steps can be used to produce a series of strains with different gene alterations.

design a PCR strategy to screen 6-TG^R clones for the second replacement event. Look hard for a restriction endonuclease site that is affected by the introduction of the alteration, which can then be detected by PCR and restriction. Failing this, PCR products can be screened by allele-specific oligonucleotide hybridization to detect the alteration.

## 5. Electroporation

The method for gene transfer and the subsequent protocols in this chapter have been developed for mouse ES cells growing, in the absence of a feeder layer, on gelatinized culture vessels in Glasgow Modification of Eagle's medium, supplemented with recombinant leukemia inhibitory factor. Further details of ES cell culture conditions are given in Chapter 6.

We prefer to use mild electroporation conditions, rather than the higher capacitance settings favored by some ES cell targeting groups, because, although the harsher conditions do give slightly higher transformation frequencies per cells electroporated, it is at the expense of the general health of the cells in the days immediately following electroporation. We prefer a protocol that minimizes the trauma to the cells. With our conditions, the transformation frequency is directly proportional to the amount of DNA used, and although we use large amounts of DNA, it is rare to find targeted clones with more than one copy of the vector integrated.

1. Add 100–200 µg of linearized targeting vector DNA in a maximum volume of 80 µL of $H_2O$ to $20–30 \times 10^6$ ES cells suspended in 0.8 mL of HEPES-buffered salt solution (HBS). To make 100 mL of 10X HBS use 8.0 g of NaCl, 0.37 g of KCl, 0.126 g of $Na_2HPO_4 \cdot 2H_2O$, 1.0 g of D-glucose, and 5.0 g of HEPES; Dissolve in 100 mL of $H_2O$ and adjust the pH to 7.05 with 1 $N$ NaOH and filter sterilize.
2. Electroporate at 850 V, 3 µF (Bio-Rad Gene Pulser) in a cuvet with a 0.4-cm electrode gap.
3. Let the cuvet stand for 5 min at room temperature before plating the cells at the required density (*see* **Subheading 6., step 1**).

## 6. Protocol for Gene Knockout

The protocol is given for the experimental design considered in **Subheading 4.1.** and illustrated in **Fig. 5**, for positive-negative selection using neo and HSVTK markers. *See* **Subheading 8.** for the equivalent protocol using HPRT and HSVTK markers.

1. After electroporation, plate cells in up to twenty 90-mm dishes in nonselective medium at the following densities: $4 \times 10^6$, $2 \times 10^6$, $1 \times 10^6$, and $0.5 \times 10^6$ cells per dish. The range of densities allows for varying efficiencies of gene transfer and negative selection between experiments and maximizes the probability that resulting colonies can be picked from dishes where the colonies are clearly individual and well separated. The selection does not work properly at densities $>5 \times 10^6$ cells per dish.
2. Twenty-four hours after electroporation, remove the nonselective medium and replace it with medium containing G418. Our G418 is purchased from Life Technologies and used at 350 µg of powder/mL. The purity of this product is 70%, so that the active [G418] is 250 µg/mL. The optimal [G418] will vary somewhat

between different ES cell lines, especially when different culture conditions are being used. The optimal concentration will not permit any surviving colony formation in untransfected ES cultures, while not seriously retarding the growth of G418$^R$ clones. This concentration should be determined empirically for your supply of G418 and your experimental conditions.

3. Change the G418 medium after a further 48 h.
4. When G418$^R$ colonies have just become visible (i.e., 4–6 d after electroporation), replace the G418 medium on most of the dishes with medium containing G418 and ganciclovir (2 μ*M*). Leave a few dishes under only G418 selection, so that the efficiency of negative selection (G418$^R$ colonies/G418$^R$ + GANC$^R$ colonies) can be calculated.
5. Eight to 11 d after electroporation, pick colonies for screening and seeding into microtiter plates in medium containing G418 and penicillin and streptomycin. Antibiotics are used for a few days only at this critical stage to minimize the risk of contamination. Do not attempt to pick from plates containing >100 colonies. At this stage, some plates can be fixed and stained for colony counting to determine the efficiency of negative selection.

## 7. Protocol for Gene Knock-in

The protocol for the example considered in **Subheading 4.2.** and illustrated in **Fig. 6** for knock-in with a replacement vector using an HPRT marker is similar to that given in **Subheading 8.** If the HPRT marker is flanked with loxP sites, then, after the first targeting step, site-specific recombination with 6-TG selection for loss of the marker would be used instead of a second targeting step. Protocols for Cre/loxP recombination are given in Chapter 9.

## 8. Protocol for Subtle Gene Alteration

The protocol is given for double-replacement gene targeting using the HPRT-deficient ES cell line, HM-1, and an HPRT marker as considered in **Subheading 4.3.** and illustrated in **Fig. 7.**

### 8.1. First Targeting Step

1. After electroporation, plate cells in up to twenty 90-mm dishes in nonselective medium at the following densities: $4 \times 10^6$, $2 \times 10^6$, $1 \times 10^6$, and $0.5 \times 10^6$ cells per dish. The range of densities allows for varying efficiencies of gene transfer and negative selection between experiments and maximizes the probability that resulting colonies can be picked from dishes where the colonies are clearly individual and well separated. The selection does not work properly at densities >5 × $10^6$ cells per dish.
2. Twenty-four hours after electroporation, remove the nonselective medium and replace it with medium containing HAT. HAT medium is made by adding 1 mL of 100X HAT concentrate to 100 mL of medium. To make 100 mL of 100X HAT use 136.1 mg of hypoxanthine, 4.4 mg of aminopterin, and 24.2 mg of thymidine.

Dissolve each component separately in ~5 mL of $H_2O$. Add ~0.5 mL of 1 $N$ NaOH to dissolve the hypoxanthine and aminopterin. Bring the individual components together in 100 mL, filter sterilize, and store in the dark at –20°C.

3. Change the HAT medium after a further 48 h.
4. When HAT$^R$ colonies have just become visible (i.e., 4–6 d after electroporation), replace the HAT medium on most of the dishes with medium containing HAT and ganciclovir (2 $\mu M$). Leave a few dishes under only HAT selection, so that the efficiency of negative selection (HAT$^R$ colonies/HAT$^R$ + GANC$^R$ colonies) can be calculated.
5. Eight to 11 d after electroporation, pick colonies for screening and seeding into microtiter plates in medium containing HAT and penicillin and streptomycin. Antibiotics are used for a few days only at this critical stage to minimize the risk of contamination. Do not attempt to pick from plates containing >100 colonies. At this stage, some plates can be fixed and stained for colony counting to determine the efficiency of negative selection.

HAT selection should be withdrawn from targeted clones prior to blastocyst injection or the second targeting step. Cultures should be weaned off HAT by growing for 3 to 4 d in HT medium prior to switching to nonselective medium. HT medium is identical to HAT, except that aminopterin is omitted from the 100X concentrate.

## 8.2. Germline Colonization by HM-1 Cells

Although the ability of the HM-1 cell line to contribute to the germline of chimeric animals has been demonstrated up to passage 35, it is prudent to proceed as rapidly as possible through the double-replacement procedure to avoid unnecessary passaging of the cells. Current experience is that at least one in three of the chimeras produced from HM-1 clones, derived from a single gene-targeting procedure, will be germline transmitters. For a second round of targeting, take on at least two independent clones that have given germline transmission after the first step and plan to generate up to 10 chimeras to obtain germline transmission.

## 8.3. Second Targeting Step

Ideally only first-step clones that are proven germline transmitters should be used in the second step. Such clones should first be screened to determine their spontaneous level of 6-TG$^R$. This is done by growing cells in the absence of selection for 7–10 d, and then plating the cells at 1 × 10$^6$ per 90-mm dish in 6-TG medium. Leave the plates for 7–10 d and then fix and stain the cells to score 6-TG$^R$ colonies. Expect less than five 6-TG$^R$ colonies per dish. If the number exceeds five per dish, it is likely that the culture of knockout cells contains a small number of parental HM-1 cells, surviving by metabolic cooperation. These parental cells will interfere with the identification of 6-TG$^R$ clones that contain the gene alteration after the second targeting step. Choose knockout clones for the second targeting step in which the spontaneous frequency of 6-TG$^R$ is <5/10$^{-6}$. If all clones tested have a high frequency of 6-TG$^R$, it is probable that there is instability of HPRT marker gene expression at this targeted locus, and it may be unwise to proceed (*see* **Subheading 3.5.** for a discussion on the stability of marker gene expression).

1. Following electroporation, distribute cells evenly into eight 75-cm$^2$ flasks in nonselective medium.
2. After 2 d, aspirate the medium and replace it with nonselective medium.
3. Flasks will require subculturing within the next 1 to 2 d. Trypsinize each flask and plate one-sixteenth or one-eighth of the cells into a fresh 75-cm$^2$ flask in nonselective medium. Plate one-sixteenth if the culture is very thick, and one-eighth if it is subconfluent.
4. Change nonselective medium daily until trypsinization.
5. Seven days after electroporation, trypsinize each flask and obtain a cell count.
6. For each flask, plate $1 \times 10^6$ cells per 90-mm dish in 6-TG medium. Plate 4 dishes per flask, 32 dishes in all. 6-TG is used at a final concentration of 5 µg/mL (made from a 2 mg/mL stock; the same instructions given in **Subheading 8.1.** for dissolving hypoxanthine apply).
7. The 6-TG$^R$ colonies should be ready for picking 6–12 d later. Expect approximately five colonies per dish. Colonies should be plated into nonselective medium.

The procedure for second-step targeting optimizes for the recovery of independently arising 6-TG$^R$ clones, while keeping the experiment manageable. In this procedure, it is essential that 6-TG selection not be imposed less than 7 d after electroporation. Time is required for the HPRT activity in targeted cells to decay below the threshold for survival in 6-TG. Plating at densities much greater than 10$^6$/90-mm dish will also reduce the recovery of HPRT-deficient clones, since at high densities, HPRT-deficient cells will be killed by metabolic cooperation with their adjacent HPRT-expressing neighbors. The frequency of targeted clones that we have observed, among colonies surviving 6-TG selection, has varied from 1 in 5 to 1 in 50, depending on the locus targeted and the nature of the alteration introduced.

## 9. Colony Isolation and Screening

### 9.1. General Considerations

Following each of the gene-targeting protocols described in **Subheadings 6.0–8.0**, we prefer to use a PCR-based screening strategy at the colony-picking stage to identify targeted clones. This is the most efficient way to manage the clones arising from a targeting experiment, enabling large numbers of clones to be screened rapidly. The nontargeted majority are then discarded, so that culture time can be devoted to expanding just the targeted clones. Pick only "average" size colonies; avoid colonies that are unusually large or small. By doing this, rather than picking completely at random, we consider that it improves the probability of isolating a correctly targeted clone and, moreover, one that will be capable of contributing to the germline. Clearly, one would have concerns about the ability of a particularly small and slow-growing colony to survive picking and be capable of the necessary expansion, especially if an additional round of targeting (or site-specific recombination) is planned. The situation with very large colonies is more complex: they may not be clonal but derived instead from two or more independent DNA transfer events. In the case of two-step targeting procedures, a very large colony at the second step could result from a carryover of nontargeted cells by metabolic cooperation, or by an early loss of marker expression in a targeted clone, preceding the second targeting step designed to lead to physical loss of the marker. In either case, a colony significantly larger than one arising from the second targeting event itself could result. Large colonies could also arise if the founding cell has undergone a karyotypic change leading to more rapid growth in vitro. Such a change could interfere with meiosis and thus render the clone incapable of contributing to the germline of any resulting chimeras. In one particular ES cell line, J1, trisomy 8 was present at a high frequency in low-passage cultures *(18)*. Cells containing this change generated larger colonies than euploid cells, with a much lower frequency of germline transmission. We have not found trisomy to be a problem in nontargeted HM-1 cultures but have occasionally noted targeted clones growing faster than normal, particularly after two-step targeting procedures. Such clones were indeed found to be trisomic, hence the advice to avoid unusually large colonies.

## 9.2. Colony Picking and Plating

1. Pick individual, well-separated, average-sized colonies through the medium, using a micropipettor with a sterile, disposable 200-μL tip. Scrape the colony from the surface while applying gentle suction.
2. Transfer the colony (which may have fragmented into several pieces) in a minimum volume of medium to a microcentrifuge tube containing 0.5 mL of medium plus antibiotics.
3. Pipet vigorously to break up the cell clumps. Do not use trypsin because this adversely affects survival at low cell densities.
4. Plate half the cells into an individual well of a flat-bottomed 24-well microtiter plate in 1 mL of medium plus antibiotics. Use the remainder of the cells from each colony to prepare DNA for PCR.

### 9.3. PCR on Individual Colonies

1. Spin down the cells in a microcentrifuge tube for 30 s.
2. Remove the medium using a micropipettor.
3. Spin down again as in **step 1**.
4. Carefully remove all the residual medium.
5. Resuspend the cell pellet in 40 µL of 1X PCR buffer containing 0.1 mg/mL of proteinase K. 1X PCR buffer is made from a 10X PCR buffer concentrate. To make 50 mL of 10X PCR buffer use 8.3 mL of 3 $M$ KCl, 0.75 mL of 1 $M$ MgCl$_2$, 5.0 mL of 1% gelatin, 5.0 mL of 1 $M$ Tris-HCl (pH 8.3), 2.25 mL of Triton X-100, 2.25 mL of Tween-20, and 26.45 mL of H$_2$O.
6. Incubate for 2 h at 65°C, followed by 15 min at 90°C to heat-inactivate the proteinase K.
7. Use 10–20 µL of this DNA solution per 50-µL PCR reaction in 1X PCR buffer to identify targeted clones. Optimal conditions for PCR will depend on the primer pair used, the thermocycler, and the source of *Taq* DNA polymerase and must be determined empirically. For targeting vectors containing an HPRT marker, primer 262W (24mer 5' AGCCTACCCTCTGGTAGATTGTCG 3'; located within HPRT exon 9 [HPRT cDNA sequence, EMBL AC J00423, positions 942–965], 350 nucleotides from the 3' end of the minigene cassette), in conjunction with a primer for a sequence from the target locus lying outside the targeting vector, is particularly useful.

### 9.4. Outcome

Clones growing in microtiter plates will not require attention for 4 to 5 d, thus providing time to identify gene-targeted clones by PCR. Positive clones should be scraped out of their wells when they are just subconfluent. This stage will usually be reached about 7 d after plating into microtiter plates. During this time, the medium in positive wells will need to be changed once. Scrape out the wells using a 1000-µL tip on a micropipettor, pipet vigorously to break up cell clumps (do not use trypsin), and transfer the cells to a 25-cm² flask. At this stage, antibiotics should be withdrawn. Cells, ideally from a number of independent gene-targeted clones, can now be expanded for further analysis.

### Acknowledgments

I wish to thank The Cancer Research Campaign, The Biotechnology and Biological Sciences Research Council, and The Medical Research Council for funding gene-targeting work in my laboratory.

### References

1. Folger, K. R., Wong, E. A., Wahl, G., and Capecchi, M. R. (1982) Patterns of integration of DNA microinjected into cultured mammalian cells: evidence for

homologous recombination between injected plasmid DNA molecules. *Mol. Cell. Biol.* **2,** 1372–1387.

2. Smithies, O., Gregg, R. G., Boggs, S. S., Koralewski, M. A., and Kucherlapati, R. S. (1985) Insertion of DNA sequences into the human chromosomal β-globin gene by homologous recombination. *Nature* **315,** 230–234.

3. Thompson, S., Clarke, A. R., Pow, A. M., Hooper, M. L., and Melton, D. W. (1989) Germline transmission and expression of a corrected HPRT gene produced by gene targeting in embryonic stem cells. *Cell* **56,** 313–321.

4. Deng, C. and Capecchi, M. R. (1992) Reexamination of gene targeting frequency as a function of the extent of homology between the targeting vector and the target locus. *Mol. Cell. Biol.* **12,** 3365–3371.

5. Adra, C. N., Boer, P. H., and McBurney, M. W. (1987) Cloning and expression of the mouse pgk-1 gene and the nucleotide sequence of its promoter. *Gene* **60,** 65–74.

6. Thomas, K. R. and Capecchi, M. R. (1987) Site-directed mutagenesis by gene targeting in mouse embryo-derived stem cells. *Cell* **51,** 503–512.

7. Selfridge, J., Pow, A. M., McWhir, J., Magin, T. M., and Melton, D. W. (1992) Gene targeting using a mouse HPRT minigene/HPRT-deficient embryonic stem cell system: inactivation of the mouse ERCC-1 gene. *Somat. Cell. Mol. Genet.* **18,** 325–336.

8. Mountford, P. and Smith, A. G. (1995) Internal ribosome entry sites and dicistronic RNAs in mammalian transgenesis. *Trends Genet.* **11,** 179–184.

9. Mansour, S. L., Thomas, K. R., and Capecchi, M. R. (1988) Disruption of the proto-oncogene int-2 in mouse embryo-derived stem cells: a general strategy for targeting mutations to non-selectable genes. *Nature* **336,** 348–352.

10. Wu, H., Liu, X., and Jaenisch, R. (1994) Double replacement: strategy for efficient introduction of subtle mutations into the murine Col1a-1 gene by homologous recombination in embryonic stem cells. *Proc. Natl. Acad. Sci. USA* **91,** 2819–2823.

11. Magin, T. M., McEwan, C., Milne, M., Pow, A. M., Selfridge, J., and Melton, D. W. (1992) A position—and orientation—dependent element in the first intron is required for expression of the mouse HPRT gene in embryonic stem cells. *Gene* **122,** 289–296.

12. Littlefield, J. W. (1964) Selection of hybrids from matings of fibroblasts in vitro and their presumed recombinants. *Science* **145,** 709, 710.

13. Melton, D. W. (1997) Double replacement gene targeting in embryonic stem cells for the introduction of subtle alterations into endogenous mouse genes, in *Gene Cloning and Analysis: Current Innovations* (Schaefer, B. C., ed.), Horizon Scientific, Wymondham, Norfolk, UK.

14. Melton, D. W., Ketchen, A.-M., and Selfridge, J. (1997) Stability of HPRT marker gene expression at different gene-targeted loci: observing and overcoming a position effect. *Nucleic Acids Res.* **25,** 3937–3943.

15. Simpson, E. M., Linder, C. C., Sargent, E. E., Davisson, M. T., Mobraaten, L. E., and Sharp, J. J. (1997) Genetic variation among 129 substrains and its importance for targeted mutagenesis in mice. *Nat. Genet.* **16,** 19–27.

16. Olson, E. N., Arnold, H. H., Rigby, P. W., and Wold, B. J. (1996) Know your neighbours: three phenotypes in null mutants of the myogenic bHLH gene MRF4. *Cell* **86,** 1–4.
17. Hasty, P., Ramirez-Solis, R., Krumlauf, R., and Bradley, A. (1991) Introduction of a subtle mutation into the Hox 2. 6 locus in embryonic stem cells. *Nature* **350,** 243–246.
18. Liu, X., Wu, H., Loring, J., Harmudzi, S., Disteche, C. M., Bornstein, P., and Jaenisch, R. (1997) Trisomy eight in ES cells is a common potential problem in gene targeting and interferes with germ line transmission. *Dev. Dynam.* **209,** 85–91.

# 9

## Cre/*loxP* Recombination System
## and Gene Targeting

**Ralf Kühn and Raul M. Torres**

## 1. Introduction

Embryonic stem (ES) cell technology has clearly established itself as a powerful technique for the examination of gene function in vivo. The vast majority of gene-targeting experiments to date have been designed simply to inactivate the function of the gene of interest by the targeted insertion of a selectable marker into the ES genome. Homologous recombinant ES cells are used without further modification for the generation of mice that bear a permanently modified allele in all cells from the onset of development. In contrast to this "conventional" gene-targeting strategy, in recent years, the use of the Cre/*loxP* recombination system in conjunction with gene targeting has greatly expanded the versatility and avenues with which biologic questions can be addressed in the mouse. In addition to the generation of subtle mutations, this system allows for a number of other genotypic options in ES cells or mice by strategically incorporating Cre recombinase recognition (*loxP*) sites into the genome and the subsequent expression of recombinase in vitro or in vivo. In particular, when Cre is expressed in mice harboring a *loxP*-containing target gene, the desired gene modification can be restricted to certain cell types or developmental stages of the mouse (conditional gene targeting) depending on the tissue specificity and timing of recombinase expression. There is no definitive rule to decide whether, for a particular experiment, conventional or conditional gene targeting is more appropriate since this depends on the specific biologic question and the peculiarities of the gene studied. However, following one of the strategies described in **Subheading 4.**, both conventional and conditional gene targeting can be applied to a particular gene requiring only a single targeting vector and germline transmission of only one mutant allele.

From: *Methods in Molecular Biology, vol. 180: Transgenesis Techniques, 2nd ed.: Principles and Protocols*
Edited by: A. R. Clarke © Humana Press Inc., Totowa, NJ

This chapter gives an overview of the various applications of Cre-mediated recombination for gene-targeting purposes and helps the newcomer to plan and perform second-generation gene-targeting experiments using site-specific recombination. **Subheading 2.** gives an introduction of the Cre/*loxP* recombination system, followed by general considerations on the design of *loxP*-containing gene-targeting vectors and their applications for gene modification in ES cells (*see* **Subheading 3.**). Finally, the last section summarizes the use of Cre/*loxP* for conditional gene targeting.

## 2. Cre Recombinase and *loxP* Sequences

The Cre (<u>c</u>auses <u>re</u>combination) recombinase is a 38-kDa protein that recognizes and mediates site-specific recombination between 34-bp sequences referred to as *loxP* (<u>lo</u>cus of crossover [<u>x</u>] in <u>P</u>1 bacteriophage) sites *(1)*. Both the Cre recombinase and *loxP* sequences are derived from the P1 bacteriophage, which utilizes this recombination system in its life cycle to maintain the phage genome as a unit copy plasmid in the lysogenic state *(2,3)*. As a member of the integrase superfamily of site-specific recombinases, the Cre recombinase does not require any host cofactors or accessory proteins to mediate *loxP*-specific recombination. This fact, coupled with its optimum operating temperature of 37°C *(4)*, has made the Cre/*loxP* recombination system an ideal system for genetic engineering in mammalian systems. Successful genetic manipulation of both endogenous loci and transgenes has been successfully accomplished with this recombination system in mammalian cells, mice, plants, yeast, and bacteria.

A *loxP* site is composed of two 13-bp inverted repeats interrupted by an 8-bp nonpalindromic sequence that provides the orientation of the overall sequence (**Fig. 1A**) *(5,6)*. Cre-mediated recombination between two *loxP* sites results in

---

Fig. 1. *(opposite page)* Structure of a *loxP* site and Cre-mediated *loxP*-specific recombination events. (**A**) *loxP* sequence is schematically represented as a solid triangle with a (relative) orientation as defined by the nonpalindromic spacer sequence. Below, the canonical *loxP* sequence is given with arrows depicting inverted repeats and the spacer region underlined. (**B**) Cre-mediated excision between two directly repeated *loxP* sites, in an intramolecular reaction, results in the circularization of the *loxP*-flanked sequence and a linear molecule, each retaining a single *loxP* site. Cre/*loxP*-mediated integration is the reverse reaction that proceeds via an intermolecular reaction. (**C**) Cre-mediated recombination between two linear molecules, each harboring a single *loxP* site in the same orientation, is resolved through an intermolecular reaction. The reaction products have reciprocally exchanged flanking sequences. (**D**) Through an intramolecular reaction, Cre mediates an inversion of DNA sequence that is flanked by *loxP* sites in opposing orientation relative to each other.

**A**  Canonical *loxP* sequence

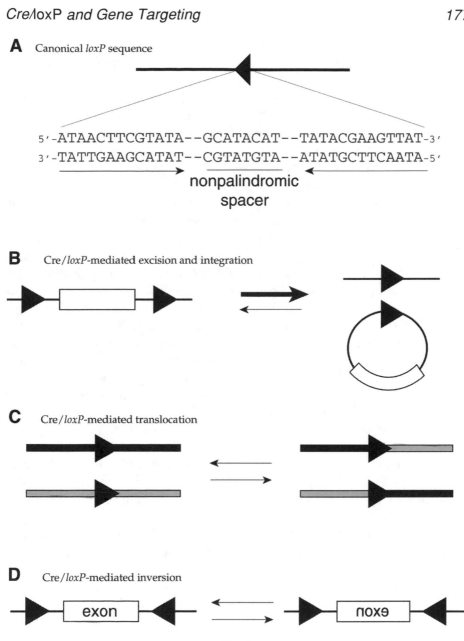

5′-ATAACTTCGTATA--GCATACAT--TATACGAAGTTAT-3′
3′-TATTGAAGCATAT--CGTATGTA--ATATGCTTCAATA-5′

nonpalindromic
spacer

**B**  Cre/*loxP*-mediated excision and integration

**C**  Cre/*loxP*-mediated translocation

**D**  Cre/*loxP*-mediated inversion

Fig. 1.

the reciprocal exchange of DNA strands between these sites but yields different overall products that depend on the orientation and location of the two sequences. When two *loxP* sites are in the same orientation on a linear DNA molecule, a Cre-mediated intramolecular recombination event resolves with the excision of the *loxP*-flanked, or "floxed," DNA sequence as a circular molecule and one *loxP* site remaining on each reaction product (**Fig. 1B**). This is a reversible reaction in which, through an intermolecular recombination event, a *loxP*-containing circular DNA molecule is introduced at a *loxP* site in a linear sequence resulting in the previously circularized DNA sequence now flanked by *loxP* sites (**Fig. 1B**). If two *loxP* sites reside on separate linear DNA sequences, then, through an intermolecular recombination event, Cre-mediates the reciprocal exchange (translocation) of the flanking sequences (**Fig. 1C**). Finally, when two *loxP* sites are on the same DNA molecule in opposite orientation relative to each other, through an intramolecular recombination event, Cre-mediates the inversion of the floxed sequence (**Fig. 1D**) *(7–9)*.

Mutational analyses of *loxP* sites have demonstrated that single nucleotide exchanges within the 8-bp spacer region can still recombine efficiently with each other but not with wild-type *loxP* sites *(10)*. This property would appear to add further versatility to the use of the Cre/*loxP* recombination system by allowing Cre to recombine *loxP* sites independently of each other and without the possibility of interchromosomal recombination. Moreover, the equilibrium of a reaction may be shifted by the incorporation of different point mutations into the two halves of a pair of *loxP* sites such that a Cre-mediated recombination will yield a wild-type and mutant *loxP* site that recombines much less efficiently than the starting *lox* sites. This has proven particularly useful in a Cre-mediated integration event (e.g., **Fig. 1B**) in ES cells in which the intramolecular excision event would be favored over an intermolecular integration event *(11)*.

The crystal structure of a Cre/lox synapse has been resolved to 2.4 Å *(12)* and has supported a model of Cre-mediated recombination in which four Cre molecules are bound to two *loxP* sites (each Cre molecule contacting a *loxP* half-site) forming a four-way Holliday-junction intermediate. Either during or immediately preceding this intermediate, tyrosine 324 cleaves the parental DNA strand and forms a phosphodiester linkage as a covalent 3'-phosphotyrosine intermediate. The resulting free 5'-hydroxyl groups attack the phosphotyrosines on the substrate of the opposing strand and form a Holliday-junction intermediate. Resolution of this intermediate is accomplished by a second round of tyrosine cleavage, linkage, and strand exchange. While Cre-mediated recombination between *loxP* sites does not appear to have a maximum distance (as evidenced by megabase deletions and interchromosomal translocations; *see* **Subheadings 3.3.2.** and **3.5.**), the minimum distance required between directly repeated *loxP* sites has been determined to be 82 bp for efficient recombination *(13)*.

For the 1029-bp coding sequence of Cre, we refer the reader to Sternberg et al. *(14)* and the EMBL database accession number X03453.

## 3. Applications of Cre/*loxP* in ES Cells

While the details may vary considerably, the use of Cre/*loxP* with gene targeting involves two basic steps. The initial stage is the introduction of *loxP* sites into the genome of ES cells using a targeting construct and standard gene-targeting technology followed by the identification and isolation of appropriate homologous recombinants. The actual genetic modification occurs on expression of the Cre recombinase in cells harboring these *loxP* sites. This recombination event can occur in ES cells in vitro on transient Cre expression after which these cells would be injected into blastocysts and transmitted to the germline as a permanent modification in the developing embryo. Alternatively, or in addition, the genetic modification can occur conditionally in vivo when Cre, expressed as a transgene, is introduced to *loxP*-containing cells in a cell-type-specific or temporally restricted manner (*see* **Subheading 4.**). Even if the initial goal is simply to ablate gene function, the incorporation of Cre/*loxP* into a gene-targeting experiment offers clear advantages to conventional targeting in that, should the experimental results validate the need, the potential approaches to examine gene function are expanded beyond that of simple gene inactivation.

### 3.1. Deletion of a Selection Marker

The removal of the selection marker used for the isolation of homologous recombinants is one of the most useful and straightforward applications of Cre/*loxP* technology. Even within the context of gene ablation studies, it is important that the selection marker be removed from the endogenous locus because its presence may affect the regulation of not only the targeted locus, but loci >100 kb away from the targeted locus (*15–19*). In experiments aiming only to modify gene function or to examine gene regulation, this issue is of even more importance. Use of the Cre recombinase to delete the selection marker results in, at a minimum, a 34-bp heterologous (*loxP*) sequence that remains in the endogenous locus.

Excision of the selection marker gene can be accomplished in both ES cells in vitro, before mice are generated, or in vivo after the selectable marker (and mutation) has been transmitted to the germline. With either route, the initial targeting construct must incorporate a *loxP*-flanked, or floxed, selection marker, and homologous recombinants must be identified and isolated after selection. For deletion in vitro, the Cre recombinase is transiently expressed in ES cells, and those clones that have undergone Cre-mediated deletion of the selection marker are identified based on their sensitivity to the selecting agent (*20*). In vivo deletion of the floxed selection marker is accomplished by breeding mice harboring the floxed selection marker on one allele to transgenic mice that express the Cre recombinase as a transgene very early in embryonic devel-

**Table 1**
**Cre-Expressing Mouse Strains**

| Cre expression | Control element | References |
|---|---|---|
| Constitutive | | |
| Thymocytes | Proximal lck promoter | *55,69* |
| Eye lens | A-crystallin promoter | *54* |
| B-lymphocytes | Insertion into CD19 gene | *68* |
| Forebrain neurons | CamKIIα promoter | *56* |
| | | R. Kühn, unpublished data |
| Schwann cells | P0 promoter | *46* |
| Pituitary gland | POMC promoter | *46* |
| Retina | IRBP promoter | *46* |
| Adipocytes | aP2 promoter | *73* |
| Cardiac muscle | αMyHC promoter | *74* |
| Epidermis | Keratin-5 promoter | *70* |
| Macrophages | Insertion into lysozyme M gene | B. Clausen and I. Förster, unpublished data |
| Mammary gland | WAP promoter | *75* |
| Deleter strains | | |
| Oocyte | ZP3 promoter | *50* |
| Embryo, mosaic | Nestin promoter/enhancer | *49* |
| Early embryo | Human CMV promoter | *21* |
| Zygote | Adenovirus EIIa promoter | *22* |
| Oocyte/zygote | CMV-IE/β-actin | *51* |
| Germline | PECAM-1 promoter | *52* |
| Inducible | | |
| Many tissues | Mx1 promoter | *62* |
| Ubiquitous | CMV promoter | *66* |
| B-lymphocytes | SV40 promoter/Ig enhancer | F. Schwenk, unpublished data |
| Viral vectors | | |
| Brain | HSV-IE promoter/herpesvirus | *47* |
| Liver | CMV-IE/adenovirus | *44* |
| Liver/brain | HSV-tk promoter/adenovirus | *45* |
| Colorectal epithelium | SRα promoter/adenovirus | *48* |
| Many tissues | hCMV-IE promoter/adenovirus | *46* |

opment *(21,22)* (*see* **Table 1**) so that the progeny of all such crosses have deleted the floxed selection marker in all tissues including germ cells. An important advantage to the in vivo deletion of the selection marker is that as soon as homologous recombinants have been identified, they may be injected into blastocysts for germline transmission and abrogate the need for continual manipulation and culturing of the ES cells in vitro.

Deletion of the selection marker in vitro results in an ES cell clone that can once again be placed under the same selective pressure as that used initially for the isolation of the homologous recombinant. That is, the original targeting construct used for isolation of homologous recombinants can again be transfected into the homologous recombinant clones to obtain ES cells homozygous (or hemizygous if the Cre-mediated deletion event also results in a null allele) for the mutation. Thus, the need to generate additional constructs harboring different selection markers, and the embryonic feeder cells that must also be resistant to the different selective pressures, is avoided. Homozygous ES cells can be used as an alternative to germline transmission for the study of gene function in fetal development through the use of tetraploid embryo aggregation *(23,24)* or for the examination of gene function in cell lineages through the use of blastocyst complementation *(25)*.

Using a phosphoglyceratekinase- or MC1-driven Cre expression vector, deletion of a floxed neo gene should occur with a frequency of at least 50% that of the efficiency of transient transfection. Therefore, if a homologous recombinant ES cell clone is transfected with 10% efficiency, 5 of 100 colonies would be expected to have undergone Cre-mediated deletion of the neo gene. Such cells can easily be identified on the basis of their G418 sensitivity in duplicate cultures or even as fluorescent cells when a green fluorescent protein (GFP)-Cre fusion protein is used *(26)*. The frequency of Cre-mediated selection marker deletion is sufficiently high such that the isolation of cells having undergone a deletion event need not be aided by the inclusion of a thymidine kinase (tk) gene and subsequent ganciclovir selection.

## 3.2. Introduction of Nonselectable (Subtle) Modifications

Introducing a (subtle) nonselectable modification by Cre/*loxP* gene targeting is accomplished with a replacement-type vector that harbors the mutation and a selectable marker flanked by *loxP* sites. For example, a point mutation can be introduced into an exon in a targeting vector and cotransferred to the endogenous locus along with the floxed selection marker. Following Cre-mediated excision of the selection marker, only the mutated exon and 34-bp *loxP* site remain. This type of strategy has been used to introduce a premature stop codon into the cytoplasmic domain of a signaling molecule used by a lymphocyte antigen receptor *(27)*. As already discussed, the removal of the selection marker may be accomplished in vitro or in vivo.

### *3.3. Genomic Deletions*

The generation of genomic deletions also has been successfully used with Cre/*loxP* technology. This involves the introduction of two *loxP* sites into ES cells, and, in a subsequent step, the transient expression of Cre deletes the intervening sequence. Depending on the length of the desired deletion, the insertion of *loxP* sequences can be accomplished with either one or two targeting constructs and homologous recombination events.

### *3.3.1. Small Genomic Deletions*

For Cre-mediated deletion of relatively short sequences (e.g., ≤10 kb), both *loxP* sites may be incorporated in a single targeting construct and, after homologous recombination, recombinants identified that have integrated both *loxP* sites into the endogenous locus (see also flox-and-delete strategy in **Subheading 3.7.**). Transient Cre expression in these homologous recombinants excises the sequence between the outer two *loxP* sites. Cre-mediated deletion of longer sequences may be aided by the use of the tk gene as a *loxP*-neo-tk cassette to enrich for clones having undergone the deletion event, although Cre appears to be quite efficient in excising lengths of DNA in the range of 1–10 kb. The limiting factor in this single vector approach is the decreasing frequency with which both the selectable marker and isolated *loxP* sites cointegrate in the endogenous locus of homologous recombinants. However, the successful cotransfer of *loxP* sites separated by 11 kb on one vector has been reported *(28)*.

### *3.3.2. Large Genomic Deletions*

Perhaps unique to the use of Cre/*loxP* gene targeting is the potential to create very large genomic deletions. While the upper limit of Cre-mediated deletion has not yet been characterized, in initial reports, DNA segments ranging from 200 kb *(29)* to 3 to 4 cM *(30)* have been successfully deleted in ES cells and transmitted through the germline in the mouse. The potential applications of such deletions would include targeted inactivation of genes spread out over a very large distance, deletion of gene clusters from the genome, and genetic screening for recessive mutations.

The generation of large deletions is accomplished by introducing the two *loxP* sites in sequential targeting events using separate targeting constructs each of which contains a different positive selection marker (e.g., neo and hygromycin) flanked by *loxP*. Homologous recombinants generated from both targeting constructs are characterized, and those that have incorporated the selection marker-*loxP* cassettes on the same chromosome (and in the same orientation) are subsequently transfected with the Cre recombinase that excises the *loxP* intervening sequence (**Fig. 2**). Instead of using two different selection markers (and embryonic feeders resistant for both selecting agents), the same *loxP*-flanked marker gene could be used in both targeting vectors if the floxed marker is deleted after the first targeting step, resulting in a single remaining *loxP* site.

Careful consideration must be applied when designing experiments for the introduction of *loxP* sites separated by a great distance since chromosomal inversions or duplications also may be generated after Cre expression in ES cells depending on the orientation of the *loxP* sequences *(30)*. In particular, it has been reported that Cre-mediated recombination of inverted *loxP* sites, when on the same chromosome, may result in chromosome loss if recombination occurs at the time of chromosome replication regardless of the distance separating the two *loxP* sites *(31)*.

To excise large genomic sequences in which the gene order and orientations are not known, a strategy must be devised such that the molecular analysis of clones that have undergone Cre-mediated recombination will reveal the relative orientations of the *loxP* sites. More important, to avoid all but the desired chromosomal rearrangements, it should be confirmed that both targeting events occurred on the same chromosome. Because after a certain distance the efficiency of Cre-mediated deletion likely declines with increasing distance between *loxP* sites, a tk gene as a negative selection marker should be included within at least one of the targeting vectors (**Fig. 2**; also *see* **ref. 29**). As another strategy for generating large genomic deletions, the reconstitution of a functional positive selection marker may be employed *(30)*. In the cited examples of Cre-mediated large deletions *(29,30)*, the frequency of Cre-mediated deletion on transient transfection of ES cells was found to be in the range of $10^{-5}$–$10^{-7}$.

Similarly as already described for Cre-mediated deletion of the selectable marker, homologous recombinant clones harboring the *loxP*-flanked chromosome segment may be directly used for the generation of a mouse strain and subsequently bred with a Cre deleter strain (**Table 1**). Although deleter strains have proven quite reliable in deleting floxed DNA segments on the order of 5 kb in vivo, their efficiency in excising longer DNA segments has not yet been characterized. Should the in vivo deletion approach be attempted, the use of the tk gene should be avoided because its presence interferes with germline transmission.

### 3.4. Gene Replacement

A targeting vector used for gene replacement is generally a conventional replacement vector that incorporates an additional isolated *loxP* site as used for the generation of genomic deletions (*see* **Subheading 3.3.1.** and **Fig. 3**).

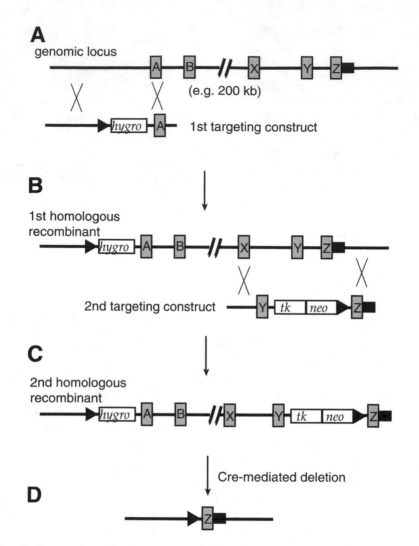

Fig. 2. Generation of a large genomic deletion via Cre-mediated *loxP*-specific recombination. (**A**) A generic genomic locus is depicted separated over 200 kb with exons represented as shaded rectangles. The first targeting construct inserts a hygromycin-resistant gene (open rectangle), flanked on one side by the *loxP* site (solid triangle), at one end of the genomic locus. (**B**) After homologous recombinants are identified, a second targeting construct is used in a second targeting event to insert a second *loxP* site at the other end of the genomic segment to be deleted. The second targeting construct harbors a second selection marker for the isolation of homologous recombinants (e.g., neo) and a tk gene that is used to select for those recombinants that have undergone Cre-mediated *loxP*-specific deletion. (**C**) Homologous recombinants are identified in which both targeting events have occurred on the same chromosome. (**D**) Transient expression of the Cre recombinase in appropriately targeted homologous recombinant clones excises the *loxP*-flanked DNA and can be enriched for by ganciclovir selection.

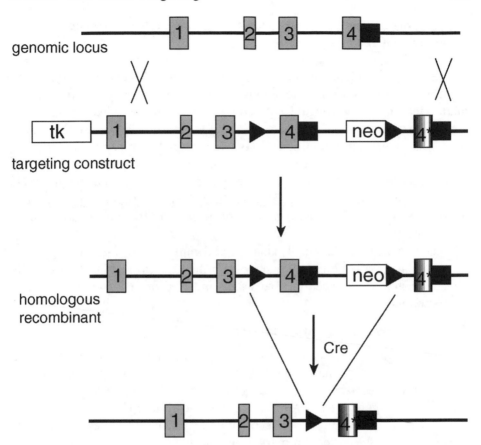

Fig. 3. Replacement of an exon using Cre/*loxP* with gene targeting. A generic genomic locus encompassing a gene of four exons is shown. Exons are depicted as shaded rectangles, 3'-untranslated regions as solid rectangles, and *loxP* sites as solid triangles. A targeting construct designed to replace the last (fourth) exon with a mutated version is shown. A selection marker (e.g., neo), flanked on one side by *loxP* followed by the mutated version of the exon, is inserted downstream of the poly-adenylation site. A second isolated *loxP* site is also incorporated in the targeting vec-tor, which on cotransfer will reside in the third intron. Cre expression in homologous recombinants that have incorporated both *loxP* sites will deleted the wild-type exon, replacing it with the mutated version.

In contrast to simple insertions by conventional targeting, this approach allows the option of removing endogenous sequence, in addition to the selectable marker, in a Cre-mediated deletion event. Thus, for relatively small genomic replacements such as an exon, a gene replacement vector encloses a region of nonhomology (i.e., the DNA sequence used for replacement) juxtaposed to a selection marker flanked on one side by *loxP* and a second isolated *loxP* site incorporated in the endogenous sequence at a site that results in the excision of the desired sequence (**Fig. 3**).

Cre-mediated deletion in homologous recombinant ES cell clones yields the deletion of the relevant endogenous sequence together with the selection marker leaving the surrogate DNA sequence. Examples of Cre/*loxP*-mediated gene replacement include the replacement of the murine *Cγ1* gene by the human homolog in the immunoglobulin (Ig) locus *(32)*, the replacement at the *engrailed* locus of a mutant *en1* gene by *en2 (33)*, and the replacement of the T-cell receptor β-chain enhancer by the Ig heavy chain intronic enhancer *(18)*. For the reasons discussed in **Subheading 3.3.**, the replacement of large genomic sequences is perhaps best accomplished in two consecutive targeting events.

Finally, gene replacement also may be accomplished by other strategies and, in particular, in a conditional manner such that a region of DNA (e.g., wild-type sequence) is replaced in vivo with a mutated or heterologous sequence (*see* **Subheading 4.**).

### 3.5. Chromosomal Translocations

Chromosomal translocations also can be generated by Cre/*loxP*-mediated recombination events, including those that mimic naturally occurring translocations. However, two targeting constructs and consecutive targeting events are required to generate a chromosomal translocation in ES cells by the use of Cre/*loxP*. The first homologous recombination event introduces, on one chromosome, a positive selection marker gene flanked on one side by *loxP*. The subsequent targeting event introduces a second selectable marker gene, again flanked on one side by *loxP*, into the second chromosome. The same selectable marker can be used in both targeting constructs provided that it is flanked on both sides in the first construct and deleted after the identification and isolation of homologous recombinants. If the two *loxP* sites are properly targeted into respective introns, a fusion protein results on translocation, whereas if the *loxP* sites are juxtaposed to respective transcription units, only regulatory regions will be exchanged. Regardless of their placement, both *loxP* sites must maintain the same orientation, with respect to the centromeres, to retain chromosomal integrity on exchange. Once a homologous recombinant ES cell clone harboring both targeting events is identified, the translocation is generated after the transient expression of the Cre recombinase in these cells (**Fig. 4**).

Despite the proven efficiency of Cre-mediated *loxP*-specific intrachromosomal deletions, in those reports in which this recombination system has been used for translocation events in ES cells, a recombination frequency on the order of $10^{-7}$–$10^{-5}$ has been observed *(30,34)*. Consequently, to facilitate the

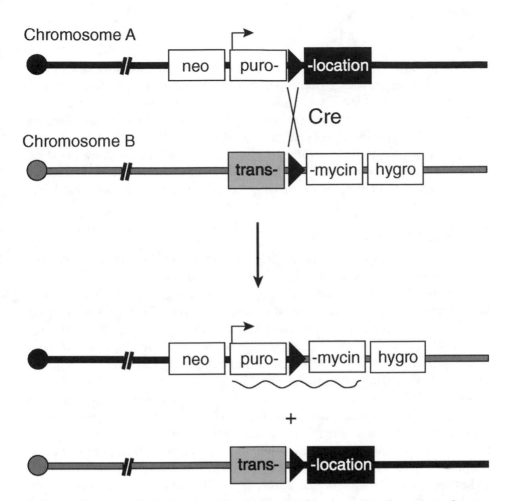

Fig. 4. Using the Cre/*loxP* recombination system for generating chromosomal trans-locations in ES cells. Different chromosomes are shown with centromeres depicted as solid and shaded circles. Distinct endogenous genetic elements are represented as "-location" on chromosome A and "trans-" on chromosome B. Homologous recombination on chromosome A introduces a selection marker, used for the identification of recombinants, and the promoter region (arrows) of second selectable marker gene (e.g., puromycin). A single *loxP* site follows these heterologous sequences. On chromosome B, another homologous recombination event introduces a second *loxP* site immediately preceding the coding sequence of the second selectable marker and a third selection marker (e.g., hygromycin) that is used for the identification of these second homologous recombinants. Cre expression and subsequent *loxP*-specific recombination in appropriately targeted clones results in a functional puromycin resistance gene (wavy line represents transcription) and the juxtaposition of the "trans-" and "-location" genetic elements on the reciprocal chromosome. *loxP* sites must be in the same orientation relative to the centromere.

identification of those ES cells that have undergone a translocation event, perhaps it is necessary to design the experiment such that a third positive selection
marker is generated on interchromosomal recombination. This might be
accomplished, e.g., by introducing in the first targeting event a promoter region
and in the second targeting event the coding region for a selectable marker
gene. Moreover, careful forethought in experimental design allows that on
recombination all three selectable markers will not reside on the desired trans-
located chromosome but, rather, on the recombination partner that would segregate by breeding (**Fig. 4**).

## *3.6. Targeted Integrations*

Besides Cre-mediated deletion of a floxed allele, the introduction of a *loxP*-
harboring circular DNA sequence (e.g., a vector harboring a transgene or
mutated endogenous sequence) into a genome that contains a *loxP* site can be
integrated by a Cre-mediated recombination event (**Fig. 1**). This feature of the
Cre/*loxP* system for site-specific integration could be used as an alternative for
homologous recombination to generate secondary modifications in *loxP* containing loci in cultured cells. Potential application of this integration event
might include analysis of a number of different promoter regions at a particular
locus in gene regulation studies or replacement of various mutated forms of a
particular exon in structure/function studies. Moreover, the generation of *loxP*-
containing loci that have proven useful for transgenic expression, or otherwise,
may be used repeatedly for the introduction of different transgenes into the
characterized locus.

Practical considerations require that the equilibrium of Cre-mediated exci-
sion/integration be shifted toward the integration event using mutant *loxP* sites
(**Fig. 1B**). Mutant *loxP*-mediated integration of nonhomologous DNA has been
successfully accomplished in ES cells with frequencies of up to 16% *(11)*.

For targeted integration, the *loxP*-containing integration vector must be
introduced together with a vector for transient Cre expression, ensuring only
limited recombinase activity, to avoid the subsequent excision of the integrated,
*loxP*-flanked DNA segment.

## *3.7. Conditional Modification: Flox-and-Delete*

A clear advantage provided by using the Cre/*loxP* system together with gene
targeting is the potential avenues of approach that exist on the generation of
homologous recombinant ES clones that harbor a floxed allele. Hence, gene
function will be ablated or modified (depending on the experimental design)
on the transient expression of the Cre recombinase in these ES cell clones.
Mice derived from these in vitro recombined ES cells will have the desired
genetic perturbation throughout development similar to those generated by

conventional targeting strategies. Alternatively, such mice also may be generated as the progeny of crosses between mice harboring the floxed allele and one of the various deleter strains that express Cre ubiquitously in the early embryo (**Table 1**). More important, homologous recombinant ES clones harboring a floxed allele also can be used for conditional gene targeting. This would obviously be useful should the targeted mutation have lethal or detrimental consequences on development. In this case, mice derived from ES cells with a floxed allele can be crossed to Cre transgenic mice for conditional gene modification or inactivation in specific cell types or on induction (*see* **Subheading 4.**). This flox-and-delete strategy has been used for conditional gene inactivation and has proven especially advantageous when the mutation has resulted in embryonic lethality.

A flox-and-delete gene-targeting vector will have three *loxP* sites in the same orientation. Two of these flank the selectable marker and the third is an isolated site a variable distance away from the selection marker within an arm of homology. Thus, the construct is configured like a replacement-type vector containing an isolated *loxP* site analogous to a nonselectable mutation, as discussed in **Subheading 3.2.** (**Fig. 5**). The isolated *loxP* site is often inserted as an oligonucleotide at a restriction site in one of the homology arms of the vector or can be recovered from various *loxP*-containing vectors *(20)*.

Once homologous recombinant ES clones that have cointegrated the isolated *loxP* site have been identified, the transient expression of Cre will lead to three different recombination products. Recombination between the outer two *loxP* sites (type I deletion) results in a large deletion that, depending on the experimental design, should inactivate or modify the gene. If recombination occurs between the two *loxP* sites flanking the selectable marker gene (type II deletion), then a single *loxP* site remains that generates, together with the isolated (outer) *loxP* site, the floxed version of the gene (**Fig. 5**).

A third recombination product will also be generated, deleting the sequence between the neo and the isolated *loxP* site; however, this is generally of no practical value and will not be found if ES colonies are screened based on G418 sensitivity after transient Cre expression. Although both type I and II deletions have been found with equal frequencies among G418-sensitive colonies in most cases, we have observed, on occasion, instances in which type I deletions were favored over type II deletions, perhaps suggesting a differential accessibility of *loxP* sites for Cre recombinase at some genomic locations. Inclusion of multiple *loxP* sites, in addition to those flanking the selection marker, into the targeting construct will permit the generation of ES clones that harbor different genomic deletions (e.g., deleting different exons) after Cre expression into appropriately targeted ES cells.

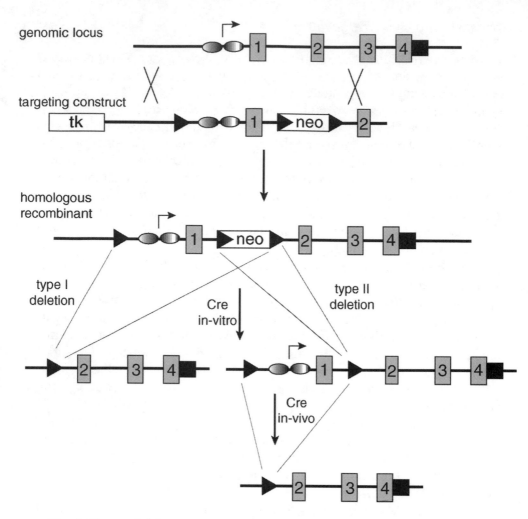

Fig. 5. Flox-and-delete gene-targeting strategy to generate ES cell clones harboring an inactive or active (floxed) allele of a gene. A generic gene is shown composed of a promoter region (shaded ellipses and arrow), four exons (shaded rectangles), and a 3'-untranslated region (solid rectangle). The targeting construct is designed to incorporate a *loxP*-flanked neo gene in the first intron and a third isolated loxP site upstream of the characterized promoter region. All three *loxP* sites are in the same orientation with respect to each other. Transient expression of the Cre recombinase in homologous recombinants will result in either recombination between the outer two *loxP* sites, which excises a critical element for gene function (type I deletion), or between the two *loxP* sites flanking the neo gene (type II deletion). The latter recombination event yields a floxed but otherwise normal allele that can be used for conditional gene targeting. The third type of recombination event is not depicted (*see* **Subheading 3.7.**).

Mice harboring a nonfunctional gene of interest can be generated from ES cell clones with a type I deletion, and ES clones with a type II deletion can give rise to a mouse strain with a floxed allele for subsequent conditional gene inactivation in vivo (*see* **Subheading 4.**). As discussed in **Subheading 3.1.**, ES cell clones with either a type I (allele) or II (neo gene) deletion also provide the option to target the second (wild-type) locus using the same targeting vector as in the first step (or a modified version if needed) because the selection marker is no longer present. The insertion site of the floxed selection marker, in addition to the isolated *loxP* site, must be carefully considered such that after the deletion of the selectable marker, neither remaining *loxP* site will affect gene function, and, furthermore, subsequent Cre-mediated recombination in vivo modifies or inactivates gene function. Note that *loxP* sites have been inserted within an intron in both orientations, with respect to transcription, and homozygous floxed mice have not exhibited any overt detrimental effects with respect to RNA splicing. The canonical *loxP* sequence has one open reading frame.

The Cre/*loxP* recombination system was first used together with gene targeting, via a flox-and-delete strategy, for the inactivation of the DNA polymerase β gene in which the promoter and first exon were flanked by *loxP* sites *(35)*. Since then, the expanding number of studies that has successfully used this strategy for the inactivation or modification of gene function and regulation attests to its usefulness in examining in vivo gene function *(15,16,18,19,36–43)*.

## 4. Conditional Gene Targeting

Conditional gene targeting can be defined as a gene modification that is restricted to certain cell types or developmental stages of the mouse. The use of the Cre/*loxP* recombination system for conditional gene targeting requires the generation both of a mouse strain harboring a *loxP*-flanked segment of a target gene and of a second strain expressing Cre recombinase constitutively, or on induction, in specific cell types. A conditional mutant is generated by crossing these two strains such that the modification of the *loxP*-flanked target gene is restricted in a spatial and temporal manner according to the pattern of Cre expression in the particular strain used (**Fig. 6**). These double transgenic strains must be bred homozygously for the chromosome containing the *loxP*-flanked locus. As an alternative to Cre-transgenic mice, Cre also can be delivered to somatic tissues through infection with viral Cre expression vectors *(44–48)*. In addition to conditional mutagenesis, strains harboring a *loxP*-flanked allele can be converted to a conventional mutant, bearing the deleted allele in its germline, by crossing to one of the deleter strains (**Fig. 6**, **Table 1**) *(21,22,49–52)* that expresses Cre in germ cells or the early embryo.

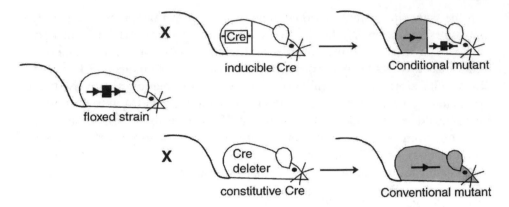

Fig. 6. Strategy for conditional gene inactivation in vivo. A mouse strain harboring a *loxP*-flanked (triangles) gene segment (square) is crossed to mice expressing Cre recombinase constitutively or in an inducible fashion in specific cell types or organs. The *loxP*-containing strain can be converted to a conventional, complete mutant by a single cross to one of the deleter strains expressing Cre in germ cells or the early embryo (bottom cross). In double transgenic animals, target gene modification occurs by Cre-mediated recombination according to the expression pattern of recombinase (shaded area).

The restricted specificity of conditional gene targeting can be used to test the function of widely expressed molecules in a particular cell lineage and to investigate gene function postnatally if conventional gene inactivation leads to a severe or lethal phenotype during embryonic development. The use of an inducible system for Cre expression has the additional advantage that the wild-type gene product under investigation can be present throughout ontogeny until the time of induction. Inducible gene targeting should be especially helpful in analyzing gene function in adults since it allows gene modification after the normal establishment of adaptational responses in mice such as immunologic or spatial memory. In both cases, the reduction in the wild-type gene product on Cre-mediated inactivation of its gene will be gradual, depending on the half-life of its mRNA and protein in the investigated cell type.

In addition to the modification of endogenous genes by gene targeting, Cre recombinase can be used to activate (or inactivate) in vivo *loxP*-interrupted, randomly integrated transgenes introduced into the mouse germline by either pronucleus injection or the ES cell route. Like Cre-mediated conditional gene targeting, transgene expression can be controlled in a tissue-specific or temporal manner by the pattern of Cre expression in vivo (**Fig. 6**). As an alternative to the use of Cre transgenic strains, a Cre expression vector can be injected directly into fertilized eggs *(53)* derived from the strain with the *loxP*-containing transgene to generate animals harboring the Cre-mediated deletion in all cells, or Cre can be delivered to somatic tissues using viral expression vectors *(44–48)*. Compared to the conventional way of controlling transgene expression directly from a cell type–specific promoter, this appears advantageous only if transgene expression is lethal or otherwise deleterious to the animal since the strain harboring the *loxP*-interrupted (inactive) transgene could be maintained normally. Possible applications of this method include the expression of oncogenes to study tumorigenesis, toxin expression for the ablation of cell lineages, and activation of reporter genes to mark the progeny of individual cells or cell lineages.

## 4.1. loxP-*Flanked Alleles*

A requirement for Cre/*loxP*-based conditional gene targeting is the generation of a mouse strain that harbors a *loxP*-flanked DNA region of interest. This is achieved by gene targeting in ES cells following one of the strategies outlined in **Subheading 3.** if an endogenous gene should be conditionally inactivated or modified. Expression of the *loxP*-modified allele should not be disturbed by the presence of the *loxP* sites, but the gene should be modified in the desired manner on Cre-mediated recombination. Cell type–specific Cre expression was first applied to the activation or inactivation of *loxP*-containing transgenes. Cre expression from the αA-crystallin promoter was used to activate a *loxP*-interrupted oncogene construct in the eye lens *(54)* and to inactivate a β-galactosidase gene specifically in T-lymphocytes using a Cre transgene driven by the thymocyte-specific *lck* proximal promoter *(55)*. Brooks et al. *(47)* have used a Cre-expressing adenovirus to locally activate an expression vector for nerve growth factor in the brain. In other transgenic recombination substrates, *loxP* sites, flanking an element interfering with transcription, have been placed between a constitutively active promoter and the coding region of β-galactosidase *(54)*. These types of strains can be used as reporters to test for the tissue specificity of Cre expression in transgenic mice by histochemical staining. Four different lacZ indicator strains have been described (**Table 2**) *(46,51,53,56)* and were confirmed for lacZ expression in specific tissues, but in none of the lines has the pattern of lacZ expression been fully characterized.

Conditional inactivation of an endogenous gene was first described in 1994 *(35)* for the DNA polymerase β gene, in which case a 1.5-kb region containing the promoter and first exon of the *pol*β gene was flanked by *loxP* sites using the flox-and-delete strategy described in **Subheading 3.7.** Presently, conditional gene targeting has been applied to 11 genes (**Table 2**), in most cases to study gene function in adult mice by circumvention of embryonic or perinatal lethality associated with the complete inactivation of the target gene.

**Table 2**
***LoxP*-Flanked Alleles**

| Target gene | *LoxP*-flanked segment (kb) | Deletion/cell type/delivery[a] | Refs. |
|---|---|---|---|
| Endogenous | | | |
| DNA polymerase β | 1.5 | 84%/thymocytes/Tg | *35* |
| | | 95%/B-lymphocytes/Tg | *68* |
| | | 100%/hepatocytes/Tg | *62* |
| IL-2Rγ | 2.5 | Up to 100%/many tissues/Tg | *49* |
| GalNac-T | 4 | 100%/thymocytes/Tg | *69* |
| LRP | 4 | 100%/hepatocytes/virus | *44* |
| NMDAR1 | 11 | ND/neurons/Tg | *28* |
| Pig-a | 7 | 100%/skin/Tg | *70* |
| APC | 3 | ND/intestine/virus | *48* |
| IgH V gene | 4.6 | >60%/B-lymphocytes/Tg | *57* |
| | | >90%/T-lymphocytes/Tg | *57* |
| VCAM-1 | — | ND | *52* |
| Mre11 | 4 | ES cells | *71* |
| C/EBPα | 4 | 80%/hepatocytes/virus | *72* |
| Transgenic | | | |
| SV40 T | 1.3 | ND/lens/Tg | *54* |
| LacZ | 3.5 | 99%/thymocytes/Tg | *22,54* |
| NGF | 2.4 | ND/neurons/virus | *47* |
| Reporter genes | | | |
| LacZ | 2 | 100%/embryo/plasmid | *11* |
| LacZ | 1.3 | 100%/neurons/Tg | *56* |
| LacZ | 2 | 78%/hepatocytes/virus | *46* |
| LacZ | 2 | 100%/embryo/Tg | *51* |

[a]Tg, transgene; ND, not determined.

For example, conditional mutagenesis allowed to inactivate the adenomatous polyposis coli (APC) cell tumor suppressor gene in the intestine of adults *(48)* or to study spatial learning in mice deficient for the *N*-methyl-D-aspartate receptor in hippocampal neurons *(28)*. Lam et al. *(57)* have used inducible gene inactivation to study the function of the surface antigen receptor in mature B-lymphocytes since its removal early in development prevents the maturation of this cell lineage. Although in some experiments the selection marker gene was left in the *loxP*-flanked loci without inhibiting target gene expression, we strongly recommend the use of a *loxP*-flanked selection marker cassette that can be removed by transient Cre expression in ES cells to avoid its potential interference with the target gene (*see* **Subheading 4.**). In the reported cases in **Table 2**, the size of the *loxP*-flanked gene segments varied between 1.5 and 11 kb including one or many exons in either the 5' or 3' region of the target gene. For Cre-mediated deletion, most of the published reports have relied on cell type–specific Cre expression in transgenic mice, while in other cases Cre was delivered by viral expression vectors (**Tables 1** and **2**). As shown in **Table 2**, the efficiency of Cre-mediated deletion in the respective cell type was complete in some but not all cases. The various efficiencies of in vivo deletion in these animals is most likely attributed to the different expression levels of Cre in the individual transgenic lines as a result of differences in transgene copy number or genomic integration site. For viral vectors, complete recombination can be obtained in liver but requires high viral titers *(44,45)*. Indeed, it is our experience that complete deletion of a *loxP*-flanked allele is certainly feasible but requires the identification of an appropriate Cre transgenic line, indicating that often a number of founder lines should be examined when attempting to express Cre as a transgene. However, additional work is required to determine whether the efficiency of Cre-mediated deletion also varies among different target loci, e.g., in relation to chromatin structure, transcription rate, or DNA methylation. In cell lines, the efficiency of Cre-mediated insertion was shown to vary depending on the chromosomal position of the target *(9)*.

## 4.2. Cre Transgenic Mice

Cre transgenic mice can be produced either by pronuclear microinjection of randomly integrating transgene constructs; by targeted introduction of the Cre gene in frame with the start codon of an endogenous gene; or, in principle, by targeted transgenesis *(58,59)*. The first approach is more straightforward considering the effort involved in vector construction and its introduction into the germline. However, a promoter region tested for transgenic expression must be available, and the number of different founder lines that must be generated in order to identify a useful strain is somewhat unpredictable because the level and pattern of transgene expression often vary greatly depending on the transgene's copy number and integration site(s). In any case, the expression pattern of the promoter region used will determine the onset and cell type specificity of Cre-mediated gene modification, while the expression level determines the efficiency of gene modification in a given cell type. However, the exact relationship between the number of Cre molecules in a cell and the efficiency of deletion of a *loxP*-flanked gene segment has not been established. Furthermore, the extent of Cre-mediated recombination can be expected to decrease as the distance between two *loxP* sites is increased and seems also to depend on the chromosomal position of the target as shown for integrative recombination in cell lines.

For the identification of a Cre transgenic strain suitable for deletion experiments, it is necessary to produce a series of transgenic mice with one type of construct and to compare the deletion efficiency among these lines empirically by crossing them to an indicator strain possessing a *loxP*-flanked gene segment. The extent of deletion in tissues or cell types can be directly assessed either at the DNA level by Southern blot analysis or at the level of single cells by detecting β-galactosidase activity in histologic sections if mice transgenic for a β-galactosidase recombination vector are used as the indicator strain (**Table 2**). In addition, monoclonal antibodies specific for Cre recombinase *(60)* (available from Babco and Eurogentec) can be used for the characterization of Cre transgenic lines, and antisera can be purchased from Novagen and Eurogentec. Transgene constructs for expression of Cre recombinase can be designed like common cDNA expression vectors for transgenic mice, i.e., a promoter region coupled to the Cre coding sequence, preceded by a region providing splice donor/acceptor sites to increase cDNA expression and followed by a polyadenylation signal sequence (**Fig. 7**). We do not recommend placing splice donor/acceptor sites 3' of the Cre gene because we have found that codon 145 can be used as a cryptic splice donor site in combination with an SV40 intron located downstream of the coding region (unpublished data). The addition of a nuclear localization signal (NLS) to the N-terminus of Cre has not been thoroughly examined for its increased efficiency in Cre-mediated deletion, and, in theory, the 38-kDa Cre protein should pass through the nuclear membrane even without this localization signal. A GFP-Cre fusion protein was rapidly targeted to the nucleus of mammalian cells in the absence of an NLS *(26)*. However, NLS-Cre has been used for the derivation of several of the published Cre transgenic strains.

In some cases, the Cre gene has been inserted downstream of the promoter of an endogenous gene by homologous recombination in ES cells as an alternative to the transgenic approach (**Table 1**). By this strategy, Cre expression becomes optimally regulated since all control elements of the targeted gene are present at their natural chromosomal position, avoiding the problems of inappropriate transgene expression often encountered with conventional transgenic mice. Furthermore, the targeting approach can be applied to any gene for which genomic clones are available without the need for characterized promoter regions necessary for transgenic expression.

Gene targeting is, however, more laborious as compared to the construction of conventional transgenic mice, and the generation of a nonfunctional allele of the gene used for targeted Cre expression might be another disadvantage in certain cases. A replacement-type vector for targeted Cre insertion should be constructed such that the initiation codon of the targeted gene is replaced by the coding region of Cre (**Fig. 7**). We suggest not adding a polyadenylation

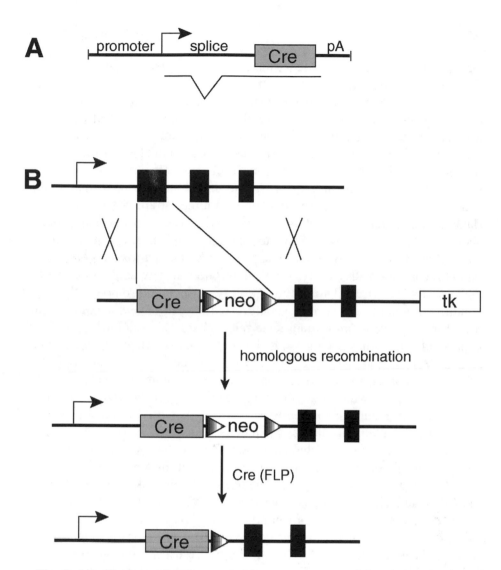

Fig. 7. (**A**) Diagram of a Cre expression cassette as used for the production of transgenic mice by pronucleus injection. A promoter region for expression in transgenic mice directs transcription of the coding region of Cre recombinase (shaded rectangle). Splice donor and acceptor sites are placed in the 5'-untranslated region for efficient transgene expression followed by a polyadenylation signal sequence (pA). The resulting transcript is drawn as a thin line. (**B**) Diagram of a vector for the targeted insertion of the Cre gene. The first exon of the target gene is replaced in the targeting vector by the coding region of Cre such that the position of the translation initiation codon is unchanged. A neo gene flanked by mutant *lox* (or *FRT*) sites (shaded triangles) is inserted downstream of Cre. The *lox* (*FRT*)-flanked neo gene must be deleted from the targeted locus by transient expression of the Cre (FLP) recombinase.

*197*

signal at the end of the Cre gene, thereby allowing transcription to proceed through the intron/exon region(s) of the targeted gene to improve Cre expression and preserve proper gene regulation. A selection marker gene must be included within the targeting vector for the identification of targeted ES clones and later should be removed from the genome to minimize the disturbance of the targeted locus. We recommend not using a canonical *loxP*-flanked selection gene for this purpose to avoid potential chromosomal rearrangements that may occur if three *loxP* sites are present in the genome by crossing the *loxP*-containing Cre-expressing mouse strain to another strain possessing a *loxP*-flanked gene. Thus, for targeting vectors that insert Cre, the selection marker should be flanked either with *FRT* sites to remove the marker gene in recombinant ES clones using FLP recombinase *(61)* or by mutant lox sites *(10)*, which do not recombine with wild-type *loxP* sites, and transient expression of Cre.

Presently, 12 different constitutive, cell type–specific promoters that allow conditional gene targeting in a variety of cell types have been successfully used for Cre expression in transgenic mice (**Table 2**). Another type of Cre-expressing strains has been generated to delete a *loxP*-flanked gene segment from the mouse genome in the germline, e.g., to remove a selection marker or to convert a conditional into a conventional, complete mutant. These so-called deleter strains, of which six different lines have been described (**Table 1**), express Cre in germ cells or during early embryogenesis so that all or part of the offspring contain the Cre-mediated deletion in germ cells.

In the first example of inducible gene targeting, Cre was placed under the control of the interferon-α/β-inducible promoter of the Mx1 gene *(62)*. In one generated transgenic line, complete deletion of a *loxP*-flanked allele is achieved in the liver and lymphocytes 2 d after treatment with interferon, indicating that induced Cre-mediated deletion can proceed rapidly and also efficiently in an organ composed mainly of resting cells. Since many tissues respond to interferon, induced deletion in Mx Cre mice is not restricted to a single cell type. Cell type–specific inducible deletion can be achieved by expression of a fusion protein of Cre and the ligand-binding domain (LBD) of steroid receptors under the control of a constitutively active cell type–specific promoter. In the absence of hormone, the steroid receptor LBDs are bound by heat-shock proteins that inactivate the recombinase, presumably by steric hindrance. The recombinase can be activated by the addition of hormone, which releases the heat-shock proteins from the fusion protein. To derive a system in mice that is unresponsive to natural steroids, Cre was fused to mutant LBDs of the estrogen *(63,64)* and progesterone receptors *(65)*, which are unresponsive to their natural ligands but can be activated by synthetic hormone antagonists. By expressing a fusion protein of Cre and a mutant estrogen receptor under the control of the cytomegalovirus promoter, Cre-mediated deletion can be induced in many tissues of transgenic mice on administration of inducer *(66)*. Induced deletion can be targeted to lymphocytes by expression of a similar protein in B-cells (unpublished results). Cell type–specific transcriptional control can be achieved by one of the systems that regulate the activity of engineered minimal promoter regions by specific transactivating proteins, which, in turn, are regulated by the presence of an inducer molecule such as tetracycline, RU486, or ecdysone *(67)*. These regulatory systems have been shown to efficiently control the activity of reporter genes in transgenic mice but have not yet been applied to the control of site-specific recombinases.

As an alternative to inducible transgenic expression vectors, Cre can be delivered to somatic tissues through infection of mice with viral Cre expression vectors (**Table 1**). These viruses can be applied either locally, infecting the cells around the injection site, or by iv injection, reaching many cells, mainly in the liver and spleen. Viral vectors can be applied at a given time point in adult mice, but they do not act in a cell type–specific manner and elicit a strong immune response against infected cells.

Presently, as summarized in **Table 1**, a growing number of Cre transgenic and *loxP*-containing strains are available, allowing conditional gene targeting in specific or many cell types, either constitutively or on induction. Because of the potential offered by the Cre/*loxP* recombination system, we assume that many tissue-specific Cre transgenic mouse strains are being developed. With respect to the substantial energy required to generate and characterize Cre-expressing strains, it would be useful to establish an international network for the exchange of such strains and information.

## Acknowledgments

We thank B. Clausen, I. Förster, F. Schwenk, and K. Rajewsky for sharing unpublished results. This work was supported by the Volkswagen Foundation.

## References

1. Sternberg, N. and Hamilton, D. (1981) Bacteriophage P1 site-specific recombination. I. Recombination between loxP sites. *J. Mol. Biol.* **150,** 467–486.
2. Austin, S., Ziese, M., and Sternberg, N. (1981) A novel role for site-specific recombination in maintenance of bacterial replicons. *Cell* **25,** 729–736.
3. Hochman, L., Segev, N., Sternberg, N., and Cohen, G. (1983) Site-specific recombinational circularization of bacteriophage P1 DNA. *Virology* **131,** 11–17.
4. Buchholz, F., Ringrose, L., Angrand, P.-O., Rossi, F., and Stewart, A. F. (1996) Different thermostabilities of FLP and Cre recombinases: implications for applied site-specific recombination. *Nucleic Acids Res.* **24,** 4256–4262.
5. Hoess, R. H., Ziese, M., and Sternberg, N. (1982) P1 site-specific recombination: nucleotide sequence of the recombining sites. *Proc. Natl. Acad. Sci. USA* **79,** 3398–3402.

6. Hoess, R. H. and Abremski, K. (1984) Interaction of the bacteriophage P1 recombinase Cre with the recombining site loxP. *Proc. Natl. Acad. Sci. USA* **81,** 1026–1029.

7. Abremski, K., Hoess, R., and Sternberg, N. (1983) Studies on the properties of P1 site-specific recombination: evidence for topologically unlinked products following recombination. *Cell* **32,** 1301–1311.

8. Hamilton, D. L. and Abremski, K. (1984) Site-specific recombination by the bacteriophage P1 lox-Cre system: Cre-mediated synapsis of two lox sites. *J. Mol. Biol.* **178,** 481–486.

9. Sauer, B. and Henderson, N. (1990) Targeted insertion of exogenous DNA into the eukaryotic genome by the Cre recombinase. *New Biol.* **2,** 441–449.

10. Hoess, R. H., Wierzbicki, A., and Abremski, K. (1986) The role of the loxP spacer region in P1 site-specific recombination. *Nucleic Acids Res.* **14,** 2287–2300.

11. Araki, K., Araki, M., and Yamamura, K. (1997) Targeted integration of DNA using mutant lox sites in embryonic stem cells. *Nucleic Acids Res.* **25,** 868–872.

12. Guo, F., Gopaul, D. N., and van Duyne, G. D. (1997) Structure of Cre recombinase complexed with DNA in a site-specific recombination synapse. *Nature* **389,** 40–46.

13. Hoess, R., Wierzbicki, A., and Abremski, K. (1985) Formation of small circular DNA molecules via an in vitro site-specific recombination system. *Gene* **40,** 325–329.

14. Sternberg, N., Sauer, B., Hoess, R., and Abremski, K. (1986) Bacteriophage P1 cre gene and its regulatory region: evidence for multiple promoters and for regulation by DNA methylation. *J. Mol. Biol.* **187,** 197–212.

15. Fiering, S., Epner, E., Robinson, K., Zhuang, Y., Telling, A., Hu, M., Martin, D. I. K., Enver, T., Ley, T. J., and Groundine, M. (1995) Targeted deletion of 5'HS2 of the murine beta-globin LCR reveals that it is not essential for proper regulation of the beta-globin locus. *Genes Dev.* **9,** 2203–2213.

16. Xu, Y., Davidson, L., Alt, F. W., and Baltimore, D. (1996) Deletion of the Ig kappa light chain intronic enhancer/matrix attachment region impairs but does not abolish V kappa J kappa rearrangement. *Immunity* **4,** 377–385.

17. Pham, C. T. N., MacIvor, D. M., Hug, B. A., Heusel, J. W., and Ley, T. J. (1996) Long-range disruption of gene expression by a selectable marker cassette. *Proc. Natl. Acad. Sci. USA* **93,** 13,090–13,095.

18. Bories, J. C., Demengeot, J., Davidson, L., and Alt, F. W. (1996) Gene-targeted deletion and replacement mutations of the T-cell receptor beta-chain enhancer: the role of enhancer elements in controlling V(D)J recombination accessibility. *Proc. Natl. Acad. Sci. USA* **93,** 7871–7876.

19. Gorman, J. R., van der Stoep, N., Monroe, R., Cogne, M., Davidson, L., and Alt, F. W. (1996) The Ig(kappa) enhancer influences the ratio of Ig(kappa) versus Ig(lambda) B lymphocytes. *Immunity* **5,** 241–252.

20. Torres, R. M. and Kühn, R. (1997) *Laboratory Protocols for Conditional Gene Targeting*, Oxford University Press, Oxford, UK.

21. Schwenk, F., Baron, U., and Rajewsky, K. (1995) A cre-transgenic mouse strain for the ubiquitous deletion of loxP-flanked gene segments including deletion in germ cells. *Nucleic Acids Res.* **23,** 5080–5081.
22. Lakso, M., Pichel, J. G., Gorman, J. R., et al. (1996) Efficient in vivo manipulation of mouse genomic sequences at the zygote stage. *Proc. Natl. Acad. Sci. USA* **93,** 5860–5865.
23. Wood, S. A., Allen, N. D., Rossant, J., Auerbach, A., and Nagy, A. (1993) Non-injection methods for the production of embryonic stem cell-embryo chimaeras. *Nature* **365,** 87–89.
24. Carmeliet, P., Ferreira, V., Breier, G., et al. (1996) Abnormal blood vessel development and lethality in embryos lacking a single VEGF allele. *Nature* **380,** 435–439.
25. Chen, J., Lansford, R., Stewart, V., Young, F., and Alt, F. W. (1993) RAG-2-deficient blastocyst complementation: an assay of gene function in lymphocyte development. *Proc. Natl. Acad. Sci. USA* **90,** 4528–4532.
26. Gagneten, S., Le, Y., Miller, J., and Sauer, B. (1997) Brief expression of a GFP cre fusion gene in embryonic stem cells allows rapid retrieval of site-specific genomic deletions. *Nucleic Acids Res.* **25,** 3326–3331.
27. Torres, R. M., Flaswinkel, H., Reth, M., and Rajewsky, K. (1996) Aberrant B cell development and immune response in mice with a compromised BCR complex. *Science* **272,** 1804–1808.
28. Tsien, J. Z., Huerta, P. T., and Tonegawa, S. (1996) The essential role of hippocampal CA1 NMDA receptor-dependent synaptic plasticity in spatial memory. *Cell* **87,** 1327–1338.
29. Li, Z.-W., Stark, G., Götz, J., Rülicke, T., Müller, U., and Weissmann, C. (1996) Generation of mice with a 200-kb amyloid precursor protein gene deletion by Cre recombinase-mediated site-specific recombination in embryonic stem cells. *Proc. Natl. Acad. Sci. USA* **93,** 6158–6162.
30. Ramirez-Solis, R., Liu, P., and Bradley, A. (1995) Chromosome engineering in mice. *Nature* **378,** 720–724.
31. Lewandoski, M. and Martin, G. R. (1997) Cre-mediated chromosome loss in mice. *Nat. Genet.* **17,** 223–225.
32. Zou, Y. R., Muller, W., Gu, H., and Rajewsky, K. (1994) Cre-loxP-mediated gene replacement: a mouse strain producing humanized antibodies. *Curr. Biol.* **4,** 1099–1103.
33. Hanks, M., Wurst, W., Anson-Cartwright, L., Auerbach, A. B., and Joyner, A. L. (1995) Rescue of the En-1 mutant phenotype by replacement of En-1 with En-2. *Science* **269,** 679–682.
34. Smith, A. J. H., De Sousa, M. A., Kwabi-Addo, B., Heppell-Parton, A., Impey, H., and Rabbits, P. (1995) A site-directed chromosomal translocation induced in embryonic stem cells by Cre-*lox*P recombination. *Nat. Genet.* **9,** 376–385.
35. Gu, H., Marth, J. D., Orban, P. C., Mossmann, H., and Rajewsky, K. (1994) Deletion of a DNA polymerase beta gene segment in T cells using cell type-specific gene targeting. *Science* **265,** 103–106.
36. Achatz, G., Nitschke, L., and Lamers, M. C. (1997) Effect of transmembrane and cytoplasmic domains of IgE on the IgE response. *Science* **276,** 409–411.

37. Alimzhanov, M. B., Kuprash, D. V., Kosco-Vilbois, M. H., et al. (1997) Abnormal development of secondary lymphoid tissues in lymphotoxin beta-deficient mice. *Proc. Natl. Acad. Sci. USA* **94,** 9302–9307.

38. DiSanto, J. P., Muller, W., Guy-Grand, D., Fischer, A., and Rajewsky, K. (1995) Lymphoid development in mice with a targeted deletion of the interleukin 2 receptor gamma chain. *Proc. Natl. Acad. Sci. USA* **92,** 377–381.

39. Dupe, V., Davenne, M., Brocard, J., Dolle, P., Mark, M., Dierich, A., Chambon, P., and Rijli, F. M. (1997) In vivo functional analysis of the Hoxa-1 3' retinoic acid response element (3'RARE). *Development* **124,** 399–410.

40. Kaisho, T., Schwenk, F., and Rajewsky, K. (1997) The roles of gamma1 heavy chain membrane expression and cytoplasmic tail in IgG1 responses. *Science* **276,** 412–415.

41. Sleckman, B. P., Bardon, C. G., Ferrini, R., Davidson, L., and Alt, F. W. (1997) Function of the TCR alpha enhancer in alphabeta and gammadelta T cells. *Immunity* **7,** 505–515.

42. Sunaga, S., Maki, K., Komagata, Y., Miyazaki, J., and Ikuta, K. (1997) Developmentally ordered V-J recombination in mouse T cell receptor gamma locus is not perturbed by targeted deletion of the Vgamma4 gene. *J. Immunol.* **158,** 4223–4228.

43. Zakany, J. and Duboule, D. (1996) Synpolydactyly in mice with a targeted deficiency in the HoxD complex. *Nature* **384,** 69–71.

44. Rohlmann, A., Gotthardt, M., Willnow, T. E., Hammer, R. E., and Herz, J. (1996) Sustained somatic gene inactivation by viral transfer of Cre recombinase. *Nat. Biotechnol.* **14,** 1562–1565.

45. Wang, Y., Krushel, L. A., and Edelman, G. M. (1996) Targeted DNA recombination in vivo using an adenovirus carrying the cre recombinase gene. *Proc. Natl. Acad. Sci. USA* **93,** 3932–3936.

46. Akagi, K., Sandig, V., Vooijs, M., Van der Valk, M., Giovannini, M., Strauss, M., and Berns, A. (1997) Cre-mediated somatic site-specific recombination in mice. *Nucleic Acids Res.* **25,** 1766–1773.

47. Brooks, A. I., Muhkerjee, B., Panahian, N., Cory-Slechta, D., and Federoff, H. J. (1997) Nerve growth factor somatic mosaicism produced by herpes virus-directed expression of cre recombinase. *Nat. Biotechnol.* **15,** 57–62.

48. Shibata, H., Toyama, K., Shioya, H., et al. (1997) Rapid colorectal adenoma formation initiated by conditional targeting of the Apc gene. *Science* **278,** 120–123.

49. Betz, U. A., Vosshenrich, C. A., Rajewsky, K., and Muller, W. (1996) Bypass of lethality with mosaic mice generated by Cre-loxP-mediated recombination. *Curr. Biol.* **6,** 1307–1316.

50. Lewandoski, M., Wassarman, K. M., and Martin, G. R. (1997) Zp3-cre, a transgenic mouse line for the activation or inactivation of loxP-flanked target genes specifically in the female germ line. *Curr. Biol.* **7,** 148–151.

51. Sakai, K. and Miyazaki, J. (1997) A transgenic mouse line that retains Cre recombinase activity in mature oocytes irrespective of the cre transgene transmission. *Biochem. Biophys. Res. Commun.* **237,** 318–324.

52. Terry, R. W., Kwee, L., Baldwin, H. S., and Labow, M. A. (1997) Cre-mediated generation of a VCAM-1 null allele in transgenic mice. *Transgenic Res.* **6,** 349–356.

53. Araki, K., Araki, M., Miyazaki, J., and Vassalli, P. (1995) Site-specific recombination of a transgene in fertilized eggs by transient expression of Cre recombinase. *Proc. Natl. Acad. Sci. USA* **92,** 160–164.

54. Lasko, M., Sauer, B., Mosinger, B. Jr., Lee, E. J., Manning, R. W., Yu, S. H., Mulder, K. L., and Westphal, H. (1992) Targeted oncogene activation by site-specific recombination in transgenic mice. *Proc. Natl. Acad. Sci. USA* **89,** 6232–6236.

55. Orban, P. C., Chui, D., and Marth, J. D. (1992) Tissue- and site-specific DNA recombination in transgenic mice. *Proc. Natl. Acad. Sci. USA* **89,** 6861–6865.

56. Tsien, J. Z., Chen, D. F., Gerber, D., Tom, C., Mercer, E. H., Anderson, D. J., Mayford, M., Kandel, E. R., and Tonegawa, S. (1996) Subregion- and cell type-restricted gene knockout in mouse brain. *Cell* **87,** 1317–1326 (see comments).

57. Lam, K. P., Kuhn, R., and Rajewsky, K. (1997) In vivo ablation of surface immunoglobulin on mature B cells by inducible gene targeting results in rapid cell death. *Cell* **90,** 1073–1083 (see comments).

58. Bronson, S. K., Plaehn, E. G., Kluckman, K. D., Hagaman, J. R., Maeda, N., and Smithies, O. (1996) Single-copy transgenic mice with chosen-site integration. *Proc. Natl. Acad. Sci. USA* **93,** 9067–9072 (see comments).

59. Jasin, M., Moynahan, M. E., and Richardson, C. (1996) Targeted transgenesis. *Proc. Natl. Acad. Sci. USA* **93,** 8804–8808.

60. Schwenk, F., Sauer, B., Kukoc, N., Hoess, R., Muller, W., Kocks, C., Kuhn, R., and Rajewsky, K. (1997) Generation of Cre recombinase-specific monoclonal antibodies, able to characterize the pattern of Cre expression in cre-transgenic mouse strains. *J. Immunol. Methods* **207,** 203–212.

61. Dymecki, S. M. (1996) Flp recombinase promotes site-specific DNA recombination in embryonic stem cells and transgenic mice. *Proc. Natl. Acad. Sci. USA* **93,** 6191–6196.

62. Kühn, R., Schwenk, F., Aguet, M., and Rajewsky, K. (1995) Inducible gene targeting in mice. *Science* **269,** 1427–1429.

63. Feil, R., Wagner, J., Metzger, D., and Chambon, P. (1997) Regulation of Cre recombinase activity by mutated estrogen receptor ligand-binding domains. *Biochem. Biophys. Res. Commun.* **237,** 752–757.

64. Zhang, Y., Riesterer, C., Ayrall, A. M., Sablitzky, F., Littlewood, T. D., and Reth, M. (1996) Inducible site-directed recombination in mouse embryonic stem cells. *Nucleic Acids Res.* **24,** 543–548.

65. Kellendonk, C., Tronche, F., Monaghan, A. P., Angrand, P. O., Stewart, F., and Schutz, G. (1996) Regulation of Cre recombinase activity by the synthetic steroid RU 486. *Nucleic Acids Res.* **24,** 1404–1411.

66. Feil, R., Brocard, J., Mascrez, B., LeMeur, M., Metzger, D., and Chambon, P. (1996) Ligand-activated site-specific recombination in mice. *Proc. Natl. Acad. Sci. USA* **93,** 10,887–10,890.

67. Saez, E., No, D., West, A., and Evans, R. M. (1997) Inducible gene expression in mammalian cells and transgenic mice. *Curr. Opin. Biotechnol.* **8,** 608–616.

68. Rickert, R. C., Roes, J., and Rajewsky, K. (1997) B lymphocyte-specific, Cre-mediated mutagenesis in mice. *Nucleic Acids Res.* **25,** 1317, 1318.

69. Hennet, T., Hagen, F. K., Tabak, L. A., and Marth, J. D. (1995) T-cell-specific deletion of a polypeptide N-acetylgalactosaminyl-transferase gene by site-directed recombination. *Proc. Natl. Acad. Sci. USA* **92,** 12,070–12,074.

70. Tarutani, M., Itami, S., Okabe, M., Ikawa, M., Tezuka, T., Yoshikawa, K., Kinoshita, T., and Takeda, J. (1997) Tissue-specific knockout of the mouse Pig-a gene reveals important roles for GPI-anchored proteins in skin development. *Proc. Natl. Acad. Sci. USA* **94,** 7400–7405.

71. Xiao, Y. and Weaver, D. T. (1997) Conditional gene targeted deletion by Cre recombinase demonstrates the requirement for the double-strand break repair Mre11 protein in murine embryonic stem cells. *Nucleic Acids Res.* **25,** 2985–2991.

72. Lee, Y. H., Sauer, B., Johnson, P. F., and Gonzalez, F. J. (1997) Disruption of the c/ebp alpha gene in adult mouse liver. *Mol. Cell Biol.* **17,** 6014–6022.

73. Barlow, C., Schroeder, M., Lekstrom-Himes, J., Kylefjord, H., Deng, C. X., Wynshaw-Boris, A., Spiegelman, B. M., and Xanthopoulos, K. G. (1997) Targeted expression of Cre recombinase to adipose tissue of transgenic mice directs adipose-specific excision of loxP-flanked gene segments. *Nucleic Acids Res.* **25,** 2543–2545.

74. Agah, R., Frenkel, P. A., French, B. A., Michael, L. H., Overbeek, P. A., and Schneider, M. D. (1997) Gene recombination in postmitotic cells: targeted expression of Cre recombinase provokes cardiac-restricted, site-specific rearrangement in adult ventricular muscle in vivo. *J. Clin. Invest.* **100,** 169–179.

75. Wagner, K. U., Wall, R. J., St-Onge, L., Gruss, P., Wynshaw-Boris, A., Garrett, L., Li, M., Furth, P. A., and Hennighausen, L. (1997) Cre-mediated gene deletion in the mammary gland. *Nucleic Acids Res.* **25,** 4323–4330.

# VI

## CRYOPRESERVATION OF MOUSE LINES

# 10

## Cryopreservation of Transgenic Mouse Lines

### Jillian M. Shaw and Naomi Nakagata

## 1. Introduction

Cryopreservation can help minimize waste and thereby increase the efficiency with which transgenic mouse colonies can be used *(1,2)*. The main applications for cryopreservation are firstly the longterm storage of transgenic mouse lines and secondly the shortterm storage of embryos when these cannot be used fresh owing to a shortage of cells, recipients, or DNA for microinjection.

No research group that is actively generating large numbers of transgenic mice has sufficient funds or space to maintain all the transgenic lines that they create. The result is that most active groups have a rapid turnover of mouse lines as all mice, except those which appear most promising and interesting, are culled. Unfortunately, some culled lines may subsequently be found to have been of potential interest to others or to the same group. Such irreversible losses would not arise if embryos (and/or eggs, sperm, and ovarian tissue) were collected and cryopreserved at, or before, the time that the animals were culled. Cryopreserved stocks can also be used as a safety backup in case of diseases or disasters affecting the remaining colony.

Researchers wishing to convert their living mouse lines into frozen stocks have many options open to them, because cryopreservation procedures have progressed further for the mouse than for any other species. The mouse is the only mammal for which there currently are highly effective cryopreservation protocols for embryos at all stages of preimplantation development, as well as protocols for mature and immature oocytes, primordial follicles, ovarian tissue, and sperm (**Table 1**). The most reliable strategy for creating frozen stocks from which the strain(s) can be recovered is to freeze sperm and embryos. In some situations, alternatives such as ovary, egg, or spermatid freezing may

From: *Methods in Molecular Biology, vol. 180: Transgenesis Techniques, 2nd ed.: Principles and Protocols*
Edited by: A. R. Clarke © Humana Press Inc., Totowa, NJ

**Table 1**
**Some Publications Examining the Viability of Frozen-Thawed Mouse Eggs, Embryos, and Ovarian Tissue as Assessed by Fetus Formation or Live Births**

| Stage | Freezing method | Fetuses per embryo transferred % (reference) | Live pups per embryo transferred % (reference) | Other |
|---|---|---|---|---|
| Immature (GV) oocytes | Slow cooling | 46 (10) | 52 (10) | 9% two-cell (12) |
| | Rapid cooling | 31–36 (11) | | |
| | Vitrification | | | 36–39% morphologically normal oocytes (33) |
| Mature MII oocytes | Slow cooling | 42–80 (13) 30–36 (14) | 26 (12) | 55% two-cell (20) |
| | Rapid cooling | 11 (15), 37–48 (11) | 18–46 (17,18) 13 (19) | 24% two-cell (20) |
| | Vitrification | 49 (16) | 38 (16) | 66% two-cell (20) |
| One–two cell embryos | Slow cooling | 22–36 (21), 53 (22), 33–44 (23) | 13 (23), 45–61 (28,29), | |
| | Rapid cooling | 9–78 (22), 43–75 (24), 17 (23), 60–72 (25) 37 (26), 44 (27), | 7–60 (30) | |
| | Vitrification | 49 (31) (may not be true vitrification) | 43–57 (32), 32 (33) | |
| Four-cell embryos | Slow cooling | 46–53 (23) | 47 (34), 15 (23) | |
| | Rapid cooling | 23 (23) | 23 (33) | |
| | Vitrification | | | |

| Eight to 16-cell embryo | Slow cooling | 57–67 (*35*), 19–64 (*36*), 81 (*37*) | 68 (*38*), 54 (*37*) | |
| | Rapid cooling | 75–88 (*24*), 36–46 (*35*) | | |
| | Vitrification | 65–73 (*37*), 18–71 (*36*) | 29 (*33*), 41 (*37*) | |
| Morulae | Slow cooling | 54–72 (*24*), 30–33 (*39*) | 45 (*43*) | |
| | Rapid cooling | 74 (*40*), 1–33 (*41*), 31–46 (*42*) | 51 (*44*), 21 (*33*) | |
| | Vitrification | | | |
| Blastocysts | Slow cooling | 10–56 (*45*) | | |
| | Rapid cooling | 52–71 (*24*) | | |
| | Vitrification | 0–36 (*41*) | 58 (*46*), 26 (*33*) | |
| Ovarian tissue | Slow cooling | | | 3/5 recipients pregnant (*47*) 25–86% pregnant (*48*), 1/4 (*49*), 1/9 (*50*) |
| | Rapid cooling | | | 2/5 recipients pregnant (*47*), 2/5 recipients pregnant (*47*) |
| | Vitrification | | | |
| Primordial follicles | Slow cooling | | | 4/9 recipients pregnant (*51*) |
| | Rapid cooling | | | ND |
| | Vitrification | | | ND |

ND, not done.

be more useful (**Tables 1**, **2**, and *3*). Several papers have recently emphasized the value of genome banking *(4–9)*.

This paper aims to provide a realistic overview of the usefulness of cryopreservation for groups aiming to set up a storage bank for a transgenic mouse colony. The same procedures can be used for short term storage of embryos if for any reason there is a shortage of constructs for pronuclear injection, cells for fusion or blastocyst injection, or pseudopregnant females to serve as foster recipients.

## 2. Choice of Cryopreservation Method

A research unit intending to bank their mouse strains will need to budget both on initial expenses to buy equipment and on subsequent running costs, namely liquid nitrogen and technician time to collect and freeze (or thaw) the materials. The unit should also expect to replace the storage tanks every 10–20 yr. The setup cost will depend on whether the unit chooses to use slow-cooling or rapid methods (rapid cooling and vitrification). Slow-cooling methods usually require specialized equipment to give controlled cooling rates to low subzero temperatures *(64)*.

Slow-cooling procedures were the first cryopreservation methods to be applied successfully to mouse embryos and oocytes. Slow cooling is well characterized *(65)* and has not been associated with severe genetic alterations in embryos *(66,67)* or sperm, but may perturb embryos in S phase *(68)* and those just entering mitosis *(21)*. It is suitable for GV and MII oocytes, but must be used correctly *(13,69,70–75)*. It is relatively versatile in that the same protocol can be used for oocytes, embryos, and tissues (**Table 1**) (earlier references are given in Friedler et al. *[76]*); and effective or at least partially effective for a wide range of mouse strains *(36,77)*, including transgenic mouse strains *(78)*. The disadvantages of slow cooling are that it is time consuming and that it usually requires a biological freezer. The cooling rate required for sperm (**Subheading 4.**) is relatively fast and simple to achieve with improvised equipment. The slower cooling rate required for oocytes, embryos, and ovarian tissue (0.3°C/min) can be achieved with improvised equipment, but the temperature gradient needs to be followed accurately for consistent results. Research groups intending to freeze oocytes or embryos by slow cooling are therefore advised to buy a biologic freezing machine. Those built for veterinary purposes are robust, portable, and reliable at relatively modest cost (under US $5,000 new).

Successful rapid methods for mouse embryos were first developed by Rall and Fahy in 1985 *(79)*. Developments since that time have resulted in many highly effective and "user friendly" freezing methods for both mouse eggs and embryos *(16,24,32,44)*. Their major advantages over slow cooling are that they do not need any specialized equipment and that most take only a few minutes to complete. Rapid methods can be divided into two broad categories: *rapid cooling* and *vitrification*. The main difference between the two is the composition of the cryoprotectant used. Vitrification solutions contain very high concentrations of penetrating cryoprotectant and additives (usually sugars and/or polymers) to ensure that ice does not form at any stage during cooling or warming. If ice forms at some stage during the cooling and warming steps, the procedure is, by definition, not a vitrification procedure. To distinguish between the two, **Table 1** gives separate information for vitrifying and nonvitrifying (rapid-cooling) procedures. Solutions used for rapid-cooling procedures tend to have a lower total concentration of penetrating cryoprotectant and/or additives. When used correctly, both vitrification and rapid-cooling methods can be very effective, giving fertilization and/or pregnancy rates comparable to those of fresh, nonfrozen controls (sperm, eggs, embryos). With any of these rapid methods, it is important that well-established protocols, that are known to work, are followed closely because both the vitrification and rapid-cooling methods can be detrimental to cells if used incorrectly or inappropriately (*see* **refs. *15,16,22,80,81***).

Irrespective of the cryopreservation method that is used or the type of material that is stored, it is vital that the material is stored under optimal conditions. The material should always be held at low subzero temperatures and never be partially or wholly thawed. Once the temperature rises above –140°C, molecules begin to move and changes will start to occur in and around the specimens. Temperatures above –90°C can damage embryos within seconds *(45,82)*. For straws, this critical temperature can be reached after as little as 5 sec in air. It is therefore important to train and educate all persons with access to the material in correct handling and thawing procedures. If used correctly, cryopreservation should improve the efficiency with which groups using or creating transgenic mice can operate *(83)*, and the material that has been frozen should have a very long shelf life *(84)*. The mouse embryos from the first batches that were successfully frozen in the early 1970s still produce the same proportion of fetuses when thawed and transferred as they did 25 yr ago *(85)* (D. Whittingham, personal communication). If infectious material is to be stored, the user should be aware that in human blood and tissue banks infective agents such as viruses have been transmitted between specimens with the liquid nitrogen *(86)*, and that potentially infectious material is therefore best stored in the vapor phase of liquid nitrogen.

## 3. Preservation of the Female Germline

Several approaches allow just the female germline (unfertilized oocytes) to be frozen for future use, including cryopreservation of the ovary *(87–93)*, primordial follicles, fully developed GV oocytes (immature oocytes), and mature

MII oocytes collected at or after ovulation. Which one is most appropriate depends on the fertility, health, and age of the animal (**Tables 2** and **3**), and the ease with which that line can be superovulated. The most versatile is cryopreservation of the ovary. This method has recently received considerable attention because it can be used for a wider range of donors than egg or embryo freezing *(49,50,90,94,95)* and appears relatively effective for mouse ovarian tissue *(90)*. Ovarian tissue freezing can be combined with oocyte or embryo freezing simply by removing and freezing the tissue at the time of egg/embryo collection. Ovarian tissue can be harvested from either live or recently deceased animals, including animals that have died unexpectedly. Its principal advantage is that the age of the donor is relatively unimportant. Live young have been obtained from grafts of nonfrozen fetal, newborn and adult tissue. Ovarian tissue is relatively easy to freeze. Slowly cooled (0.3°C/min to –40°C), rapidly cooled, and vitrified mouse ovaries have all given rise to live young. The disadvantage of either oocyte or ovarian tissue cryopreservation is that in-vitro maturation and culture protocols still need optimization *(13,51,96–100)*. As a result ovaries are most commonly grafted. The disadvantages of this approach are that the grafts need to be placed in histocompatible, or immunologically compromised, recipients and that tissues from a sick donor animal may transmit the disease to the recipient (viral or bacterial infections, cancers *[101]*, etc.). A guide to the current protocol for ovarian grafting and cryopreservation is given in this volume (*see* Chapter 11). It is not known how much the success of the ovarian grafting and cryopreservation procedures is strain or operator dependent, but several groups have used ovarian grafting of fresh tissue as a strategy to propagate transgenic mouse lines *(102,103)*.

Oocytes (fully developed GV or MII) cannot normally be obtained from mice until they are sexually mature (older than 5–6 wk). However, developmentally competent oocytes can be obtained from mice as young as 3 wk by treating them with gonadotrophins (superovulation). There are several effective cryopreservation procedures for mouse oocytes *(104*; **Table 1**). Unfortunately, good results cannot be expected unless the best methods are used and steps are taken to minimize zona hardening *(13)*, and the user has reliable and consistent in vitro fertilization results or, in the case of immature oocytes, reliable in vitro maturation and in vitro fertilization results. If oocytes are to be fertilized by sperm or spermatid injection (Intracytoplasmic sperm injection [ICSI]), it is best to ascertain in advance that the oocytes survive the injection procedure and handling steps. Embryos generated from frozen oocytes are usually transferred to day-1 pseudopregnant recipients at the one- or two-cell stage. It must be recognized that these published results are in most cases generated from strains or hybrids that have oocytes that tolerate handling and in-vitro conditions very well (e.g., the C57BL x CBA F1 hybrid).

## 4. Preservation of the Male Germline

An important problem associated with cryobanking of transgenic mice is to ensure that enough material is stored to allow the line to be reestablished, if it is required. Sperm cryopreservation can be more economical than oocyte and embryo cryopreservation in achieving this aim, particularly if only a few strains are subsequently thawed for later use *(8,105–114)*.

Oocyte freezing is costly because each donor female has to carry the transgene. Embryo freezing is slightly less costly because embryos can be collected from either transgenic females or from normal (nontransgenic) females mated to transgenic male studs. Both oocyte and embryo freezing therefore require a number of female donors whether or not the strain is ever used in the future. If, on the other hand, sperm are frozen, females are only required (as egg donors) once it has been decided to reestablish the transgenic line. Sperm freezing is relatively simple and requires no specialized equipment, but in order for it to work the males of the transgenic mouse must produce freezable spermatozoa and the personnel must have (or develop) expertise with both sperm freezing and in-vitro fertilization procedures.

If the transgenic males produce freezable sperm, relatively few donors are required since normally about 10 million cells can be collected from the epididymis of each mouse. With current methods, about 10 (0.25 mL) straws can normally be prepared for each male, and usually only one straw needs to be thawed for each in-vitro fertilization procedure. Each straw generally contains 1,000,000 cells/10 μL. For IVF, a small volume (~1–2 μL, ~100,000–200,000 cells) of the sperm suspension is added to 200 μL insemination drops giving a final sperm concentration of 500–1000/μL. Most researchers place ~40–60 eggs per drop and achieve a fertilization rate of between 26 and 91% (**Table 4**). From this it can be seen that each male mouse can be used to fertilize many oocytes, on more than one occasion.

Mouse sperm can be frozen by slow or by rapid methods. With rapid cooling the spermatozoa are frozen as pellets in holes on a block of dry ice. This brings the temperature from about +20°C to ~ –79°C at a rate of around 200° to 250°C/min. Each frozen pellet is then cooled in LN2 (cooling rate from –79° to –196°C at around 7000–8000°C/min), placed in a precooled cryotube and finally stored at –196°C. In slow-cooling methods, the mouse spermatozoa can be frozen in freezing vials or straws. In these methods, the temperature falls from +20°C to ~ –170°C at 20–60°C/min but more slowly from –170° to ~ –196°C (2 to ~5°C/min). With these cooling rates, the straws can be moved from the LN2 gas phase to LN2 around 10 min after the start of the cooling. The slow-cooling method for mouse spermatozoa is well established except for certain strains such as C57BL/6 *(57)*. If the motility of the

**Table 2**
**Summary of the Advantages and Disadvantages**
**of the Different Materials When Aiming to Implement Cryobanking**

| | Advantages | Disadvantages |
|---|---|---|
| Immature (GV) oocytes | Unfertilized allows female line only to be stored. The source of sperm can be decided at the time of thawing. | Minimum donor age ~20 days. Strain must have oocytes that are suited to in vitro maturation, in vitro fertilization and culture after thawing. Only oocytes that can be punctured can be used in ICSI. Donor must carry the transgene. |
| Mature MII oocytes | Unfertilized allows female line only to be stored. The source of sperm can be decided at the time of thawing. | Minimum donor age ~21 days. Strain must have oocytes that are suited to in vitro fertilization and culture after thawing. Only oocytes that can be punctured can be fertilized by ICSI. The donor must carry the transgene. |
| One- to two-cell embryos | Easy to collect from the ampulla of the oviduct. Either the donor or the male can be transgenic. | Minimum donor age ~22 days. Many strains sensitive to handling and freezing |
| Morulae | Robust and tolerate handling freezing and transfer[a]. Either the donor or the male can be transgenic. | Minimum donor age ~23 days. May be difficult to flush from tract |
| Blastocysts | Robust and tolerate handling freezing and transfer[a]. Either the donor or the male can be transgenic. | Minimum donor age ~24 days. May be difficult to recognize in uterine flushing, quality may be reduced in superovulated mice |

| | | |
|---|---|---|
| Sperm | Male line only. Sperm can be collected from mated females. | Needs to be sexually mature. Unpredictable and variable post-thaw survival, requires IVF, ICSI or insemination after thawing |
| Ovarian tissue | Can be collected from fetal, newborn, immature, or adult oocytes present in large numbers (~3000 per ovary in young mice). Should still be viable 24 h after death. Graft recipients may have many litters. | Need to be grafted into histocompatible, or immunologically compromised recipients, may transmit disease. Donor must carry the transgene. |
| Primordial follicles | Can be collected from fetal, newborn, immature, or adults. Easy to cryopreserve. | Difficult: requires extraction and manipulation of tiny follicles (30–40 μm). Requires grafting or long-term in vitro culture. If follicles are grafted they have the potential to transmit disease. Donor must carry the transgene. |

[a]Embryos obtained by in-vitro culture may be less robust than those grown in vivo.

## Table 3
### Strategies for Preserving Lines of Transgenic Mice, in Relation to Their Sex, Health, Fertility, and Age[a]

| Sex of donor | Health of donor | Options if the donor is fertile | Options if the donor is infertile | Options if the donor is immature or sterile |
|---|---|---|---|---|
| Male | Good | 1. Sperm freezing without killing the donor: Mate then cull the female, freeze the sperm flushed from the female tract. Thaw for IVF (±PZD) or AI (*52,126*). | If sperm are present in ejaculate attempt options 1 and 2 as for fertile male; if the sperm fail to fertilize, apply ICSI. | Attempt options 4 and 5 as for infertile donors. |
| | | 2. Embryo freezing without killing the donor: Mate and then collect and freeze embryos from the female tract. Thaw and transfer to pseudopregnant recipients. | 4. No sperm in ejaculate. Anesthetize or kill; if sperm are in tract, attempt IVF (±PZD) ICSI, AI, and/or freeze for subsequent IVF, ICSI, or AI (*52,126*). | Immature: let grow to puberty then try options 1–3 as for fertile males. |
| | | 3. Kill mouse, collect and cryopreserve sperm. Thaw for IVF (±PZD), ICSI or AI (*52,126*). | 5. Anesthetize or kill. If only spermatids are present attempt ICSI, or freeze for subsequent ICSI. | |
| | Poor | If options 1 and 2 fail, use option 3. | Attempt options 4 and 5. | Options are as above. |
| | Dead | If sperm are motile try IVF (±PZD) or AI, if not try ICSI and/or cryopreserve sperm for subsequent IVF (±PZD), ICSI or AI. | If sperm are motile, try IVF or AI; if not, try ICSI and/or cryopreserve sperm for subsequent IVF (±PZD), ICSI, or AI (*52,126*). | Attempt options 4 and 5. |

| Female | Good | 6. Mate and then collect and freeze embryos and ovaries.<br>7. Superovulate, mate, and then kill to collect and freeze embryos and ovaries.<br>8. Graft and/or cryopreserve one or both ovaries. | Attempt options 6–8 as for fertile female. | Graft and/or cryopreserve ovaries. |
|--------|------|------|------|------|
| | Poor | Attempt options 6–8 as for fertile female. | Graft and/or cryopreserve one or both ovaries. | Graft and/or cryopreserve ovaries. |
| | Dead | Graft and/or cryopreserve ovaries. | Graft and/or cryopreserve ovaries. | Graft and/or cryopreserve ovaries. |

[a]ICSI, intracytoplasmic sperm injection; AI, artificial insemination; IVF, in vitro fertilization; PZD, partial zona dissection (*61*).

**Table 4**
**Results for Frozen Mouse Spermatozoa**

| Method and reference | Fertilization rate (%) | Fetuses (%) | Live young (%) |
|---|---|---|---|
| Slow *(52)* | 42 | | 14 |
| Rapid *(53)* | 13–64 | 19 | |
| Slow *(54)* | 19–37 | | 17–75 |
| Slow *(55)* | 36 | | 45 |
| Slow *(56)* | 84 | | 51 |
| Slow *(57)* | 26–89 | | 35–62 |
| Slow *(58)* | 50 | 16 | 17 |
| Slow *(59)* | 29 | | 24–47 |
| Slow *(60)* | 91 | 37 | 38 |
| Slow *(61)* | 73–76[a] | | |
| Slow *(62,63)* | 9–83 | | 31–42 |

[a]Partial zona dissection oocytes were inseminated with frozen-thawed transgenic spermatozoa with low motility.

cryopreserved spermatozoa is low after thawing, the fertilization rate can be increased by partial zona dissection, that is, in vitro fertilization with oocytes that have the zona pellucida partially dissected *(61)*. In extreme cases, intracytoplasmic sperm injection (ICSI) may be required *(115)*. This technique allows sperm heads from immotile or dead sperm, or even spermatids, *(116)* to be injected directly into each egg to achieve fertilization. In cases where a male does not produce mature sperm, spermatogonial transplantation could be used in an attempt to obtain mature sperm *(117)*.

In the hands of experienced workers, ICSI results in 60–70% of eggs being fertilized *(115)*. However, ICSI requires expensive equipment, experienced operators, and is only suited to strains that have oocytes that tolerate handling and injection. Current research aims to establish which protocol(s) are the easiest, least damaging, and most versatile for mouse sperm cryopreservation.

## 5. Embryo Cryopreservation

Excellent results (fetuses/live young) have been obtained with cryopreserved mouse embryos at all stages of preimplantation development (**Table 1**). Studies have shown that transgenic mouse embryos can be frozen *(78)* and that frozen-thawed pronuclear-stage embryos are suitable for subsequent DNA microinjection *(118,119)*. The transfer of frozen-thawed embryos is likely to establish pregnancies, but how many live young are born will depend on the genetic background of the embryo, the stage at which it was collected, and possibly the transgene. Groups working with transgenic embryos may not get as high a proportion of fetuses or live young as shown in **Table 1**, because many of these results were obtained with hybrid embryos or strains that tolerate handling and culture very well (e.g., the C57BL x CBA F1 hybrid). Even though certain mouse strains are known to be difficult to freeze *(36,77,84)*, and the overall survival per embryo collected may be low *(83,120)*, there are no known strains of laboratory mice with unfreezable embryos. Even damaged *(121)*, or micromanipulated *(122,123)* embryos can produce viable fetuses. To maximize viability, it is generally best to use later stage embryos (eight-cell or morula) collected directly from the female tract as these stages are less likely to be damaged by handling or freezing than earlier embryos or embryos previously cultured in vitro. While slow methods can be used, rapid methods may be as or more effective and are significantly cheaper, quicker, and easier. The protocol developed by Kasai et al. 1990 *(44)* for mouse morulae is currently one of the easiest to use and has proved very successful for eight-cell– to blastocyst–stage mouse embryos. Details of this technique have been given in several recent publications *(124,125)*.

The biggest disadvantage with oocyte or embryo freezing is, as already mentioned, the cost. Each donor produces relatively few eggs or embryos, and these may not all be transgenic. If a normal donor female is mated to a homozygous male, or if a homozygous donor female is used, all the embryos will carry the transgene (providing that the transgene is located on an autosome). If a heterozygous donor female is used, no more than 50% of the oocytes and embryos she produces will carry the transgene. If the transgene is located on a sex chromosome, its presence could depend on the sex of the embryo. This indicates that a combination of embryo and sperm freezing may be the best strategy for ensuring that a cryobanked mouse line can be reestablished for future users.

## 6. Alternative Preservation Strategies

In cases in which the transgene(s) is incorporated into embryonic stem (ES) cell lines, it may be more economic to freeze the cell line rather than the gametes or zygotes. The ES cells can be thawed, cultured, and injected as required into fresh or thawed recipient embryos. An alternative approach, cloning from somatic or embryonic cells of the transgenic line, may be possible but is still experimental.

## 7. Summary

Cryopreservation will, if used appropriately, lower maintenance costs while creating embryo or germline banks from which transgenic lines can be restored. To maximize the likelihood of restoring a transgenic line, it is best to establish that the gametes or zygotes of that strain do survive cryopreservation before all of the animals have been culled. Furthermore, it is advisable to store more than one type of material (e.g., sperm and embryos) and to ensure that the material is stored under optimal conditions (i.e., at low subzero temperatures) and never partially or wholly thawed, until it is removed for use.

## Acknowledgments

Dr. J. Shaw wishes to acknowledge funding from Monash Research Fund and Monash IVF. We thank Dr. Stan Leibo for helpful criticisms on the manuscript.

## References

1. Mobraaten, L. (1997) Cryopreservation and strain re-derivation. *Lab. Animal* **26,** 21–25.
2. Sharp, J. J. and Mobraaten L. (1997) To save or not to save: the role of repositories in a period of rapidly expanding development of genetically engineered strains of mice, in *Transgenic Animals: Generation and Use* (Houdebine, L. M., ed.), Harwood Academic, Switzerland, pp. 525–532.
3. Nakagata, N. (1995) Studies on cryopreservation of embryos and gametes in mice. *Exp. Anim.* **44,** 1–8.
4. Critser, J. K. and Russell, R. J. (2000) Genome resource banking of laboratory animal models, in Cryobiology of Embryos, Germ Cells, and Ovaries. *ILAR J.* **41,** 183–186.
5. Glenister, P. H. and Thornton, C. E. (2000) Cryoconservation—archiving for the future. *Mamm. Genome.* **11,** 565–571.
6. Pinkert, C. A. (1998) Mouse sperm cryopreservation: a legacy in the making. *Lab. Anim. Sci.* **48,** 224.
7. Rall, W. F. Schmidt, P. M., Lin, X., Brown, S. S., Ward, A. C., and Hansen, C. T. (2000) Factors affecting the efficiency of embryo cryopreservation and rederivation of rat and mouse models. *ILAR J.* **41,** 221–227.
8. Thornton, C. E., Brown, S. D., and Glenister, P. H. (1999) Large numbers of mice established by in vitro fertilization with cryopreserved spermatozoa: implications and applications for genetic resource banks, mutagenesis screens, and mouse backcrosses. *Mamm. Genome* **10,** 987–992.
9. Wildt, D. E. (2000) Genome resource banking for wildlife research, management, and conservation. *ILAR J.* **41,** 228–234.
10. Candy, C. J., Wood, M. J., Whittingham, D. G., Merriman, J. A., and Choudhury, N. (1994) Cryopreservation of immature mouse oocytes. *Human Reprod.* **9,** 1738–1742.
11. Shaw, J. M., Oranratnachai, A., and Trounson, A. O. (2000) Fundamental cryobiology of mammalian oocytes and ovarian tissues. *Theriogenology* **53,** 59–72.
12. Schroeder, A. C., Champlin, A. K., Mobraaten, L. E., and Eppig, J. J. (1990) Developmental capacity of mouse oocytes cryopreserved before and after maturation in vitro. *J. Reprod. Fert.* **89,** 43–50.

13. Carroll, J., Wood, M. J., and Whittingham, D. G. (1993) Normal fertilization and development of frozen-thawed mouse oocytes: Protective action of certain macromolecules. *Biol. Reprod.* **48**, 606–612.

14. George, M. A., Johnson, M. H., and Howlett, S. K. (1994) Assessment of the developmental potential of frozen-thawed mouse oocytes. *Human Reprod.* **9**, 130–136.

15. Van der Elst, J., Neinckx, S., and Van Steirteghem, A. C. (1993) Association of ultrarapid freezing of mouse oocytes with increased polyploidy at the pronucleate stage, reduced cell numbers in blastocysts and impaired fetal development. *J. Reprod. Fert.* **99**, 25–32.

16. Bos-Mikich, A., Wood, M. J., Candy, C. J., and Whittingham, D. G. (1995) Cytogenetical analysis and developmental potential of vitrified mouse oocytes. *Biol. Reprod.* **53**, 780–785.

17. Nakagata, N. (1989) High survival rate of unfertilized mouse oocytes after vitrification. *J. Reprod. Fertil.* **87**, 479–483.

18. Nakagata, N. (1990) Cryopreservation of unfertilized mouse oocytes from inbred strains by ultrarapid freezing. *Exp. Anim.* **39**, 303–305.

19. Anzai, M., Nakagata, N., Matsumoto, K., Ishikawa, T., Takahashi, Y., and Miyata, K. (1994) Production of transgenic mice from in vitro fertilized eggs cryopreserved by ultrarapid freezing. *Jikken Dobutsu. Exp. Anim.* **43**, 445–448.

20. Men, H. S., Chen, J. C., Ji, W. Z., Shang, E. Y., Yang, S. C., and Zou, R. J. (1997) Cryopreservation of kunming mouse oocytes using slow cooling, ultrarapid cooling and vitrification protocols. *Theriogenology* **47**, 1423–1431.

21. Chedid, S., Van den Abbeel, E., and Van Steirteghem, A. C. (1992) Effects of cryopreservation on survival and development of interphase- and mitotic stage 1-cell mouse embryos. *Human Reprod.* **7**, 1451–1456.

22. Shaw, J. M., Kola, I., MacFarlane, D. R., and Trounson, A. (1991) An association between chromosomal abnormalities in rapidly frozen 2-cell mouse embryos and the ice-forming properties of the cryoprotective solution. *J. Reprod. Fert.* **91**, 9–18.

23. Liu, J., Van den Abbeel, E., and Van Steirteghem, A. C. (1993) Assessment of ultrarapid and slow freezing procedures for 1-cell and 4-cell mouse embryos. *Human Reprod.* **8**, 1115–1119.

24. Shaw, J. M., Diotallevi, L., and Trounson, A. O. (1991). A simple rapid dimethyl sulphoxide freezing technique for the cryopreservation of one-cell to blastocyst stage preimplantation mouse embryos. *Reprod. Fertil. Devel.* **3**, 621–626.

25. Vasuthevan, S., Ng, S.-C., Bongso, A., and Ratnam, S. S. (1992) Embryonic behaviour of two-cell mouse embryos frozen by the one- and two-step ultrarapid techniques. *J. Assist. Reprod. Genet.* **9**, 545–550.

26. Nowshari, M. A., Nayudu, P. L., and Hodges, J. K. (1995) Effect of cryoprotectants and their concentration on post thaw survival and development of rapid frozen-thawed pronuclear stage mouse embryos. *Human Reprod.* **10**, 3237–3242.

27. Van den Abbeel, E., Van der Elst, J., Van der Linden, M., and Van Steirteghem, A. C. (1997) High survival rate of one-cell mouse embryos cooled rapidly to −196°C after exposure to a propylene glycol-dimethylsulfoxide-sucrose solution. *Cryobiology* **34,** 1–12.

28. Nakagata, N. (1989) High survival rate of pronuclear mouse oocytes derived from in vitro fertilization following ultrarapid freezing and thawing. *Jpn. J. Fert. Steril.* **34,** 757–760.

29. Nakagata, N. (1989) Survival of 2-cell mouse embryos derived from fertilization in vitro after ultrarapid freezing and thawing. *Jpn. J. Fert. Steril.* **34,** 470–473.

30. Nakagata, N. (1990) Cryopreservation of mouse strains by ultrarapid freezing. *Exp. Anim.* **39,** 299–301.

31. Tada, N., Sato, M., Amann, E., and Ogawa, S. (1993) A simple and rapid method for cryopreservation of mouse 2-cell embryos by vitrification: beneficial effect of sucrose and raffinose on their cryosurvival rate. *Theriogenology* **40,** 333–344.

32. Nakao, K., Nakagata, N., and Katsuki, M. (1997) Simple and efficient vitrification procedure for cryopreservation of mouse embryos. *Exp. Animals.* **46,** 231–234.

33. Miyake, T., Kasai, M., Zhu, S. E., Sakurai, T., and Machida, T. (1993) Vitrification of mouse oocytes and embryos at various stages in an ethylene glycol based solution by a simple method. *Theriogenology* **40,** 121–134.

34. Nakagata, N. (1989) Survival of 4-cell mouse embryos derived from fertilization in vitro after ultrarapid freezing and thawing. *Exp. Anim.* **38,** 279–282.

35. Tarin, J. J. and Trounson, A. O. (1993) Effects of stimulation or inhibition of lipid peroxidation on freezing-thawing of mouse embryos. *Biol. Reprod.* **49,** 1362–1368.

36. Dinnyes, A., Wallace, G. A., and Rall, W. F. (1995) Effect of genotype on the efficiency of mouse embryo cryopreservation by vitrification or slow freezing methods. *Mol. Reprod. Devel.* **40,** 429–435.

37. Rall, W. F. and Wood, M. J. (1994) High *in vitro* and *in vivo* survival of day 3 mouse embryos vitrified or frozen in a non-toxic solution of glycerol and albumin. *J. Reprod. Fert.* **101,** 681–688.

38. Liu, J., Van den Abbeel, E., and Van Steirteghem, A. C. (1993) The in-vitro and in-vivo developmental potential of frozen and non-frozen biopsied mouse embryos. *Human Reprod.* **8,** 1481–1486.

39. Mezzalira, A. and Rubin, M. I. B. (1992) Ultrarapid freezing and transfer of mouse morulae. *Theriogenology* **37(abst),** 257.

40. Kasai, M., Nishimori, M., Zhu, S. E., Sakurai, T., and Mashida, T. (1992) Survival of mouse morulae vitrified in an ethylene glycol-based solution after exposure to the solution at various temperatures. *Biol. Reprod.* **47,** 1134–1139.

41. Ali, J. and Shelton, J. N. (1993) Design of vitrification solutions for the cryopreservation of embryos. *J. Reprod. Fertil.* **99,** 471–477.

42. Barichello, E. M. M. R., Silvera, D. R., Theisen, F. A., Medina, F. T., and Rubin, M. I. B. (1992) Effects of equilibration time and pre cooling on the "in vivo"

survival of mouse morulae cryopreserved by vitrification. *Theriogenology.* **37(abst),** 190.

43. Nakagata, N. (1993) Survival of mouse morulae and blastocysts derived from in vitro fertilization after ultrarapid freezing. *Exp. Anim.* **42,** 229–231.

44. Kasai, M., Komi, J. H., Takakamo, A., Tsudera, H., Sakurai, T., and Machida, T. (1990) A simple method for mouse embryo cryopreservation in a low toxicity vitrification solution, without appreciable loss of viability. *J. Reprod. Fert.* **89,** 91–97.

45. Shaw, J. M., Ward, C., and Trounson, A. O. (1995) Survival of mouse blastocysts slow cooled in propanediol or ethylene glycol is influenced by the thawing procedure, sucrose and antifreeze proteins. *Theriogenology* **43,** 1289–1300.

46. Zhu, S. E., Kasai, M., Otoge, H., Sakurai, T., and Machida, T. (1993) Cryopreservation of expanded mouse blastocysts by vitrification in ethylene glycol based solutions. *J. Reprod. Fert.* **98,** 139–145.

47. Schneider-Kolsky, M., Shaw, J., Jenkin, G., and Trounson, A. (1995) Cryopreservation of fresh and frozen mouse ovaries. Proceedings of the 14th conference of the Fertility Society of Australia (abstract 68).

48. Cox, S.-L., Jenkin, G., and Shaw, J. (1996) Transplantation of cryopreserved fetal ovarian tissue to adult recipients. *J. Reprod. Fert.* **107,** 315–322.

49. Gunasena, K. T., Lakey, J. R. T., Villines, P. M., Critser, E. S., and Critser, J. K. (1997) Allogeneic and xenogeneic transplantation of cryopreserved ovarian tissue to athymic mice. *Biol. Reprod.* **57,** 226–231.

50. Gunasena, K. T., Villines, P. M., Critser, E. S., and Critser, J. K. (1997) Live births after autologous transplant of cryopreserved mouse ovaries. *Human Reprod.* **12,** 101–106.

51. Carroll, J. and Gosden, R. G. (1993) Transplantation of frozen-thawed mouse primordial follicles. *Hum. Reprod.* **8,** 1163–1167.

52. Okuyama, M., Isogai, S., Saga, M., Hamada, H., and Ogawa, S. (1990) In vitro fertilization (IVF) and artificial insemination (AI) by cryopreserved spermatozoa in mice. *J. Fertil. Implant.* **7,** 116–119.

53. Tada, N., Sato, M., Yamanoi, J., Mizorogi, J., Kasai, K., and Ogawa, S. (1990) Cryopreservation of mouse spermatozoa in the presence of raffinose and glycerol. *J. Reprod. Fert.* **89,** 511–516.

54. Yokoyama, M., Akiba, H., Katsuki, M., and Nomura, T. (1990) Production of normal young following transfer of mouse embryos obtained by in vitro fertilization using cryopreserved spermatozoa. *Exp. Anim.* **39,** 125–128.

55. Takeshima, T., Nakagata, N., and Ogawa, S. (1991) Cryopreservation of mouse spermatozoa. *Exp. Anim.* **40,** 493–497.

56. Nakagata, N. and Takeshima, T. (1992) High fertilizing ability of mouse spermatozoa diluted slowly after cryopreservation. *Theriogenology* **37,** 1283–1291.

57. Nakagata, N. and Takeshima, T. (1993) Cryopreservation of mouse spermatozoa from inbred and F1 hybrid strains. *Exp. Anim.* **42,** 317–320.

58. Penfold, L. M. and Moore, H. D. M. (1993) A new method for cryopreservation of mouse spermatozoa. *J Reprod. Fert.* **99,** 131–134.

59. Songsasen, N., Betteridge, K. J., and Leibo, S. P. (1997) Birth of live mice resulting from oocytes fertilized in vitro with cryopreserved spermatozoa. *Biol. Reprod.* **56,** 143–152.

60. Sztein, J. M., Farley, J. S., Young, A. F., and Mobraaten, L. E. (1997) Motility of cryopreserved mouse spermatozoa affected by temperature of collection and rate of thawing. *Cryobiology* **35,** 46–52.

61. Nakagata, N., Okamoto, M., Ueda, O., and Suzuki, H. (1997) The positive effect of partial zona-pellucida dissection on the in vitro fertilizing capacity of cryopreserved C57BL/6J transgenic mouse spermatozoa of low motility. *Biol. Reprod.* **57,** 1050–1055.

62. Songsasen, N. and Leibo, S. P. (1997) Cryopreservation of mouse spermatozoa: I. effect of seeding on fertilizing ability of cryopreserved spermatozoa. *Cryobiology* **35,** 240–254.

63. Songsasen, N. and Leibo, S. P. (1997) Cryopreservation of mouse spermatozoa: II. Relationship between survival after cryopreservation and osmotic tolerance of spermatozoa from three strains of mice. *Cryobiology* **35,** 255–269.

64. Elliot, K. and Whelan, J., eds. (1977) CIBA foundation symposium 52, *The freezing of mammalian embryos.* Elsevier/North Holland, Amsterdam.

65. Mazur, P. (1990) Equilibrium, quasi-equilibrium, and nonequilibrium freezing of mammalian embryos. *Cell Biophys.* **17,** 53–92.

66. Bongso, A., Ng, S. C., Sathananthan, H., Lee, M.-L., Mok, H., Wong, P. C., and Ratnam, S. (1988) Chromosome analysis of two-cell mouse embryos frozen by slow and ultrarapid methods using two different cryoprotectants. *Fertil. Steril.* **49,** 908–912.

67. Dulioust, E., Toyama, K., Busnel, M. C., Moutier, R., Carlier, M., Marchaland, C., et al. (1995) Long-term effects of embryo freezing in mice. *Proc. Natl. Acad. Sci.* **92,** 589–593.

68. Balakier, H., Zenzes, M., Wang, P., MacLusky, and Casper, R. F. (1991) The effect of cryopreservation on the development of S- and $G_2$ phase mouse embryos. *J. In Vitro Fert. ET* **8,** 89–95.

69. Bos-Mikich, A. and Whittingham, D. G. (1995) Analysis of the chromosome complement of frozen thawed mouse oocytes after parthenogenetic activation. *Mol. Reprod. Devel.* **42,** 254–260.

70. Glenister, P. H., Wood, M. J., Kirby, C., and Whittingham, D. G. (1987) The incidence of chromosome anomalies in first-cleavage mouse embryos obtained from frozen-thawed oocytes fertilized in vitro. *Gamete Res.* **16,** 205–216.

71. Gook, D. A., Osborn, S. M., and Johnston, W. I. H. (1993) Cryopreservation of mouse and human oocytes using 1,2-propanediol and the configuration of the meiotic spindle. *Human Reprod.* **8,** 1101–1109.

72. Pickering, S. J., Braude, P. R., Johnson, M. H., Cant, A., and Currie, J. (1990) Transient cooling to room temperature can cause irreversible disruption of the meiotic spindle in the human oocyte. *Fertil. Steril.* **54,** 102–108.

73. Pickering, S. J., Braude, P. R., and Johnson, M. H. (1991) Cryoprotection of human oocytes: inappropriate exposure to DMSO reduces fertilization rates. *Human Reprod.* **6,** 142–143.

74. Shaw, J. M. and Trounson, A. O. (1989) Parthenogenetic activation of unfertilized mouse oocytes by exposure to 1,2-propanediol is influenced by temperature, oocyte age and cumulus removal. *Gamete Res.* **24,** 269–279.

75. Vanblerkom, J. and Davis, P. W. (1994) Cytogenetic, cellular, and developmental consequences of cryopreservation of immature and mature mouse and human oocytes. *Microsc. Res. Tech.* **27,** 165–193.

76. Friedler, S., Giudice, L. C., and Lamb, E. J. (1988) Cryopreservation of embryos and ova. *Fertil. Steril.* **49,** 743–764.

77. Schmidt, P. M., Schiewe, M. C., and Wildt, D. E. (1987) the genotypic response of mouse embryos to multiple freezing variables. *Biol. Reprod.* **37,** 1121–1128.

78. Pomeroy, K. O. (1991) Cryopreservation of transgenic mice. *Genet. Anal. Techn. Applicn.* **8,** 95–101.

79. Rall, W. F. and Fahy, G. M. (1985) Ice-free cryopreservation of mouse embryos at −196°C by vitrification. *Nature* **313,** 573–575.

80. Ishida, G. M., Saito, H., Ohta, N., Takahashi, T., Ito, M. M., Siato, T., Nakahara, K., and Hiroi, M. (1997) The optimal equilibration time for mouse embryos frozen by vitrification with trehalose. *Hum. Reprod.* **12,** 1259–1262.

81. Kola, I., Kirby, C., Shaw, J., Davey, A., and Trounson, A. (1988) Vitrification of mouse oocytes results in aneuploid zygotes and malformed fetuses. *Teratology* **38,** 467–474.

82. Shaw, J. M., Ward, C., and Trounson, A. O. (1995) Evaluation of propanediol, ethylene glycol, sucrose and antifreeze proteins on the survival of slow cooled mouse pronuclear and 4-cell embryos. *Hum. Reprod.* **10,** 396–402.

83. Suzuki, H. (1996) Recent advances in cryopreservation facilitate transgenic mouse technology. *Low Temp. Med.* **22,** 128–136.

84. Whittingham, D. G., Lyon, M. F., and Glenister, P. H. (1977) Re-establishment of breeding stocks of mutant and inbred strains of mice from embryos stored at −196°C for prolonged periods. *Genet. Res.* **30,** 287–299.

85. Whittingham, D. G., Leibo, S. P., and Mazur, P. (1972). Survival of mouse embryos frozen to −196°C and −269°C. *Science* **178,** 411–414.

86. Tedder, R. S., Zukerman, M. A., Goldstone, A. H., Hawkins, A. E., Fielding, A., Briggs, E. M., et al. (1995) Hepatitis transmission from contaminated cryopreservation tank. *Lancet* **346,** 137–140.

87. Sztein, J. M., McGregor, T. E., Bedigian, H. J., and Mobraaten, L. E. (1999) Transgenic mouse strain rescue by frozen ovaries. *Lab. Anim. Sci.* **49,** 99–100.

88. Sztein, J., Sweet, H., Farley, J., and Mobraaten, L. (1998) Cryopreservation and orthotopic transplantation of mouse ovaries: new approach in gamete banking. *Biol. Reprod.* **58,** 1071–1074.

89. Sztein, J. M., O'Brien, M. J., Farley, J. S., Mobraaten, L. E., and Eppig, J. J. (2000) Rescue of oocytes from antral follicles of cryopreserved mouse ovaries:

competence to undergo maturation, embryogenesis, and development to term. *Hum. Reprod.* **15,** 567–571.

90. Candy, C. J., Wood, M. J., and Whittingham, D. G. (1997) Effect of cryoprotectants on the survival of follicles in frozen mouse ovaries. *J. Reprod. Fert.* **110,** 11–19.

91. Candy, C. J., Wood, M. J., and Whittingham, D. G. (2000) Restoration of a normal reproductive lifespan after grafting of cryopreserved mouse ovaries. *Hum Reprod.* **15,** 1300–1304.

92. Shaw, J. M., Cox, S. L., Trounson, A. O., and Jenkin, G. (2000) Evaluation of the long-term function of cryopreserved ovarian grafts in the mouse, implications for human applications. *Mol. Cell Endocrinol.* **161,** 103–110.

93. Agca, Y. (2000) Cryopreservation of murine oocyte and ovarian tissue. *ILAR J.* **41,** 207–220.

94. Karow, A. M. and Critser, J. K., eds. (1997) *Reproductive tissue banking: scientific principles*. Academic, San Diego, CA.

95. Shaw, J. M., Dawson, K. J., and Trounson, A. O. (1997) A critical evaluation of ovarian tissue cryopreservation and grafting as a strategy for preserving the human female germline. *Reprod. Med. Rev.* **6,** 163–183.

96. Carroll, J., Whittingham, D. G., Wood, M. J., Telfer, E., and Gosden, R. G. (1990) Extra-ovarian production of mature viable mouse oocytes from frozen primary follicles. *J. Reprod. Fertil.* **90,** 321–327.

97. Eppig, J. J. and O'Brien, M. (1995) In vitro maturation and fertilization of oocytes isolated from aged mice: a strategy to rescue valuable genetic resources. *J. Assist. Reprod. Genet.* **12,** 269–273.

98. Eppig, J. J. and O'Brien, M. J. (1998) Comparison of preimplantation developmental competence after mouse oocyte growth and development in vitro and in vivo. *Theriogenology* **49,** 415–422.

99. Eppig, J. J., Hosoe, M., O'Brien, M. J., Pendola, F. M., Requena, A., and Watanabe, S. (2000) Conditions that affect acquisition of developmental competence by mouse oocytes in vitro: FSH, insulin, glucose and ascorbic acid. *Mol. Cell Endocrinol.* **163,** 109–116.

100. Liu, J., Van Der Elst, J., Van Den Broecke, R., Dumortier, F., and Dhont, M. (2000) Maturation of mouse primordial follicles by combination of grafting and in vitro culture. *Biol. Reprod.* **62,** 1218–1223.

101. Shaw, J. M., Bowles, J., Koopman, P., Wood, C., and Trounson, A. O. (1996) Fresh and cryopreserved ovarian tissue from donors with lymphoma, transmit the cancer to graft recipients. *Hum. Reprod.* **11,** 1668–1673.

102. Brem, G., Buanack, E., Muller, M., and Winnacker, E. L. (1990) Transgenic offspring by transcaryotic implantation of transgenic ovaries into normal mice. *Mol. Reprod. Dev.* **25,** 42–44.

103. Harari, D., Bernard, O., and Shaw, J. (1997) Rescue of an infertile transgenic line by ovarian transplantation. *Transgenics* **2,** 143–151.

104. Shaw, J. M., Oranratnachai, A., and Trounson, A. O. (2000) Cryopreservation of oocytes and embryos, in *Handbook of in vitro fertilization*, 2nd ed. (Trounson, A. O. and Gardner, D., eds.), CRC, Boca Raton, FL, pp. 373–412.

105. An, T. Z., Iwakiri, M., Edashige, K., Sakurai, T., and Kasai, M. (2000) Factors affecting the survival of frozen-thawed mouse spermatozoa. *Cryobiology.* **40,** 237–249.
106. Critser, J. K. and Mobraaten, L. E. (2000) Cryopreservation of murine spermatozoa. *ILAR J.* **41,** 197–206.
107. Devireddy, R. V., Swanlund, D. J., Roberts, K. P., and Bischof, J. C. (1999) Subzero water permeability parameters of mouse spermatozoa in the presence of extracellular ice and cryoprotective agents. *Biol. Reprod.* **61,** 764–775.
108. Dewit, M., Marley, W. S., and Graham, J. K. (2000) Fertilizing potential of mouse spermatozoa cryopreserved in a medium containing whole eggs. *Cryobiology* **40,** 36–45.
109. Katkov, I. I., Katkova, N., Critser, J. K., and Mazur, P. (1998) Mouse spermatozoa in high concentrations of glycerol: chemical toxicity vs osmotic shock at normal and reduced oxygen concentrations. *Cryobiology* **37,** 325–338.
110. Kishikawa, H., Tateno, H., and Yanagimachi, R. (1999) Fertility of mouse spermatozoa retrieved from cadavers and maintained at 4 degrees C. *J. Reprod. Fertil.* **116,** 217–222.
111. Mazur, P., Katkov, I. I., Katkova, N., and Critser, J. K. (2000) The enhancement of the ability of mouse sperm to survive freezing and thawing by the use of high concentrations of glycerol and the presence of an *Escherichia coli* membrane preparation (Oxyrase) to lower the oxygen concentration. *Cryobiology* **40,** 187–209.
112. Nakagata, N. (2000) Cryopreservation of mouse spermatozoa. *Mamm. Genome* **11,** 572–576.
113. Phelps, M. J., Liu, J., Benson, J. D., Willoughby, C. E., Gilmore, J. A., and Critser, J. K. (1999) Effects of Percoll separation, cryoprotective agents, and temperature on plasma membrane permeability characteristics of murine spermatozoa and their relevance to cryopreservation. *Biol. Reprod.* **61,** 1031–1041.
114. Songsasen, N. and Leibo, S. P. (1998) Live mice from cryopreserved embryos derived in vitro with cryopreserved ejaculated spermatozoa. *Lab. Anim. Sci.* **48,** 275–281.
115. Lacham-Kaplan, O. and Trounson, A. (1995) Intracytoplasmic sperm injection in mice: increased fertilization and development to term after induction of the acrosome reaction. *Human Reprod.* **10,** 2642–2649.
116. Ogura, A., Matsuda, J., Asano, T., Suzuki, O., and Yanagimachi, R. (1996) Mouse oocytes injected with cryopreserved round spermatids can develop into normal offspring. *J. Assisted Reprod. Genetics* **13,** 431–434.
117. Ogawa, T. (2000) Spermatogonial transplantation technique in spermatogenesis research. *Int. J. Androl.* **23(Suppl 2),** 57–59.
118. Nakagata, N. (1996) Use of cryopreservation techniques of embryos and spermatozoa for production of transgenic (Tg) mice and for maintenance of Tg mouse lines. *Lab. Anim. Sci.* **46,** 236–238.
119. Leibo, S. P., Francesco, J. D., and O'Malley, B. (1991) Production of transgenic mice from cryopreserved fertilized ova. *Mol. Reprod. Dev.* **30,** 313–319.
120. Boubelik, M. and Cerna, Z. (1993) A modified 2-step method for cryopreservation of mouse embryos for purposes of embryo banking. *Folia Biologica* **39,** 211–219.

121. Rulicke, T. and Autenried, P. (1995) Potential of two-cell mouse embryos to develop to term despite partial damage after cryopreservation. *Lab. Anim.* **29,** 320–326.

122. Wilton, L., Shaw, J., and Trounson, A. O. (1989) Successful single-cell biopsy and cryopreservation of preimplantation mouse embryos. *Fertil. Steril.* **51,** 513–517,

123. Snabes, M. C., Cota, J., and Hughes, M. R. (1993) Cryopreserved mouse embryos can successfully survive biopsy and refreezing. *J. Assisted Reprod. Genetics* **10,** 513–516.

124. Kasai, M. (1995) Cryopreservation of mammalian embryos: vitrification, in *Methods in Molecular Biology* **38,** 211–219 (Day, J. G. and McLellan, M. R., eds.), Humana Press, Totowa NJ.

125. Shaw, J. and Kasai, M. (2001) Embryo cryopreservation for transgenic mouse lines. Chapter 25 in *Methods in Molecular Biology*, vol: 158 (Tymms, M. J. and Kola, I., eds.), Humana Press, Totowa, NJ, pp. 397–419.

126. Nakagata, N. (1992) Production of normal young following insemination of frozen-thawed mouse spermatozoa into Fallopian tubes of pseudopregnant females. *Exp. Anim.* **41,** 519–522.

# 11

## Ovarian Tissue Transplantation and Cryopreservation

*Application to Maintenance and Recovery of Transgenic and Inbred Mouse Lines*

### Jillian M. Shaw and Alan O. Trounson

## 1. Introduction

A major problem for research groups breeding inbred, mutant, and transgenic mouse lines is that many are poor breeders, or do not breed at all. This problem is often difficult to alleviate. Sperm banking can be used for mice *(1–3)*, but in some instances it is also important to bank the female genome *(4)*. Superovulation and embryo transfer procedures are of little value for lines that produce few eggs or embryos, lines that die soon after birth or whose animals that die unexpectedly. In these cases, ovarian tissue grafting may provide a solution *(5,6)*. Ovarian tissue grafting (transfer) is not a widely used technique, but it has been used successfully to propagate subfertile, infertile, and sterile mouse lines and is used by commercial breeders such as the Jackson Laboratories. In cases in which the mice carry transgenes that interfere with reproduction or shorten the animal's lifespan *(7–10)* ovarian transfer may be the only way of propagating the female line. Ovarian grafting has several significant advantages over embryo transfer *(11)*. One advantage is that it is less age dependent. Tissue collected from fetal, newborn, juvenile, or adult female mice can all produce functional grafts *(12–18)*. A second advantage is that viable tissue can be "rescued" from recently deceased animals. If each "donor" ovary is cut into small pieces and each is grafted into a recipient, several breeding animals can be generated from each donor. Successful ovarian grafts develop quickly and usually restore normal estrous cycles in the recipient (her own ovaries are removed) within 2–6 wk. These grafted recipients may,

From: *Methods in Molecular Biology, vol. 180: Transgenesis Techniques, 2nd ed.: Principles and Protocols*
Edited by: A. R. Clarke © Humana Press Inc., Totowa, NJ

**Table 1**
**Summary of Current Protocols**
**for Cryopreservation of Ovarian Tissue** *(5,12,15,32,36,44–46)*

|                      | Usual protocol           | Variations on usual protocol                    |
| -------------------- | ------------------------ | ----------------------------------------------- |
| Cryoprotectant type  | DMSO                     | Propanediol, ethylene glycol, glycerol          |
| Concentration        | 1.5 *M*                  | 1.4 *M*                                         |
| Equilibration        | 0°C, 30 min              | Room temperature                                |
| Ramp 1               | 2°C/min                  | Direct to seeding temperature                   |
| Seeding temperature  | 7°C                      | 6 or 8°C                                        |
| Ramp 2               | 0.3°C/min to −40°C       |                                                 |
| Ramp 3               | 10°C/min to −140°C       | Plunge                                          |
| Thaw                 | Water bath               | Air, air + water bath                           |

if placed with a male, conceive, bear, and raise one or more litters and may produce litters for 10 mo or longer after grafting *(16,17)*.

Ovarian grafting techniques for mice and rats were established more than 40 yr ago *(13,14,16–20)*, Grafting provides a useful tool in breeding *(7,8)* and rescuing *(9)* mouse lines because it can be combined with embryo collection from naturally mated or superovulated donors as well as cryopreservation (oocyte *[21,22]*, embryo *[22]*, or ovarian tissue *[23,24]*). However, the success of ovarian grafting is dependent on minimizing graft rejection and on good surgical technique. In the case of infertile or sterile donors, the etiology of their infertility will influence the success of the grafting procedure. This chapter aims to help the reader establish an ovarian grafting program.

Frozen-thawed ovarian tissue can be as effective as fresh tissue in restoring fertility to graft recipients. Ovarian tissue cryopreservation can therefore be used for longterm storage of the female germline *(25–28)*. The best characterized cryopreservation protocol for ovarian tissue is based on the slow-cooling procedures developed for mouse eggs and embryos in the 1970s *(12,15,21,29–32)*. This process (**Table 1**) requires access to a controlled rate biological freezer or equivalent equipment. It is now documented that in the mouse both fresh and frozen thawed grafts have the potential to restore longterm fertility (i.e., for 1 yr) to the graft recipient *(9,12,15,17,26,27,33–35)*. This chapter details this cryopreservation technique and how it can be used to complement an ovarian tissue-grafting program.

## 2. Materials

### 2.1. Animals (see Notes 1–8)

#### 2.1.1. The Donor(s)

It is best to use healthy donors. The minimum age at which ovaries can easily be collected for grafting is around d 15–16 of fetal life (3–4 d before birth). Ovaries collected on d 16 of gestation or later in fetal or postnatal life are known to restore cycles and produce normal live young *(12,16,35)*. The upper age of the donor is generally about 6–7 mo *(16,35)*. Older animals can be used, but these grafts tend to be less productive than those from younger donors because the number of primordial follicles in the graft is maximal (about 3000/ovary) at 1–3 d after birth and falls to 300 or less in 1-yr-old mice *(14)*.

### 2.1.1.1. OVARIAN TISSUE COLLECTION FROM HEALTHY DONORS

Gonads can be collected from fetuses before d 15, but this is not ideal because the gonads are not fully differentiated and ovaries are difficult to distinguish from testes. From d 16 of gestation, the ovaries cannot be confused with testes because ovaries are located higher in the abdominal cavity (just under the kidneys), and they have a relatively uniform, spotty appearance. Testes are located closer to the pelvis and look stripy. After birth, the testes descend into the scrotum.

It is best to collect the donor tissue from freshly killed animals or by surgery. Surgery is useful if the donor needs to be kept alive for subsequent studies. If only one ovary or part of an ovary is removed at surgery the donor animal can subsequently be used for breeding (*see* **Notes 2–8**).

### 2.1.1.2. OVARIAN TISSUE COLLECTION FROM DECEASED OR DISEASED DONORS

If ovarian tissue is collected for use as a graft to generate live offspring, it is best if the donor is healthy. This minimizes the risk of pathogen/disease transmission from the donor to the recipient *(36)*. It is also best if the interval between the donor's death and the time of grafting to a recipient is kept to a minimum. Two recent studies *(37,38)* have shown that the number of viable follicles within a grafted ovary is dependent on the collection protocol and storage conditions prior to grafting. Ovaries collected by surgery (or immediately after the death of the donor) and then grafted without further delay to a recipient contained twice as many viable follicles than grafts stored on ice or at room temperature in phosphate buffered saline (PBS) for 3–12 h before grafting. In cases in which the ovaries were not removed from the carcass of a donor until 3 h or longer after its death, only very few viable follicles remained within the ovaries after grafting. However, even grafts derived from ovaries collected (or held in PBS) 24 h or longer (e.g., 48 h) after the death of the donor could, in some instances, contain viable follicles *(37,38)*.

If the cause of death is not known or the donor is known to be sick (such as viral or bacterial infection, cancer), the graft may transmit this disease to the recipient. In the mouse, both fresh and cryopreserved pieces of ovarian tissue from donors with lymphoma have transmitted the lymphoma to healthy recipients and have lead to their death *(36)*. Information from humans shows that transplants (such as cornea, liver, or heart) can transmit a wide range of ailments (such as human immunodeficiency virus, hepatitis, bacterial infections, parasites, or cancers; *6,11*). For this reason, recipients of grafts from diseased or deceased donors should, if possible, be isolated from other mice and monitored closely. In vitro maturation of follicles or oocytes is also possible *(39–41)*.

### 2.1.2. Recipients (see **Notes 1–8**)

Recipients should be mature (3–4 mo) and histocompatible with the donor (*see* **Notes 2–8**). Ovarian tissue carries histocompatibility antigens, and the ovarian bursa is not an immunologically privileged site (*see* **Note 1**). The success of the graft will therefore depend on the choice of recipient because there is a risk of the graft being rejected by the recipient if it is recognized by the immune system. To overcome problems with rejection, the graft recipient should be either histocompatible (tissue matched) with the donor or immunologically compromised. If the donor has a transgene that may cause the graft to become antigenic, the graft might not succeed even if it is placed in a recipient of the same inbred strain. Tissue from these animals may be best grafted into immunologically compromised recipients (e.g., NIH III, nude, SCID, or "RAG" mice). Ovarian tissue that has been cryopreserved can be treated in the same way as fresh tissue, providing that histocompatible mouse strains are still available. If histocompatible strains are no longer available, these tissues also should be grafted to immunologically compromised recipients.

### 2.1.3. Studs and Progeny

When the graft recipients recover from surgery (~2 wk), they can be placed with fertile males (studs). It is important to remember that the progeny arise from the donor, not the recipient, female. However, complete ovariectomies are difficult to achieve, especially for beginners; therefore, a proportion of the progeny may arise from remnant ovarian tissue. This means that tests to confirm the maternal origin of the progeny are needed before the progeny are used.

### 2.2. Solutions and Equipment (see **Notes 9–17**)

All solutions should be made up with high-quality Milli-Q water (Millipore Corporation) or water of equivalent quality.

PBS can be made or purchased in powder or liquid form from a commercial supplier (e.g., Gibco). The solutions should be filter sterilized through a 0.2 µm filter (*see* **Notes 9–13**).

### 2.2.1. Grafting

1. Anesthetic solution: 0.5 mL ketamine (100 mg/mL ketamine hydrochloride), 0.5 mL of xylazine (20 mg/mL xylazine hydrochloride [sold as Rompun by Bayer AG, Germany]) and 9 mL PBS. Filter sterilize through a 0.2 µm filter.

2. Reverzine 0.5 mL (1.25 mg/mL yohimbine hydrochloride, 2 mg/mL 4-amino-pyridine) and 2 mL PBS. Filter sterilize through a 0.2 μm filter.
3. 70% alcohol.

## 2.2.2. Cryopreservation (see **Notes 9–13**)

1. Cryoprotectant solution: 1.5 $M$ dimethyl sulfoxide (DMSO, e.g., D8779 or D2650, Sigma, St. Louis, MO) and 0.1 $M$ sucrose. Place 1.1 mL of DMSO, and 0.342 g of sucrose (S 9378, Sigma, St. Louis, MO) in a 10 mL graduated tube. Make up to 10 mL with PBS. Filter sterilize. Allow 5 mL per mouse and a further 1 mL per tissue piece (assuming one piece per freezing vial).
2. First dilution solution: 0.75 $M$ DMSO and 0.25 $M$ sucrose. Place 0.55 mL of DMSO and 0.86 g of sucrose in a graduated tube and make up to the 10-mL line with PBS. Filter through a 0.2 μm filter.
3. Second dilution solution: 0.25 $M$ sucrose. Place 0.86 g of sucrose in a graduated tube and make up to 10 mL with PBS. Filter through a 0.2 μm filter.

## 2.2.3. General Equipment for Grafting

1. Stereo dissecting microscope(s). A stereo dissecting microscope with zoom (e.g., ×4–40) is helpful, especially when ovaries from fetuses or newborn mice are to be grafted. If surgery is to be carried out under the dissecting microscope, one with a long working distance is best (e.g., 6–10 cm).
2. Light source. Good lighting is needed for grafting. Although any type of lamp can be used, it is best to have a fiberoptic light source because the light it provides is "cold" and does not dehydrate the operation site. Fiberoptic units with two or more flexible arms or a ring light that attaches to the microscope are ideal.
3. Surgical instruments. Fine scissors and very fine forceps (#5 watchmaker's forceps). Although cheap instruments can be used, good quality instruments last longer. Forceps must be sharp and meet at the tip. If necessary, the forceps can be sharpened on a grinder, wet stone, or equivalent.
4. Hood. If immunologically compromised recipients are to be used, a laminar flow or biohazard hood, providing a sterile work environment, may prove beneficial. It can be difficult to operate under these conditions, particularly when a dissecting stereomicroscope is used.

## 2.2.4. Equipment for Cryopreservation (see **Notes 14–25**)

1. Freezing machine. A controlled-rate freezer that allows tubes to be held at –6°C and then cooled at 0.3°C min to –40°C is optimal.
2. Storage tank. To store ovaries efficiently for long periods, they must be held permanently at low subzero temperatures (e.g., below –140°C). Tissue is most commonly stored in liquid nitrogen (below –196°C) or in liquid nitrogen (LN) vapor. Ultra-low electrically powered freezers (with or without LN backup) that hold a temperature of –140°C or below can be used also. Discuss your needs with suppliers and distributors of cryostorage tanks because the specifications will

determine the quantity of LN that will be used and how many specimens it will hold. See notes on the safety issues associated with the use of LN/nitrogen (*see* **Notes 2–10**).

To safely dispense LN, the user should wear a face mask and protective gloves. Thick leather gloves such as the type used by welders are suitable.

3. Dewars for benchtop work. Small, approx 1 L, wide-necked containers (dewars) for LN are used for bench topwork and for transporting frozen samples short distances such as between the bench or freezing machine and the storage tank. Use stainless steel dewars or other insulated containers (e.g., thick-walled styrofoam boxes). Do not use any type of evacuated glass, thermos-style containers because they have the potential to explode without warning.

4. Freezing vials. Only use sterile vials that are specifically made for cryopreservation (e.g., Nunc, Denmark). Some suppliers provide caps or cap inserts of different colors. Many large-scale storage tanks are configured so that the vials are stored in boxes. Alternatively, the storage tank may have tube-like canisters. In the latter case, the vials should be clipped onto canes for storage. Ensure that the vials and cane match, so that the vials will seat firmly on the cane (*see* **Note 25**).

5. Straws. Very small ovaries can be frozen in 0.25 mL "insemination" straws (e.g., IMV, L'Aigle, France). If straws are used, a heat sealer or polyvinylpyrrolidone (PVP) or polyvinylalcohol (PVA) powder is required to seal the ends before freezing. Straws are usually stored in 9-, 13-, or 16-mm wide, 13-cm high goblets on canes.

6. Water bath. Stirred water baths with accurate temperature control can be obtained from scientific suppliers. Alternatively, any container filled with tap water at the correct temperature is adequate.

7. Minor equipment. Other equipment used for most freezing protocols includes
   a. Tongs with very long handles or blades.
   b. Timers, preferably with a countdown alarm.
   c. Glass pipets to handle the smallest ovaries.
   d. Large artery forceps or the equivalent to manipulate tubes within liquid nitrogen.
   e. Marker pens with fine tips to label dishes, tubes, and canes. These must be resistant to water.
   f. Sharp scissors. If straws are used, these are needed to give a clean cut when removing the ends of the straws.

# 3. Methods

## 3.1. Grafting Protocol

Because the tissue is to be grafted into the body cavity of other animals, it is important to use aseptic techniques or, in the case of immunologically compromised animals, a sterile technique throughout the collection (cryopreservation) and grafting steps.

### 3.1.1. Preparation

Before starting, check that you have all the animals (donor and/or recipient mice), solutions (70% alcohol, anesthetic, PBS) and equipment (26-gage needles and 1-mL syringe) to anesthetize the mouse, 35-mm Petri dishes, #20 or #22 scalpel blades, gloves, two pairs of scissors, two but preferably four pairs of fine #5 watchmakers forceps, tissues) needed for collecting and grafting the ovary. Equipment such as a dissecting stereomicroscope, flexible cold light source, and a prewarmed warm plate to keep the mice warm until they are returned to their cages, are helpful but not essential.

The donor ovary can be collected and prepared before anesthetizing the recipients. However, it is possible to reduce the length of time between the collection and grafting of the donor ovary by keeping the donor alive until the recipient is anesthetized, ovariectomized, and the bursal cavities are prepared to receive the graft.

If tissue is to be cryopreserved, additional equipment and solutions are also required (**Subheadings 2.2.2.** and **2.2.4.**).

### 3.1.2. Collection of the Donor Ovary

1. Kill the donor and then swab all the skin on and around the abdomen with 70% alcohol. For this to have full effect, apply the alcohol several times with a 3–4 min interval between each application.
2. Using clean instruments, sterilized in alcohol, make a transverse cut in the skin across the middle of the abdomen. Then, using your fingers, get a firm grip on the skin above and below the cut and widen the slit by firmly pulling the skin apart.
3. Keep pulling the skin back until the whole of the abdominal wall is exposed.
4. Use a second set of instruments to enter the abdominal cavity. To minimize the risk of cutting the intestines during this step, it is best to start by grasping a small amount of the abdominal wall with forceps and lifting the wall up as far as possible. Then make a very small hole immediately under the forceps to let air into the abdominal cavity. As soon as the air enters, the intestines should slip down, away from the abdominal wall. Using scissors, extend the cut from the hole all the way across the abdominal wall down to the muscles next to the spine (in both directions). Lift the abdominal wall out of the way, and push the intestines forward to expose the reproductive tract.
5. Follow each uterine horn toward the kidney and locate the fatpad/ovary/oviduct.
6. Using sharp (e.g., #5 watchmaker's) forceps, tear open the bursa to expose the ovary.
7. Slide one pair of forceps under the ovary and then pull gently upward. The ovary should detach easily.
8. Wash the ovary through three changes of sterile buffer (e.g., phosphate or HEPES-buffered media, pH 7.2–7.4, when used at room temperature in air). In mice younger than weanling age, it is best to remove the whole reproductive

tract and then dissect off the ovary with 26-gage needles under a dissection microscope. Ovaries from mice younger than 21 d are usually grafted whole whereas ovaries from larger mice are often cut into pieces. Donors which are to be kept alive should have their ovary/ovaries removed surgically as described for recipients.

### 3.1.3. Ovariectomy/Transplantation Procedure

1. Anesthetize the mature female recipient with an intraperitoneal injection (e.g., 0.3 mL Rompun ketamine for a 20-g mouse). Return it to its cage, and do not disturb it until it stops moving. Do not start the surgery until the mouse ceases to respond to a toe pinch. The whole time that the mouse is under anesthetic, it should be monitored regularly. If at any time, the breathing becomes very slow or irregular, the anesthetic may be too deep. Stop the operation and try to restore the body temperature and breathing before resuming surgery. If the mouse starts to move before the operation is over, it should be given more anesthetic (e.g., 0.05 mL). Take care not to give an overdose nor to inject any anesthetic subcutaneously or intramuscularly.
2. Swab the back of the anesthetized recipient with 70% alcohol, shave or clip the fur at the operation site, and then reswab the area with 70% alcohol to remove loose fur.
3. With a pair of sharp scissors, cut a single 1- to 1.5-cm middorsal incision through the skin running from a line level with the top of the hips to a line level with the bottom of the ribs (**Fig. 1**).
4. Pull the skin sideways over the flank until the abdominal wall can be seen. It should be possible to see the mouse's own ovary just inside the abdominal wall.
5. Make a 1-cm cut through the abdominal wall, parallel to the backbone. To minimize the risk of the intestines being cut during this step, it is safest to start by grasping a small amount of the abdominal wall with forceps and lifting the wall up as far as possible. Then make a very small hole immediately under the forceps to let air into the abdominal cavity. As soon as the air enters, the intestines should slip down, away from the abdominal wall.
6. Locate the ovarian fatpad inside the hole. Then use blunt forceps to gently grasp the fatpad and exteriorize it and the ovary through the incision, and rotate the tissue until the ovary is clearly visible. Use clips or a fine needle (26 gage) to secure the fatpad so that the ovary does not fall back into the abdomen (it is easiest to secure the fatpad to the skin).
7. Locate the junction of the transparent membrane and the portion of the fatpad that is farthest from the oviduct (**Fig. 2**).
8. Using two pairs of fine #5 forceps, very gently tear along the junction of the fatpad and the membrane (**Fig. 2**) until the opening is just big enough to allow the membrane to be pulled over the ovary (toward the oviduct). Use caution since the transparent membrane has blood vessels, and try to avoid tearing any particularly large blood vessels while the opening is being made or the ovary is being exposed. The ovary is attached along the length of its lower surface by connec-

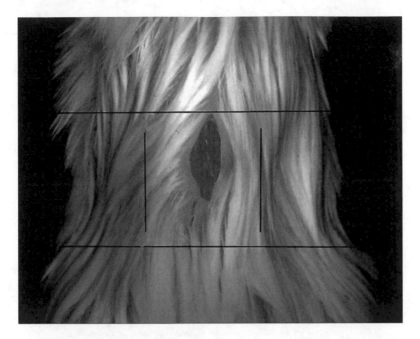

Fig. 1. Dorsal incision as used for ovarian grafting in the mouse. Make a 1–1.5-cm incision through the skin. It should be over the backbone, between the level of the hips (lower line) and the bottom of the rib cage (upper line). Do not damage the back muscles, which are in the area between the two upright short lines. The skin, with its incision, should be pulled sideways (beyond the marked lines) before trying to enter the abdominal cavity.

tive tissue (the hilum) to the back of the bursa (opposite the transparent membrane). To maximize the amount of ovarian tissue that is removed, sever or cut the hilum as near to the base of the bursa as possible. This is difficult because the hilum is rich in large blood vessels. The grip on the forceps should be adjusted to get the best grip and access to the tissue (**Fig. 3**).

9. To minimize bleeding, slide one pair of fine forceps under the entire length of the ovary with one prong on either side of the hilum. Then tightly close the forceps around the hilum to stop the blood flow.
10. Insert a second pair of forceps between the first pair of forceps and the ovary.
11. Close the forceps very tightly around the upper hilum, and then raise these forceps together with the ovary out of the bursa. Alternatively, a fine scalpel can be slid over the top of the first pair of forceps, or fine scissors can be used to cut off the ovary.
12. Keep the first pair of forceps (which are applying pressure to the hilum) in place while you try to assess whether there is any remnant ovarian tissue. Keep these forceps in place for at least 1–2 min before removing them, to try to reduce bleed-

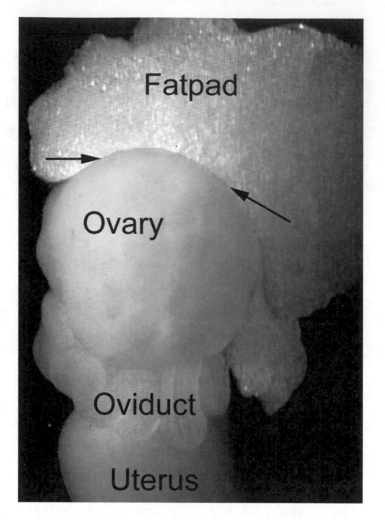

Fig. 2. Appearance of the upper reproductive tract of the female mouse. As viewed in this photo, the thick transparent bursal membrane lies over the ovary, while the hilum lies directly behind it. The hilum is the band of connective tissue that holds the ovary in place and contains all blood vessels entering and leaving the ovary. To access the ovary, use very fine forceps or very fine scissors to make a short slit along the boundary of the bursa and the fatpad (between the arrows). Stop widening the slit as soon as it is big enough for the bursal membrane to be pulled back to expose the ovary. To remove the ovary, ease the tips of the two pairs of fine forceps under the ovary (the tips of each pair should be positioned on either side of the hilum) and then close them very firmly onto the hilum to stop blood circulation to the ovary. After a 1–2 min wait, pull the upper forceps, together with the ovary, out of the bursal cavity. The donor ovary is placed in the vacated bursal cavity and the bursal membrane pulled up towards the fatpad to help hold the new ovary in position.

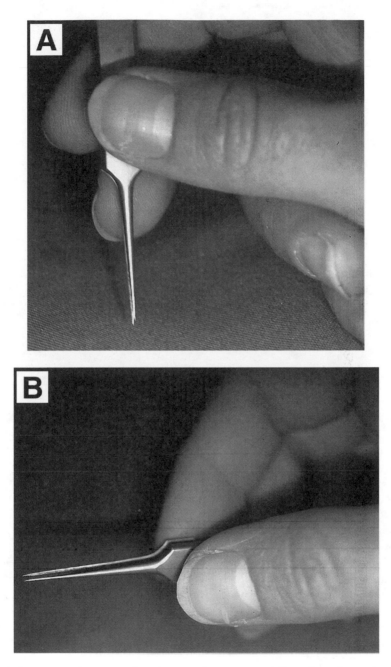

Fig. 3. Methods of holding fine forceps for surgery. Adjust the grip on the forceps to maximize their usefulness. The vertical position (**A**) can be used to grip skin, abdominal wall, and bursa. The horizontal position (**B**) makes it easier to slide the tips under the ovary to grip the hilum.

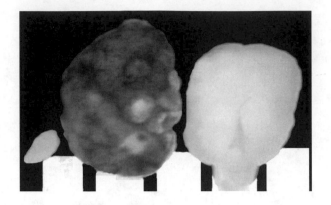

Fig. 4. Appearance of mouse ovaries. One ovary from a newborn mouse (left) and two ovaries from an adult mouse (right). The adult ovaries had been grafted 24 h earlier from the same donor. The dark graft has established a new blood supply and is filled with erythrocytes (successful graft); the light graft has not established a blood supply (unsuccessful graft). The scale is the millimeters.

     ing from the severed blood vessels. Bleeding is common and will, in many instances, flood the operation site. Trying to irrigate the area or removing the blood can exacerbate the problem.

13. Insert the donor ovary into the bursal cavity. Adjust the grafting method to suit the size of the ovary (**Fig. 4**). If the donor ovary is large, place it in the blood-filled cavity (as near the hilum as possible), and gently pull the thin bursal membrane over it. If the graft is too large, cut a portion off and try again. Small ovaries or small pieces of ovary can be treated the same way unless there is a risk that the blood flow will wash the tissue out of the bursa. In this case wait for the blood flow to stop (this may take 10 min), remove the clot, and insert the donor ovary. If possible, place the hilum against a cut surface of the graft (preferably the hilar region).

14. Pulling the bursal membrane back to its original position should be sufficient to hold the graft in place. However, some research groups reclose the bursa with one or two stitches (9-0 or 10-0 suture) at the junction of the bursa and fatpad.

15. A small amount of pressure on the bursa may cause the ovary to slip out; it is therefore best to hold the hole in the abdominal wall wide open with forceps while very gently lowering the oviduct, bursa, and fatpad back into the abdominal cavity.

16. Repeat the same ovariectomy/grafting procedure on the opposite ovary. The same middorsal skin incision can be used, but a new incision needs to be made in the abdominal wall on the opposite side of the mouse. Both of the recipient's own ovaries must be removed (or they may inhibit ovulation by the graft), but donor ovarian tissue can be placed in either (or both) bursal cavities. Large ovaries can be divided into two or more pieces and implanted in one or more recipients.

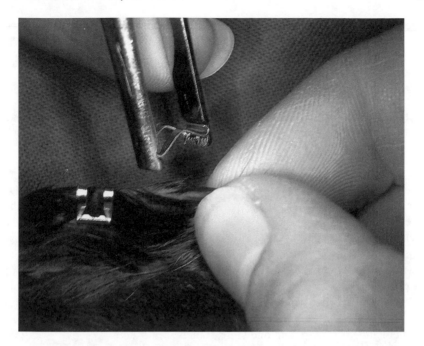

Fig. 5. Example of how to apply Michel clips with a Michel clip applicator.

17. Close the skin wound with two to four firmly applied Michelle clips (e.g., 9-mm MikRon Autoclip, Becton Dickinson). To ensure that the skin on both sides of the wound are clipped together, pull the sides of the wound together by taking a large pinch of skin above and below the wound (i.e., the skin along the backbone at the level of the hips and the ribs). This should stretch the skin on either side of the wound and bring the skin on both sides together. Once the skin on both sides of the wound is aligned, apply the clip (**Fig. 5**). The two skin edges should be together near the top (highest point) of the clip; properly applied clips do not fall off.

18. Keep the mice warm until they recover from the anesthetic by placing them or their cage on a warming tray (check the temperature regularly).

19. If Rompun ketamine has been used, the mice can be given Reverzine (intravenously into the tail vein) to speed recovery.

20. Monitor the recipients daily to ensure that the wounds heal properly.

21. After 2 wk, the wound should be healed and the clips can be removed (optional). To remove the clips, sedate the mouse (half dose of anesthetic, or a short exposure to $CO_2$), and then widen the clip until it falls off. Use a clip remover or insert the tips of some pointed scissors inside the clip and open the jaws of the scissors to pry the clip open. This procedure can be performed without sedation if two persons are available (one person holding the mouse and the second to remove the clip). If a clip falls off before the wound has fully healed, it must be replaced.

The ovary should, if it is correctly placed within the bursa, stay in place and show the first signs of vascular ingrowth (turning red *[42]*) within 24 h (**Fig. 4**). Small ovaries do not exhibit this redness as clearly as large ovaries.

## 3.2. Tissue Cryopreservation (see Notes 9–26)

Sterile solutions and equipment should be used throughout this procedure.

### 3.2.1. Freezing Procedure

1. Before killing the donor or starting the freezing procedure prepare all the necessary solutions and equipment: cryoprotectant solution (1.5 *M* DMSO with 0.25 *M* sucrose), ice bath, freezing machine precooled to –6°C (check that the selected cooling program is 0.3°C/min to –40°C), and timer (15 min). Label the vials (1 mL cryotubes, Nunc, Greiner) with the date, strain, and other essential information using a fine-tipped indelible marker pen. (Do not use pens with water-soluble ink because it will wash off during thawing.) Make up the cryoprotectant and dispense it into a large tube (e.g., 5 mL in each of 2 Falcon 2001 tubes) and freezing vials (e.g., 1 mL in Nunc 1.5 mL tubes). Place the labelled vials and tubes on ice to cool (allow at least 30 min). Allow for one Falcon tube per donor and one vial per piece of tissue.
2. Collect the ovary as outlined in **Subheading 3.1.2.** (*see* **Subheading 3.1.3.** for donors that have to be kept alive).
3. Place the ovary or pieces of ovary in the 10-mL tube containing precooled cryoprotectant solution, and return it to the ice bath.
4. Every 5 min, agitate the tube so that all the pieces become more uniformly exposed to the cryoprotectant.
5. After 15–20 min, tip out the contents of the tube into a sterile dish and rapidly allocate tissue to each vial. Distribute the material as evenly as possible between all vials.
6. Screw on each vial cap and return the vials to the ice.
7. After the tissue has been in the cryoprotectant on ice for 30 min, place the vials in a freezing machine precooled to –6°C (seeding temperature).
8. Seed (initiate ice formation) by touching the plastic wall overlying the upper meniscus of the cryoprotectant solution with a metal rod or equivalent metal instrument precooled in liquid nitrogen.
9. Remove the seeding rod when ice is first seen inside the tube.
10. Once ice is present in all the tubes, leave the tubes 10 min longer at –6°C and then start the cooling program: cooling at 0.3°C/min to –40°C. On reaching –40°C, the vials can be transferred into liquid nitrogen.
11. Label each cane or box with an indelible marker record the date, and strain on each one and, if possible, also have a unique label or number that can be cross-referenced to paper or computer records.
12. Place the vials onto precooled canes or in precooled boxes, and place in a storage vessel.

13. Complete the records (e.g., cards, books, sheets, or computer database).
14. If ovaries are to be frozen for more than one donor use fresh equipment, solutions, and dishes for each.
15. During storage, ensure that the liquid nitrogen (LN) levels in the storage tanks (or freezers) are maintained.

### 3.2.2. Thawing Procedure

1. Before starting to thaw any tissue, have the following equipment and solutions prepared in advance:
   a. A dewar which is safe containing LN (e.g., 1 L) for the benchtop work and a water-bath at 37°C.
   b. Label four 35-mm petri dishes for each vial to be thawed. The first dish should contain 3 mL of diluent (0.75 $M$ DMSO + 0.25 $M$ sucrose in PBS, i.e., 0.55 mL of DMSO and 0.86 g sucrose/10 mL of PBS), the second 3 mL 0.25 $M$ sucrose (0.86 g sucrose/10 mL PBS), and the other two 3 mL of PBS. All solutions should be at room temperature (20 to 22°C).
   c. You will also need fine forceps and two or more timers.
2. If the thawed material is to be grafted, all animals, equipment and solutions required for grafting should also be prepared in advance.
3. Find the stored vials and use long-handled forceps to move them from the storage vessel into liquid nitrogen in a small dewar suitable for benchtop work.
4. Take great care at this stage because the vial may be filled with LN. LN trapped inside a vial will rapidly expand when it is warmed and may cause the straw to hiss. Vials not specifically designed for LN may explode.
5. If the vial does contain nitrogen, leave or hold it in the vapor phase (just above the surface of the LN of the dewar where the temperature is lowest) for 3–5 min until the nitrogen in the vial has fully evaporated.
6. To thaw a vial, grasp it with the long-handled forceps and rapidly transfer it to a 37°C water bath. Agitate it in the water until the contents have thawed.
7. Wipe the vial with alcohol before removing the lid. Make sure that the information on the label is recorded before doing this because the alcohol could remove the writing.
8. Open the vial and locate the pieces of tissue.
9. Either empty the contents of the vial into the first thawing solution (0.75 $M$ DMSO + 0.25 $M$ sucrose) or grasp the piece of tissue with sterile forceps and move it alone to the first thawing solution.
10. After 10 min, grasp the ovarian tissue gently with fine forceps and move it to the second solution (0.25 $M$ sucrose).
11. Ten minutes later grasp the ovarian tissue gently with fine forceps and move it into the dish with PBS.
12. Wash the tissue a second time before grafting.
13. Apply sterile technique throughout. All solutions, equipment, and containers should be sterile.

## 4. Notes

1. The following definitions apply to the terminology used.

   *Inbred mice.* Mice that have been brother/sister mated for more than 30 generations to eliminate genetic variability. The result is that tissue from any individual
   will be accepted as "self" by all other individuals (of the same sex) of that same
   inbred strain, since their genetic makeup is largely identical.

   *Hybrid mice.* F1 hybrids are the offspring of a mating between two mice belonging to different strains. The progeny are typically stronger and healthier than the
   parents. This "hybrid vigor" is particularly evident in the hybrid offspring of two
   highly inbred strains. F2 hybrids are the offspring of the F1 generation.

   *Transgenic mice.* Mice with a genome that has been altered by human intervention.

   *Donor.* The mouse whose breeding potential you wish to enhance.

   *Recipient.* The mouse into which the ovarian tissue is grafted.

   *Immunologically incompetent.* Animals which have defective immune systems
   and are therefore not able to recognize and/or reject foreign tissue. Because these
   types of mice also have a compromised capacity to fight infections, they should
   be housed under aseptic or sterile conditions.

   *Histocompatibility.* The outer membrane of most cells in the body carry histocompatibility antigens. These molecules are recognized by the immune system.
   Genes belonging to the MHC govern the expression of histocompatibility antigens as well as the immune system and its responsiveness to foreign materials.
   These genes form a large part of the mouse's genome.

   *Immunologically privileged site.* Grafts to immunologically privileged sites are
   recognized less readily as foreign and are less likely to be rejected. Unfortunately, only a limited number of sites in the body (the anterior chamber of the
   eye, the brain) are immunologically privileged.

   *Bursa.* In the mouse, the ovary and the fimbrium lie together inside a bursa
   (a cavity). Approximately 40% of the wall surrounding this cavity is no more
   than a very thin transparent membrane. The ovary and fimbrium can easily be
   accessed through this membrane. The rest of the bursa is surrounded by layers of
   fat or connective tissue.

   *Primordial follicle.* A very immature oocyte (egg). Each follicle contains an egg
   closely associated with a small number of somatic cells.

   *Ischemia.* Oxygen depletion.

   *Cryopreserve/cryopreservation.* Antifreeze solutions are added to the tissue to
   allow it to be cooled to very low subzero temperatures (e.g., $-196°C$ in liquid
   nitrogen). At such temperatures all metabolic processes cease allowing tissue to
   be stored indefinitely. Tissue frozen without antifreeze cannot be stored at these
   temperatures.

2. Ovaries collected from d 16 of gestation through to d 3 after birth will start to
   ovulate approx 20 d after they are placed in an adult (ovariectomized) recipient.

Ovaries collected from mice over 2 wk of age should start to ovulate 1–2 wk after grafting.

3. Disease transmission by ovarian tissue is a greater problem than for oocytes or embryos. Mature oocytes and embryos are usually enclosed within a zona and are washed before transfer. This reduces the likelihood that eggs or embryos will transmit infections to the recipient. Ovarian tissue cannot be cleared of disease in this way. One consequence is that unwanted cells or infections may be transmitted with the graft. Thus, lymphoma can be spread from a sick donor to a healthy recipient with either fresh or frozen ovarian tissue grafts. Ovarian tissue can be dissociated into primordial follicles that can be frozen thawed and give rise to live young after grafting. It remains to be established whether this will remove infectious agents or undesirable cell types. Grafts with isolated primordial follicles can give rise to live young.

4. Histocompatibility matching is not a problem if the donor belongs to an *inbred* mouse strain, as its ovarian tissue will be accepted by all other females belonging to that same inbred strain and all F1 hybrids of that strain. If a hybrid is to be used, an F1 hybrid of two inbred strains should be a more uniform recipient than an F1 hybrid derived from a random bred or outbred line. Do not graft to females resulting from the mating of a male F1 and a female F1, as it is not possible to predict which, if any, of their progeny (F2 hybrids) will be histocompatible with the original inbred line (or with the F1 hybrid). Ovaries from donors which have two inbred lines as ancestors can usually be grafted into an F1 hybrid of the same two inbred lines. If the donor ovary is from an inbred line, it may be possible to select F1 recipients and studs with very different characteristics (e.g., coat color) to the donor. This can simplify the task of identifying the origin of the graft recipients progeny (graft vs recipient origin).

5. Ovarian tissue is significantly less immunogenic than skin, but most grafts between nonidentical individuals are rejected.

6. It is important to use an anesthetic that suits the mouse strain being used. Use whatever anesthetic you, or those you work with, routinely use for the given mouse strain. We currently use 0.4 mg of Rompun and 2 mg of ketamine in 0.3 mL of saline/20 g mouse given by ip injection. The advantage of this drug combination is that there is an antedote, Reverzine (it counteracts the Rompun). This is given intravenously at the end of the operation to aid the animal's recovery (speeds up breathing and return to consciousness). Alternative anesthetics are Avertin (e.g., 4 mg of 2,2,2-tribromoethanol-[Sigma, Aldrich], and 4 µL of tertiary amyl alcohol in 0.4 mL of saline/20 g mouse) or Hypnorm.

7. If only one ovary is removed, the remaining ovary usually compensates for this loss (ovarian hypertrophy), releasing nearly twice the normal number of oocytes. As the resulting embryos implant normally, the litter sizes of mice with only one ovary are usually comparable with those of mice with two functional ovaries. By surgically removing only one ovary and grafting it into one or more ovariectomized recipients, it may be possible to obtain more offspring from the donor animal.

8. Sutures can be used instead of clips.

9. The PBS described in this chapter can be replaced by HEPES-buffered medium (e.g., M2).

10. The buffer described in this chapter contained no protein. It is not known whether this is beneficial or detrimental to ovarian tissue. Some users add 4 mg/mL of bovine serum albumin, 10% serum, or macromolecules (e.g., PVA, Ficoll, or PVP).

11. We do not add protein to the cryoprotectant solution to avoid any possible effects protein may have on the antigenicity of the tissue at the time of grafting. Macro-molecular substitutes for protein such as Ficoll 70000 or P.V.P. at low concentrations (e.g., 5%) may be added, but it has yet to be determined whether this would have any protective effects.

12. DMSO is the most widely used cryoprotectant (antifreeze) for ovarian tissue freezing for species other than the mouse. Recent evidence shows that DMSO, propanediol, and ethylene glycol-based solutions are all equally effective for freezing mouse ovarian tissue *(44)*. Glycerol can be used, but current evidence indicates that it is less suitable than the other three cryoprotectants.

13. DMSO is a powerful solvent. Before filter sterilizing any solution containing DMSO, ascertain that the filters are resistant to (not dissolved by) DMSO. If in doubt ask the company that supplies your filters.

14. If you have to purchase a freezing machine, the following information may be useful.

   a. There are many different types of freezing machines. Small inexpensive portable machines are widely used by veterinarians. A new machine need not cost more than US $5–7,000. Important features that should be considered before purchasing a machine are: that it can cool vials (and straws) at 0.3°C/min to –40°C, how much liquid nitrogen is used for each freeze, and the ease of loading, seeding, and unloading the straws. Buy one with a large enough capacity to suit your purposes.

   b. In machines that have vials arranged in several (unsupported) tiers during freezing, you must only seed one tier of vials at a time (or else the unsupported uppermost vials will fall out). Place one vial per channel in the chamber and wait 2–3 min. Then, for each vial in turn, pull it halfway out of the chamber, and seed it (e.g., with a large pair of precooled forceps). Once all vials in a tier have been seeded, reexamine each one in turn to determine whether the solution still contains ice crystals. The next layer can then be added to the chamber. Seed this layer using the same protocol as with the first layer. Repeat this until all vials have been seeded. If you have difficulties in getting a vial into the channel, it could be because the O-ring is bulging out sideways; loosen the cap marginally and try again. The reason for this layer-by-layer approach is to minimize the risk of vials falling out of the chamber into the liquid nitrogen bath. Because there is a time delay for each layer, the tissue from the first ovary (first piece to be dissected) can be put in cryoprotectant before the dissection of the second ovary starts.

15. If you have to purchase a storage container, the following information may be of assistance:

    a. Liquid nitrogen evaporates from storage tanks at a rate determined by the efficiency of the storage tank. When purchasing a tank for liquid nitrogen, the required storage capacity should be considered as well as its weight (when full), neck width, insulation properties, and whether the materials are going to be stored in goblets or on canes. Wide-necked tanks are more convenient as storage containers, but they tend to allow faster evaporation of the liquid nitrogen. The insulation properties of a storage tank can decline quite markedly with age; therefore, old or second-hand tanks may be uneconomical.

    b. At the present time, ovaries are most commonly stored in the liquid phase of LN. If there is any risk that the samples could contain infectious agents such as viruses, they should be stored in the vapor phase of LN or in a very low temperature freezer (–140 to –150°C), because viruses can be spread by the LN between samples within a tank. There are documented cases of bags for blood transfusion acquiring hepatitis by this route *(43)*. LN tanks with only a small amount of LN at the bottom can serve as vapor phase storage tanks, but the temperature at the top of the tank needs to be monitored to ensure that it does not rise too far above –150°C. If the temperature gradient from the surface of the LN to the top of the tank is too great, then the installation of copper tubes, rods, or sheets reaching from the LN to the top of the specimens may be necessary.

    c. A spare tank can be useful when searching for lost straws or goblets. The spare tank enables the canisters with frozen materials to be placed in the backup tank while the storage tank is thoroughly searched (or emptied).

    d. It is advisable to install alarms in all tanks and freezers containing valuable materials. If the temperature rises above a set level, these should give out an audible alarm or automatically contact security or those in charge of the laboratory. An automatic phone alarm system must be organized to exclude telephone answering machines from answering on the emergency contact numbers. Storage tanks may swiftly develop problems, as exemplified by a 120-L tank in our department going from full to totally empty overnight. The only external evidence of a problem was some frost on the tank.

16. The volume of LN needed will depend on the freezing procedure chosen, the type of biologic freezer, the size and specifications of the storage tanks, and the frequency of use of the equipment. It is usual to use 20–100 L each week. LN is potentially dangerous and must be handled with care. Pure nitrogen gas asphyxiates living organisms.

17. A torch and very long-handled tongs should be available to search for vials lost in the liquid nitrogen.

18. Nitrogen at high concentrations can displace oxygen and cause asphyxia. Rooms in which LN is to be used or in which tanks are to be kept must be well ventilated.

19. No LN should accompany ovaries that are to be transported by car or plane. Specialized vapor shippers that are precooled with LN but emptied before the ovaries are placed in them are highly effective for transport. Most brands remain below $-190°C$ for 2–3 wk.

20. Contact with LN or any conductive material (e.g., metal) that is or has been in contact with $LN_2$ can cause severe frostbite. Eye and hand protection should be worn. If LN saturates a person's clothing or shoes, these items should be removed as quickly as possible or rapidly doused with a large quantity of water.

21. LN behaves like water and can splash and spill and can cause personal injury. Spills will cause most floor coverings to crack.

22. Beware of any situation in which LN is trapped within an enclosed space. Its vaporization can lead to an explosion. To reduce the risk of serious injury, store vials and straws out of the liquid phase of the LN. If the specimens cannot be stored in nitrogen vapor, the vials and straws should be sealed correctly to prevent LN from getting in. If LN does get into a container that is to be thawed, hold the container in the vapor phase until all the trapped gas has evaporated.

23. Only use materials designed for use with LN. Eppendorf tubes and silvered glass liners (such as "thermos" flasks) must be avoided. Although such evacuated glass flasks appear highly suitable because they hold their temperature well, they are dangerous; at $-196°C$ they are very fragile and can, and do, explode without warning, causing glass fragments to spread explosively in all directions.

24. Viruses can survive in and be transmitted by LN *(43)*. Tanks can also harbor viable fungi and bacteria. There are now products (e.g., straws, straw loaders, and straw sealers as supplied by IMV, France) which aim to minimize cross-contamination between specimens.

25. Cryotubes come in many types. We use 1-mL, star-shaped foot, conical base, internal-thread tubes with a frosted area on which to write (e.g., Nunc or Greiner). The advantage of the star-shaped base is that it can be opened and closed with one hand when placed in a tightly fitting polystyrene rack (or in a tailor-made tube holder, Nunc). With the conical base, it is easy to see the tissue even during thawing. In some cases, tubes with an internal thread are more uniform in width than those with an external thread. This makes them seat better on canes and gives more uniform contact with the freezing machine (in those machines in which the bore size is fixed). Do not use tubes not specifically designed for use in LN (*see* **Notes 18–24**).

26. Always perform a test run for each strain. It is possible that your specific strain cannot be frozen by this particular method. In this case, try an alternative cryoprotectant (such as ethylene glycol) for that strain.

## Acknowledgments

We wish to thank Emily Spilseth for evaluating the manuscript and helping with the photography. This chapter was prepared while in receipt of funding from Monash Research Fund, Monash IVF, and the Australian NH and MRC.

# References

1. Critser, J. K. and Mobraaten, L. E. (2000) Cryopreservation of murine spermatozoa. *ILAR J.* **41,** 197–206.
2. Sztein, J. M., Farley, J. S., Young, A. F., and Mobraaten, L. E. (1997) Motility of cryopreserved mouse spermatozoa affected by temperature of collection and rate of thawing. *Cryobiology* **35,** 46–52.
3. Thornton, C. E., Brown, S. D., and Glenister, P. H. (1999) Large numbers of mice established by in vitro fertilization with cryopreserved spermatozoa: implications and applications for genetic resource banks, mutagenesis screens, and mouse backcrosses. *Mamm. Genome* **10,** 987–992.
4. Critser, J. K. and Russell, R. J. (2000) Genome resource banking of laboratory animal models. *ILAR J.* **41,** 183–186.
5. Karow, A. M. and Critser, J. K., eds. (1997) *Reproductive tissue banking: Scientific principles.* Academic, San Diego, CA.
6. Wood, E. C., Shaw, J. M., and Trounson, A. O. (1997) Cryopreservation of ovarian tissue Potential "reproductive insurance" for women at risk of early ovarian failure. *Med. J. Aust.* **166,** 366–369.
7. Brem, G., Buanack, E., Muller, M., and Winnacker, E. L. (1990) Transgenic offspring by transcaryotic implantation of transgenic ovaries into normal mice. *Mol. Reprod. Dev.* **25,** 42–44.
8. Cecim, M., Kerr, J., and Bartke, A. (1995) Infertility in transgenic mice overexpressing the bovine growth hormone gene: Luteal failure secondary to prolactin deficiency. *Biol. Reprod.* **52,** 1162–1166.
9. Harari, D., Bernard, O., and Shaw, J. (1997) Rescue of an infertile transgenic line by ovarian transplantation. *Transgenics* **2,** 143–151.
10. Russell, W. L. and Gower, J. S. (1950) Offspring from transplanted ovaries of fetal mice homozygous for a lethal gene (sp) that kills before birth. *Genetics* **35,** 133.
11. Shaw, J. M., Dawson, K. J., and Trounson, A. O. (1997) A critical evaluation of ovarian tissue cryopreservation and grafting as a strategy for preserving the human female germline. *Reprod. Med. Review.* **6,** 163–183.
12. Cox, S.-L., Jenkin, G., and Shaw, J. (1996) Transplantation of cryopreserved fetal ovarian tissue to adult recipients. *J. Reprod. Fert.* **107,** 315–322.
13. Green, S. H., Smith, A. U., and Zuckerman, S. (1956) The numbers of oocytes in ovarian autografts after freezing and thawing. *J. Endocrinol.* **13,** 330.
14. Greenwald, G. S. and Roy, S. K. (1994) Follicular development and its control, in *The Physiology of Reproduction,* 2nd ed. (Knobil, E. and Neill, J. D., eds.), Raven Press, New York, NY, pp. 629–724.
15. Gunasena, K. T., Villines, P. M., Critser, E. S., and Critser, J. K. (1997) Live births after autologous transplant of cryopreserved mouse ovaries. *Human Reprod.* **12,** 101–106.
16. Jones, E. C. and Krohn, P. L. (1960) Orthotopic ovarian transplantation in mice. *J. Endocrin.* **20,** 135–146.

17. Krohn, P. L. (1977) Transplantation of the ovary, in *The Ovary, Vol. II, Physiology*, 2nd ed. (Zuckerman, S. and Weir, B. J., eds.), Academic Press, New York, NY, pp. 101–128.
18. Parrott, D. M. V. (1960) The fertility of mice with orthotopic ovarian grafts derived from frozen tissue. *J. Reprod. Fertil.* **1**, 230–241.
19. Deanesly, R. (1957) Egg survival in immature rat ovaries grafted after freezing and thawing. *Proc. R. Soc. Lond. B. Biol. Sci.* **147**, 412.
20. Parkes, A. S. and Smith, A. U. (1953) Regeneration of rat ovarian tissue grafted after exposure to low temperatures. *Proc. R. Soc. Lond. B. Biol. Sci.* **140**, 455–470.
21. Shaw, J. M., Oranratnachai, A., and Trounson, A. O. (2000) Fundamental cryobiology of mammalian oocytes and ovarian tissues. *Theriogenology* **53**, 59–72.
22. Shaw, J. M., Oranratnachai, A., and Trounson, A. O. (2000) Cryopreservation of oocytes and embryos, in *Handbook of in vitro fertilization*, 2nd ed. (Trounson, A. O. and Gardner, D., eds.), CRC, Boca Raton, FL, pp. 373–412.
23. Rall, W. F., Schmidt, P. M., Lin, X., Brown, S. S., Ward, A. C., and Hansen, C. T. (2000) Factors affecting the efficiency of embryo cryopreservation and rederivation of rat and mouse models. *ILAR J.* **41**, 221–227.
24. Shaw, J. M., Wood, E. C., and Trounson, A. O. (2000) Transplantation and cryopreservation of ovarian tissue, in *Handbook of in vitro fertilization*, 2nd ed. (Trounson, A. O. and Gardner, D., eds.), CRC, Boca Raton, FL, pp. 413–430.
25. Agca, Y. (2000) Cryopreservation of murine oocyte and ovarian tissue. *ILAR J.* **41**, 207–220.
26. Sztein, J., Sweet, H., Farley, J., and Mobraaten, L. (1998) Cryopreservation and orthotopic transplantation of mouse ovaries: new approach in gamete banking. *Biol. Reprod.* **58**, 1071–1074.
27. Sztein, J. M., McGregor, T. E., Bedigian, H. J., and Mobraaten, L. E. (1999) Transgenic mouse strain rescue by frozen ovaries. *Lab. Anim. Sci.* **49**, 99–100.
28. Sztein, J. M., O'Brien, M. J., Farley, J. S., Mobraaten, L. E., and Eppig, J. J. (2000) Rescue of oocytes from antral follicles of cryopreserved mouse ovaries: competence to undergo maturation, embryogenesis, and development to term. *Hum. Reprod.* **15**, 567–571.
29. Mazur, P. (1990) Equilibrium, quasi-equilibrium, and nonequilibrium freezing of mammalian embryos. *Cell Biophys.* **17**, 53–92.
30. Whittingham, D. G., Lyon, M. F., and Glenister, P. H. (1977) Re-establishment of breeding stocks of mutant and inbred strains of mice from embryos stored at –196°C for prolonged periods. *Genet. Res.* **30**, 287–299.
31. Whittingham, D. G., Leibo, S. P., and Mazur, P. (1972). Survival of mouse embryos frozen to –196°C and –269°C. *Science* **178**, 411–414.
32. Gunasena, K. T., Lakey, J. R. T., Villines, P. M., Critser, E. S., and Critser, J. K. (1997) Allogeneic and xenogeneic transplantation of cryopreserved ovarian tissue to athymic mice. *Biol. Reprod.* **57**, 226–231.
33. Shaw, J. M., Cox, S. L., Trounson, A. O., and Jenkin, G. (2000) Evaluation of the long-term function of cryopreserved ovarian grafts in the mouse, implications for human applications. *Mol. Cell. Endocrinol.* **161**, 103–110.

34. Candy, C. J., Wood, M. J., and Whittingham, D. G. (2000) Restoration of a normal reproductive lifespan after grafting of cryopreserved mouse ovaries. *Hum. Reprod.* **15,** 1300–1304.
35. Vom Saal, F., Finch, C. E., and Nelson, J. F. (1994) Natural history and mechanisms of reproductive ageing in humans, laboratory rodents and other selected vertebrates, in *The Physiology of Reproduction* (Knobil, E. and Neill, J. D., eds.), Raven Press, New York, NY, pp. 1213–1315.
36. Shaw, J. M., Bowles, J., Koopman, P., Wood, C., and Trounson, A. O. (1996) Fresh and cryopreserved ovarian tissue from donors with lymphoma, transmit the cancer to graft recipients. *Hum. Reprod.* **11,** 1668–1673.
37. Cleary, M. (1999) Cryopreservation and transplantation of wombat ovarian tissue—as strategies for the preservation of endangered species—Honours thesis, Department of Physiology, Monash University, Clayton, Victoria, Australia.
38. Snow, M. (1999) The cryopreservation and transplantation of ovarian tissue—as strategies for the preservation of endangered species. Honours thesis, Department of Physiology, Monash University, Clayton, Victoria, Australia.
39. Eppig, J. J., Hosoe, M., O'Brien, M. J., Pendola, F. M., Requena, A., and Watanabe, S. (2000) Conditions that affect acquisition of developmental competence by mouse oocytes in vitro: FSH, insulin, glucose and ascorbic acid. *Mol. Cell. Endocrinol.* **163,** 109–116.
40. Eppig, J. J. and O'Brien, M. J. (1998) Comparison of preimplantation developmental competence after mouse oocyte growth and development in vitro and in vivo. *Theriogenology* **49,** 415–422.
41. Liu, J., Van Der Elst, J., Van Den Broecke, R., Dumortier, F., and Dhont, M. (2000) Maturation of mouse primordial follicles by combination of grafting and in vitro culture. *Biol. Reprod.* **62,** 1218–1223.
42. Dissen, G. A., Lara, H. E., Fahrenbach, W. H., Costa, M. E., and Ojeda, S. R. (1994) Immature rat ovaries become revascularized rapidly after autotransplantation and show a gonadotrophin-dependent increase in angiogenic factor gene expression. *Endocrinol.* **134,** 1146–1154.
43. Tedder, R. S., Zukerman, M. A., Goldstone, A. H., Hawkins, A. E., Fielding, A., Briggs, E. M., et al. (1995) Hepatitis transmission from contaminated cryopreservation tank. *Lancet* **346,** 137–140.
44. Candy, C. J., Wood, M. J., and Whittingham, D. G. (1997) Effect of cryoprotectants on the survival of follicles in frozen mouse ovaries. *J. Reprod. Fert.* **110,** 11–19.
45. Carroll, J. and Gosden, R. G. (1993) Transplantation of frozen-thawed mouse primordial follicles. *Hum. Reprod.* **8,** 1163–1167.
46. Harp, R., Leibach, J., Black, J., Keldahl, C., and Karrow, A. (1994) Cryopreservation of murine ovarian tissue. *Cryobiology* **31,** 336–343.

# VII

## TRANSGENESIS IN THE RAT

# 12

# Transgenesis in the Rat

## Linda J. Mullins, Gillian Brooker, and John J. Mullins

## 1. Introduction

The first transgenic experiment was carried out more than 15 yr ago, and transgenesis is now a routine technique for the study of gene and cell function, and the development of animal models of disease. The mouse remains the species of choice for the vast majority of transgenic applications, reflecting the cost-effectiveness of using a small species of high fecundity and rapid generation time. An important factor is the establishment of embryonic stem cells from this species, allowing the development and exploitation of gene "knockout" technology.

For certain applications and research areas, however, the rat may be a more appropriate choice. Historically, the rat has been widely used in the fields of cardiovascular disease, aging, infectious diseases, autoimmunity, transplantation biology, cancer-risk assessment, industrial toxicology, pharmacology, behavioral studies, and neurobiology. There is therefore a wealth of background knowledge against which transgenic studies can be interpreted. Additionally, the size of the rat makes it more amenable to chronic studies, in which, e.g., sequential sampling of blood of sufficient quantities for multiple analyses may be necessary. The rat is also more suitable for microsurgery, tissue and organ sampling, tissue transplantation, and organ perfusion, although continuous technical refinements and miniaturization mean that such analyses may become equally applicable to the mouse in the future.

Another consideration is the availability of inbred and congenic rat strains, and specific strains such as the spontaneously hypertensive rat, which provide a wide variety of genetic backgrounds for transgenic experiments. Microsatellite mapping oligonucleotides are also available at a sufficient map density to allow the identification of genetic modifier loci through appropriate genetic crosses.

From: *Methods in Molecular Biology, vol. 180: Transgenesis Techniques, 2nd ed.: Principles and Protocols*
Edited by: A. R. Clarke © Humana Press Inc., Totowa, NJ

In the following sections, we briefly outline aspects of animal husbandry that must be considered at the inception of a rat transgenic program.

## 1.1. General Considerations of Animal Husbandry

### 1.1.1. Space Requirements

To run a transgenic rat program of size equivalent to that typically required for the mouse, approximately four to five times more space is required. Animals are generally housed three to a cage (16 × 9 × 8 in. [UK Home Office regulations]) or six to a cage (19 × 12 × 8.5 in.), depending on their age and sex. We use racks that can hold 24 small or 12 large cages, and a 16 × 12 ft room, which can comfortably hold six racks, if adequately ventilated (with 15–20 air changes per hour) and maintained at a temperature of 18–22°C and humidity of 50–55%. For general considerations of animal handling, such as appropriate light–dark regimes, feeding, and breeding, the reader is referred to appropriate reference manuals *(1,2)*.

A significant breeding program is required to regularly maintain the number of young donor and recipient females (*see* **Subheading 1.1.4.**), and this might be beyond the capacity of small animal establishments. Also, considerable space is required for the establishment and maintenance of the multiple transgenic rat lines that may be generated for any given construct.

### 1.1.2. Time Consideration

The gestation time of the rat is typically 20–24 d, compared with 19–21 d for the mouse. The rat reaches sexual maturity at 8–12 wk, as opposed to 5–8 wk for the mouse. Therefore, it takes a correspondingly longer period of time to derive a transgenic rat line, especially if the founder is chimeric, or if problems in breeding proficiency are encountered. This time constraint may be circumvented by the judicious superovulation and mating of young transgenic female rats, followed by embryo transfer, but this cannot be considered for valuable founder animals.

### 1.1.3. Choice of Strain

The choice of strain used for superovulation has a major bearing on the number of recoverable fertilized eggs. Generally, outbred strains such as Sprague-Dawley or F1 hybrids between two inbred strains give a higher yield of eggs than inbred strains. Another consideration is the breeding performance of the animals, since inbred strains generally produce reduced litter sizes and are less capable mothers. The choice of strain may be dictated by constraints of the experimental design, such as distinction of the exogenous transgene expression from its endogenous counterpart. Parameters such as the light/dark regime, timings of drug administration, mating, egg collection, and microinjection may need to be optimized for any new strain being considered as a transgene host.

Note that the response of one species to a gene construct may vary significantly from that of another *(3,4)*. The genetic background of the recipient strain may affect expression of the transgene or its resultant phenotype in unexpected ways. It is also noteworthy that different sources of Sprague-Dawley rats have been shown to respond differently in the presence of an identical transgene *(5)*.

## 1.1.4. Egg Production and Transfer

For a typical day of microinjection (a single operator and microinjection rig), five females should be superovulated and individually paired with five stud males on the day prior to egg recovery. If all five females are plugged, one can recover up to 200 eggs, not all of which will be fertilized. Following microinjection, the eggs are transferred to recipient females. To ensure that enough recipient females are available (one requires between four and six plugged females, to accommodate the transfer of microinjected eggs), a minimum of 12 (and up to 20) vasectomized males should be paired with two females on the day prior to transfer. To reduce the number of pairings, one can check that the females are in estrus, but care needs to be taken to avoid sending females into pseudopregnancy. Alternatively, one can synchronize estrus in the females by injecting them with luteinizing hormone–releasing hormone (LHRH) (*see* **Subheading 3.3.1.**).

If more than one rig is used, or microinjection is carried out several days a week, then the logistics of egg production and transfer are considerable, and economic and space considerations may justify or necessitate the use of commercial breeders for the production and supply of donor and recipient animals. In this case, however, the quality and exact age of the animals must be guaranteed by the supplier.

## 1.1.5. Animal Health

It is preferable that the rat colony be free of any infection before a transgenic program is undertaken, and specific pathogen-free (SPF) conditions are desirable. In addition, the rat seems to be particularly sensitive to environmental stress such as adverse humidity and overcrowding. If recipient females are stressed in any way, they may neglect their pups or even resort to cannibalism, with the mother eating some or all of her litter. If this becomes a serious problem, one should consider transferring the embryos to a recipient female from a robust strain such as Sprague-Dawley, irrespective of the genetic origin of the zygotes. It is conceivable that this approach may have repercussions owing to maternal influences *in utero*, but it ensures that valuable transgenic founders are not lost. The fostering of pups at birth (or within 1 to 2 d of age) to an experienced mother is another option, but success rate depends largely on the skill of the animal care staff and the quality of the animal husbandry.

### 1.1.6. Animal Identification

It is advisable to use a reliable identification system, such as ear tagging, with a backup of ear punching. The cost and time invested in the generation of transgenic animals is significant, and problems of identification may result in repeat testing, costly delays, and possibly even loss of founder animals through mistaken identity. For the most valuable animals, it may be prudent to invest in a system such as microchip tagging (R. S. Biotech, UK).

## 1.2. Superovulation

In this and subsequent sections, we have assumed that the reader is conversant with the general principles of transgenic production. There are several subtle changes that have to be made to the standard protocols for transgenic mouse production, which are covered in depth earlier in this book (*see* Chapter 3) and elsewhere (*6*). As with the mouse, embryo quality, DNA preparation, and optimal conditions for microinjection are crucial to high transgenic efficiency. In subsequent sections, we highlight the technical details in the generation of transgenic rats that are specific to this species.

Since eggs are recovered in relatively low numbers, approx 8–15, following natural matings, it is beneficial to use superovulation of immature females. The exact timing of superovulation is critical and has to be optimized for each strain. To achieve this, females should be monitored from about 25 d of age and checked for both vaginal opening and weight. In the production of donor females, it is advantageous to cull male offspring in the litter at birth or shortly after, leaving the mother with fewer pups (all female) that will therefore be larger at weaning. It is also advisable to delay weaning for a few days (up to 28 d) to ensure that the females are as large as possible. Ideally, one should superovulate a female at the maximum weight prior to vaginal opening. In inbred rats such as Fischer F344, the vagina opens between 33 and 35 days, at a weight of 80–90 g. Once the average age of vaginal opening has been ascertained, the superovulation protocol should be implemented 3 d before. We select the best females at 30 d old for superovulation.

Classically, ovulation of immature rodents was achieved using a mixture of the gonadotropic hormones, follicle-stimulating hormone (FSH), and luteinizing hormone, or, alternatively, pregnant mare's serum gonadotropin (PMSG). The latter has been used successfully by several groups to generate transgenic rats (*7*), despite evidence suggesting that PMSG induces a variable ovulatory response with an increased number of abnormal ova (*8,9*).

In our experience, superovulation in the rat is best achieved by the procedure of Armstrong and Opavsky *(9)*, using a highly purified FSH preparation.

### 1.3. Microinjection

The time taken for pronuclei to become visible, following removal from the oviduct, varies among different strains of rat. Rat pronuclei are less regular and uniform and take several hours longer to develop than those of the mouse; they become visible within 1–4 h of removal from the donor. Once the pronuclei are visible, the eggs are ready for microinjection of DNA.

### 1.4. Rat Oviduct Transfer

It has been reported that reimplantation of two-cell embryos into 1-d pseudo-pregnant females gives lower pregnancy rates *(10)*; thus, it is preferable to transfer the majority of microinjected embryos on the day of injection (day +1). Whether at the one- or two-cell stage, embryos are reimplanted into the oviduct of anesthetized pseudopregnant females. These are experienced mothers that have been mated the previous night with vasectomized or genetically infertile males, with copulation confirmed by the presence of a vaginal plug, which may be deep-seated. To reduce the numbers of pairings required to produce enough pseudopregnant females for the day of transfer, it is advantageous to synchronize estrus through the administration of LHRH *(11)*.

### 1.5. Animal Screening

At weaning, progeny ($G_0$ animals) can be tested for incorporation of the transgene, by polymerase chain reaction (PCR) *(12)* or Southern blot analysis *(13)* of genomic DNA isolated from tail biopsies or by PCR analysis of whole blood *(14)*. We have noted a higher degree of mosaicism and multiple insertion sites in transgenic rat founders than has been our experience with mice. Multiple transgene insertions can be segregated through breeding and individual sublines analyzed.

The selection of appropriate controls deserves careful consideration. The derivation of a transgenic line from an outbred strain such as Sprague-Dawley will, by definition, create a new "substrain." Unless the transgenic animals have been produced on a pure inbred background or careful backcrossing has been performed, the optimal controls would be nontransgenic littermates.

### 1.6. Cryopreservation of Transgenic Rat Lines

Cryopreservation of embryos is a crucial part of any transgenic breeding project. It provides the ultimate assurance that a transgenic line is preserved against loss for the foreseeable future, awaiting recovery at any time. Cryopreservation ensures that microbial contamination, loss by fire or accident, shift in genetic background, or alteration of transgenic expression will not occur. The benefits are particularly apparent for lines that have been extensively characterized.

Success varies between 30 and 90%, depending on the method used. There are two strategies for the cryopreservation of embryos: The first involves stepwise freezing of the embryos; and the second involves instant freezing of the embryos in liquid nitrogen, in the presence of a range of cryoprotective additives.

Cryopreservation also can be used to store surplus eggs following superovulation, for microinjection at a later date. This has been demonstrated for mice, with success rates approaching those for fresh embryo microinjections *(15,16)*.

In this section, we briefly outline techniques for the cryopreservation of rat embryos, either for the potential storage of microinjectable eggs or long-term storage of important transgenic lines.

## 2. Materials

### 2.1. Superovulation

#### 2.1.1. Preparation of Minipump

1. Alzet model 1003D microosmotic pump (Charles River, Kent, UK).
2. Highly purified FSH preparation from either the pig pituitary (Folltropin; Vetrepharm, Ontario, Canada) or the sheep (Ovagen; Synergy, Melksham, Wiltshire, UK).
3. 0.9% Saline.

#### 2.1.2. Minipump Implantation

1. Anesthetic (halothane; Zeneca, UK).
2. Small animal shaver.
3. 70% Ethanol.
4. Large-toothed forceps.
5. Blunt-ended forceps.
6. Surgical staples.
7. Surgical spray (Nobecutane; Astra).

#### 2.1.3. Administration of Human Chorionic Gonadotropin

1. Human chorionic gonadotropin (hCG) powder (5000 IU) (Sigma, St. Louis, MO).
2. 0.9% Saline.
3. Syringe.
4. Damp cotton buds.
5. Proven stud male rat.

#### 2.1.4. Egg Removal

All media are prepared exactly as for mouse microinjection protocols *(6)*.

1. Smooth speculum.
2. Dissection kit.
3. Petri dishes.
4. Fine watchmaker's forceps.
5. Hyaluronidase (500 U/mg) (Calbiochem) made up in 0.9% saline to give a concentration of 100 U/μL.
6. M16 culture medium: 94.66 m$M$ NaCl, 4.78 m$M$ KCl, 1.71 m$M$ $CaCl_2$, 1.19 m$M$ $KH_2PO_4$, 1.19 m$M$ $MgSO_4$, 25.00 m$M$ $NaHCO_3$, 23.28 m$M$ sodium lactate, 0.33 m$M$ sodium pyruvate, 5.56 m$M$ glucose, 100 U/mL of penicillin G, 50 μg/mL of streptomycin sulfate, 10 μg/mL of phenol red. Adjust the pH to 7.0 by bubbling $CO_2$ gas through the medium, which should be straw colored (yellow-orange). Filter sterilize the medium and store in 10-mL aliquots. Add bovine serum albumin (BSA) (final concentration of 5 mg/mL) and refilter the solution just before use.
7. M2 culture medium: As per M16 but with 4.15 m$M$ $NaHCO_3$ and 20.85 m$M$ HEPES. Adjust the pH to 7.0 with NaOH.

## 2.2. Microinjection

1. DNA solution at a concentration of approx 1 μg/mL.
2. Glass tubing for microinjection pipet (od of 1 mm, id of 0.58 mm, filament diameter of 0.133 mm) (e.g., Hilgenberg, Germany).
3. Mechanical micropipet puller (e.g., Sutter model P-87).
4. Glass tubing for holding pipets (e.g., New Brunswick; 1 mm od).
5. Light paraffin oil.
6. Micrometer syringe.
7. Microinjection rig with right- and left-hand micromanipulators.
8. Inverted microscope with ×40 and ×10 objective.
9. Pneumatic pump for delivery of DNA into nucleus.
10. Base plate.
11. Isolation table.
12. Microforge.

## 2.3. Rat Oviduct Transfer

The operating area should be prepared as sterile as possible.

1. LHRH (cat no. L-4513; Sigma) made up in phosphate-buffered saline (PBS) (pH 7.2–7.4) at a final concentration of 200 μg/mL.
2. Small animal shavers.
3. 70% Ethanol.
4. Sharp, blunt, strong scissors, to open the outside skin.
5. Long-toothed forceps, to hold outside skin.
6. Sharp pointed scissors, to separate the muscle from the skin and to make the incision in the muscle wall.
7. Fine-pointed forceps, to hold the muscle wall.

8. Long, blunt, fine forceps, to grasp the fat pad protecting the ovary.
9. Sterile moist gauze (0.9% saline), on which to rest the ovary and fat pad.
10. Small clip, to hold the fat pad in place.
11. Syringe loaded with adrenaline solution (Suprarenin; Hoechst AG).
12. Two pairs of microfine watchmaker's size 5 forceps.
13. Straight microscissors.
14. Cotton swabs.
15. Size 4 mersilk suture.
16. Wound clips.
17. Wound spray.
18. Small, heated operation pad.
19. Good surgery microscope magnification ($10 \times 10$).
20. Dissection microscope.
21. Mouth pipet, for loading microinjected eggs into transfer pipet.
22. Penibritan, made up in sterile water and administered at a dose of 5 mg/kg of body wt after surgery.
23. Injection anesthetic.
24. Transfer pipet.
25. Balance.
26. Surgical gown, face mask, hat, and gloves.
27. Eye wash.

## 2.4. Cryopreservation of Transgenic Rat Lines

1. Cryopreservation straws.
2. Wash medium (e.g., 100% sheep serum; Sigma).
3. Programmable embryo-freezing machine.
4. Storage cartridges, precooled in liquid nitrogen.
5. Tongs, precooled in liquid nitrogen.
6. Liquid nitrogen chamber.
7. Cryoprotective agents (e.g., ethylene glycol; Sigma).

# 3. Methods

## 3.1. Superovulation

### 3.1.1. Preparation of Minipump

The minipump is prepared at 8:00 AM on d −2. Since purified FSH has a short half-life, it is administered as a constant infusion using an Alzet model 1003D microosmotic pump (*see* **Note 1**).

1. Dissolve 18 mg of lyophilized hormone (Folltropin or Ovagen) in 2.1 mL of 0.9% saline.
2. Introduce 100 μL of hormone solution into the minipump, using the filler needle supplied and taking care not to introduce air bubbles.

3. Once filled, allow the minipump to equilibrate for 1 h at 37°C (or overnight at 4°C, if setting up on d –3), immersed to the neck in isotonic saline. This promotes the osmotic pumping action, prior to implantation.

### 3.1.2. Implantation of Minipump

The minipump is implanted at 9:00 AM on d –2 (*see* **Note 2**).

1. Lightly anesthetize the rat with halothane.
2. Quickly shave a small area on the back of the neck and swab with alcohol.
3. Pinch the pelt with toothed forceps and make a small incision.
4. Insert blunt-ended forceps into the incision and gently work down to the left hind leg (this separates the muscle from the pelt to make a pocket). Remove the forceps.
5. Insert the minipump into the pocket such that the top of the minipump is in an anterior position (this prevents blockage while the pump is being maneuvered into position).
6. Close the wound with sterile surgical staples and spray the area with Nobecutane (Astra) (*see* **Notes 3** and **4**).

### 3.1.3. Administration of hCG

hCG is administered at 11:00 AM on d 0. Fifty hours after implantation of the minipump, the rats receive an ip injection of hCG.

1. Reconstitute 5000 IU of hCG powder (Sigma) in 50 mL of 0.9% saline.
2. Inject intraperitoneally 0.3 mL (30 IU) of hCG into the female rat.
3. The vagina should be open at this time, but to ensure successful mating, fully open the vagina using a damp cotton bud.
4. Pair the rat with a proven stud male overnight.

### 3.1.4. Egg Removal

Egg removed is performed at 9:00 AM on d +1.

1. Check the superovulated females for a waxy vaginal mating plug using a smooth speculum.
2. Sacrifice plugged females and cleanly dissect each oviduct away from the fat pad, ovary, and oviduct.
3. Transfer the oviducts to a Petri dish containing M2 medium. Release fertilized ova from the swollen ampullae of the oviduct under a dissecting microscope, by gently tearing the ampullae with fine watchmaker's forceps and squeezing the contents out into the dish (*see* **Note 5**).
4. Remove the surrounding cumulus cells by placing the eggs in 2.5 mL of M2 medium containing 100 µL of hyaluronidase for 2 min (*see* **Note 6**). Their removal can be assisted by gently pipetting through a finely pulled Pasteur pipet.
5. Rinse the eggs in two changes of M2 medium, and maintain in M16 culture medium at 37°C under 5% $CO_2$. In these conditions, the eggs will develop to the two-cell stage if cultured overnight.

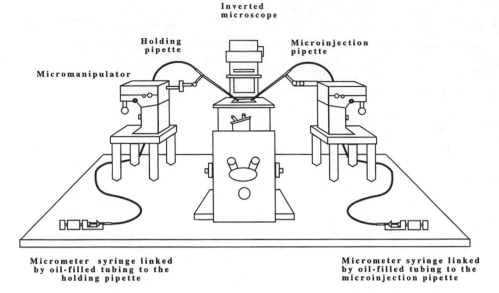

Fig. 1. Schematic diagram of the microinjection rig.

## 3.2. Microinjection

We have assumed that the reader has a microinjection rig (**Fig. 1**) and is conversant with its use. A variety of methods are available for setting up the embryos for microinjection, depending on the preference of the operator. These include a concave microscope slide, a flat slide, and a hanging droplet.

1. Place the embryos within a 50-μL drop of M2 medium covered with liquid paraffin (Sigma) (*see* **Note 7**). The embryos will remain viable for at least 20 min at room temperature. Typically, 20–25 fertilized eggs are processed at a time.
2. One by one, pick up the eggs by applying gentle suction on a holding pipet. Suction to the holding pipet is applied through oil-filled tubing via a micrometer-controlled syringe (*see* **Notes 8** and **9**). The movement of both holding and microinjection pipets is controlled by micromanipulators that are either hand or pneumatically controlled.
3. Gently but firmly manipulate the microinjection needle, which has an internal tip diameter of <1 μm and is filled with DNA at a concentration of approx 1 μg/mL, until both the zona pellucida and the nuclear membrane of one of the pronuclei have been pierced (*see* **Note 10**). **For the rat egg, the tip of the microinjection needle must be pushed through the pronucleus and then drawn back into it in order to achieve penetration of the membrane** (*see* **Note 11**).
4. Inject DNA into the pronucleus using a pneumatic pump or a hand-operated syringe/micrometer. Successful injection is indicated by swelling of the pronucleus prior to removing the pipet tip (*see* **Note 12**).

5. Return eggs that have been successfully injected to the incubator; these may be left to develop to the two-cell stage overnight. This allows one to check that the eggs are still viable following injection (a proportion of the injected eggs does not survive the ordeal), and that the DNA solution was not toxic (*see* **Note 13**).

## 3.3. Rat Oviduct Transfer

### 3.3.1. Synchronization of Rat Estrus

Synchronization of estrus takes place on d −4. Through synchronization of estrus, sufficient numbers of pseudopregnant females should be generated using eight vasectomized males and eight LHRH-treated females.

1. Take female rats (at least 7 wk old) and inject 40 µg of LHRH intraperitoneally (0.2 mL of prepared solution). This stops the current estrous cycle.
2. On the evening of d 0 the rats will be in estrus again. Pair them singly with vasectomized males.
3. Check for presence of a plug next morning (*see* **Note 14**).

### 3.3.2. Transfer of Microinjected Eggs into Oviduct

Microinjected eggs are transferred into the oviduct on d +1.

1. Weigh the rat and administer Vetalar (100 mg/mL) (Pharmacia and Upjohn) anesthetic (1.2 mL/kg of body wt) and Rompun 2% (Bayer) analgesic (0.2 mL/kg of body wt) by ip injection. Then place the anesthetized rat on a heated operation pad.
2. Under sterile surgical conditions, shave a 3-in. square in the middle of the back, and wipe with 70% ethanol.
3. Make a 1.5-in. incision in the skin from the center of the rat's back (just below the ribs), and gently separate the skin from the muscle (1-in. square). Repeat this procedure on the other side of the rat.
4. Make a 1-in. incision in the muscle wall, and locate the large fat pad beneath it.
5. Gently pull the fat pad through the opening, and secure on a moist gauze (soaked in 0.9% saline).
6. Under the operating microscope, locate the oviduct beneath the highly vascularized bursa membrane.
7. Make a fine incision in the bursa, and locate the infundibulum (opening to the oviduct) (*see* **Note 15**),
8. Meanwhile, load the transfer pipet (*see* **Note 16**) with microinjected eggs (M2 medium, air bubble, M2, air bubble, microinjected embryos, air bubble, M2) (*see* **Note 17**).
9. Place the tip of the transfer pipet gently inside the infundibulum, and hold in place with watchmaker's forceps, while the eggs are transferred. If transfer is successful, you should be able to see the air bubbles move around the oviduct.
10. Remove the pipet, taking care not to disturb the loaded oviduct.

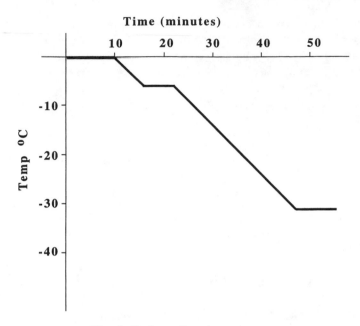

Fig. 2. Embryo-freezing program.

11. Gently place the fat pad back into the body cavity, again taking care not to displace the air bubbles/eggs from the oviduct.
12. Securely stitch the wound and repeat the procedure on the other side (*see* **Notes 18** and **19**).
13. Allow the rat to recover from the anesthetic on a heated pad, and maintain under optimal husbandry conditions. Application of eye drops will prevent eyes from becoming dry. Administration of antibiotic (Penibritan) is also beneficial.

### 3.4. Cryopreservation of Transgenic Rat Lines

### 3.4.1. Stepwise Freezing

By slowly reducing the temperature to −6°C, microcrystals are allowed to seed in the cells of the embryo. These cause a minimum amount of damage to the internal structures of the cells and prevent damage by large crystal formation.

1. Flush embryos from the oviduct using wash medium. We use 100% sheep serum.
2. Collect two-cell embryos and load into a cryopreservation straw containing wash medium + 10% ethylene glycol.
3. Place the cryostraws in the embryo-freezing machine (Haake, Germany) and freeze using the following program (shown schematically in **Fig. 2**):
   a. Step 1: Hold 0°C for 10 min.

b. Step 2: Drop –1°C every minute to –6°C.

c. Step 3: Hold at –6°C for 5 min to seed the straws.

d. Step 4: Drop –1°C every minute until the temperature reaches –32°C.

e. Step 5: Hold at –32°C for 10 min.

4. Remove the cryostraws from the embryo-freezing machine with precooled tongs and drop into prelabeled, precooled cartridges that are stored in a liquid nitrogen chamber.

### 3.4.2. Defrosting

1. Rapidly defrost embryos at room temperature.
2. Dilute the contents of the cryostraw every 30 s with two drops of the same wash medium that was used to flush out the embryos (100% sheep serum).
3. Collect viable embryos and transfer into a pseudopregnant female. We regularly attain a 60% recovery rate (*see* **Note 20**).

### 3.4.3. Vitrification

The embryos are frozen rapidly in liquid nitrogen, in the presence of a range of cryoprotective additives. By the process of vitrification, the cytoplasm is instantly turned to glass, again preventing damage by large crystal formation.

Sato et al. *(17)* demonstrated the successful vitrification and recovery of spontaneously hypertensive rat-stroke-prone (SHRSP) two-cell embryos and showed that cryopreservation did not affect phenotypic characteristics of the SHRSPs. For vitrification, they used solution containing 2.75 *M* dimethylsulfoxide, 2.75 *M* propylene glycol, and 1.0 *M* sucrose diluted in modified PBS containing 0.3% BSA. The embryos were exposed to the solution for 15 s before vitrifying in liquid nitrogen. Following rapid warming and transfer to recipient animals, 62% of the embryos were delivered normally.

## 4. Notes

1. It is advisable to test each batch of FSH hormone before buying in bulk, because there can be some variation in superovulation efficiency between batches owing to the relative purity of the preparations.
2. The timing of the minipump implantation is important. Typically, the light/dark cycle has the midpoint of the dark cycle at midnight, so the minipumps are introduced at approx 9:00 AM on d –2.
3. The whole procedure of minipump implantation should take no more than 2 min, with the rat recovering from between 1 and 3 min later. A sterile technique is advisable throughout the procedure.
4. The minipumps can be used more than once if they are carefully cleaned and refilled, making them more cost-effective, but the success rate can be as low as 50% on the second use.

5. Using the superovulation protocol outlined in **Subheading 3.1.**, between 30 and 110 fertilized eggs can be recovered from a single female, depending on the strain of rat (25–50 from an inbred rat and 70–100 from an outbred rat), with seasonal variation. The percentage of healthy eggs also varies with strain and superovulation protocol.

6. A high specific activity hyaluronidase enzyme (500 U/mg; Calbiochem) is essential for disaggregation of rat cumulus cells.

7. All media, paraffin, and so forth should be assessed for nontoxicity by testing the ability of noninjected embryos to divide into two-cell embryos following exposure and subsequent overnight culture.

8. It is imperative that the microinjection rig be vibration free. This can be achieved by using an isolation table.

9. We find that handheld manipulators are better for the microinjection of rat embryos.

10. Microinjection needles are prepared from borosilicate glass capillaries with an internal glass filament, which are drawn out mechanically using the Sutter Flame-Browning micropipet puller.

11. It has been found that the zona pellucida and pronuclear membranes are much more elastic in the rat egg than the mouse, making them more difficult to penetrate.

12. Care must be taken to ensure that the highly elastic nuclear membrane is punctured, and that the needle does not touch the nucleoli, causing blockage of the needle and damage to the egg.

13. DNA up to 100 kb is injected in TE buffer (10 m$M$ Tris, pH 7.4, and 0.1 m$M$ EDTA). DNA up to 150 kb is injected in TE containing 100 m$M$ NaCl (TEN). DNA >150 kb is injected in TEN buffer containing 30 μ$M$ spermine and 70 μ$M$ spermidine.

14. Unlike the mouse, the mature pseudopregnant female rat often does not have a visible plug. The plug may be deep-seated and only observed using a smooth speculum and miniature torch.

15. Rupture of the bursa membrane can lead to excessive bleeding, which may completely obscure the entrance to the infundibulum. This can be prevented by either cauterization or topical application of adrenaline (Suprarenin; Hoechst AG).

16. The transfer pipet is prepared from a Pasteur pipet by drawing the nose out in a Bunsen flame, and then scoring and breaking off the tip. The broken end of the transfer pipet is smoothed using a microforge, to leave an id of 80–120 μm.

17. Typically, a maximum of 30 embryos are transferred to the pseudopregnant female, 15 to each oviduct. Alternatively, up to 20 can be placed on one side. Transfer of this many embryos should result in a litter size of 8–10.

18. A sterile surgical technique is essential for high pregnancy rate, and results may be further improved by the administration of antibiotics to the recipient female, at the time of surgery. The pregnancy is then continued to term.

19. These procedures must be carried out under appropriate home office (or local equivalent) regulations (e.g., with complete project and personal licences). It is

essential to keep adequate records of all the procedures carried out, together with the number of animals used for the home office (or local equivalent) returns.

20. There are numerous variations of the basic protocols, both for stepwise freezing and vitrification, in the literature. Some claim success rates of up to 90% for recovery of embryos. Readers are advised to scan the literature before choosing the method of cryopreservation most appropriate for their needs.

## References

1. Poole, T. B., ed. (1989) *UFAW Handbook on Care and Management of Laboratory Animals*, 6th ed., Longman Group, Harlow, UK.
2. Kelly, P. J., Millican, K. G., and Organ, P. J., eds. (1988) *Principles of Animal Technology: Volume 1*, Institute of Animal Technology, UK.
3. Mullins, J. J., Peters, J., and Ganten, D. (1990) Fulminant hypertension in transgenic rats harbouring the mouse Ren-2 gene. *Nature* **344,** 541–544.
4. Hammer, R. E., Maika, S. D., Richardson, J. A., Tang, J.-P., and Taurog, J. D. (1990) Spontaneous inflammatory disease in transgenic rats expressing HLA-B27-associated human disorders. *Cell* **63,** 1099–1112.
5. Whitworth, C., Fleming, S., Kotelevtsev, Y., Manson, L., Brooker, G., Cumming, A., and Mullins, J. J. (1995) A genetic model of malignant phase hypertension in rats. *Kidney Int.* **47,** 529–535.
6. Hogan, B., Beddington, R., Costantini, F., and Lacy, E., eds. (1994) *Manipulating the Mouse Embryo: A Laboratory Manual*, 2nd ed., Cold Spring Harbor Laboratory Press, Cold Spring Harbor, NY.
7. Hochi, S., Ninomiya, T., Homma, M., and Yuki, A. (1990) Successful production of transgenic rats. *Anim. Biotechnol.* **1,** 175–184.
8. Young, W. Y., Yuen, B. H., and Moon, Y. S. (1987) Effects of superovulatory doses of pregnant mare serum gonadotropin on oocyte quality and ovulatory and steroid responses in rats. *Gamet. Res.* **16,** 109–120.
9. Armstrong, D. T. and Opavsky, M. A. (1988) Superovulation of immature rats by continuous infusion of follicle-stimulating hormone. *Biol. Reprod.* **39,** 511–518.
10. Charreau, B., Tesson, L., Soulillou, J.-P., Pourcel, C., and Anegon, I. (1996) Transgenesis in rats: technical aspects and models. *Transgen. Res.* **5,** 223–235.
11. Rouleau, A. M. J., Kovacs, P. R., Kunz, H. W., and Armstrong, D. T. (1993) Decontamination of rat embryos and transfer to specific pathogen free recipients for the production of a breeding colony. *Lab. Anim. Sci.* **43,** 611–615.
12. Mullis, K. B., Faloona, F., Scharf, S. J., Saiki, R. K., Horn, G. T., and Erlich, H. A. (1986) Specific enzymatic amplification of DNA in vitro: the polymerase chain reaction. *Cold Spring Harbor Symp. Quant. Biol.* **51,** 263–273.
13. Southern, E. M. (1975) Detection of specific sequences among DNA fragments separated by gel electrophoresis. *J. Mol. Biol.* **98,** 503–517.
14. Ivinson, A. J. and Taylor, G. R. (1991) Polymerase chain reaction in generic diagnosis, in *PCR—A Practical Approach* (McPherson, M. J., Quirke, P., and Taylor, G. R., eds.), Oxford University Press, NY, pp. 15–17.

15. Leibo, S. P., Demayo, F. J., and O'Malley, B. (1991) Production of transgenic mice from cryopreserved fertilized ova. *Mol. Reprod. Dev.* **30,** 313–319.
16. Tada, N., Sato, M., Kasai, K., and Ogawa, S. (1995) Production of transgenic mice by microinjection of DNA into vitrified pronuclear stage eggs. *Transgen. Res.* **4,** 208–211.
17. Sato, M., Yokokawa, K., Kasai, K., and Tada, N. (1996) Successful vitrification of stroke-prone spontaneously hypertensive and normal Wistar rat 2-cell embryos. *Lab. Anim.* **30,** 132–137.

# VIII

## TRANSGENESIS IN DOMESTIC SPECIES

# 13

## Generation of Transgenic Livestock by Pronuclear Injection

### A. John Clark

### 1. Introduction

The first transgenic livestock were reported in 1985 *(1)*. The techniques for producing these animals used pronuclear injection, which had been established previously in the mouse *(2)*. This technique involves the direct introduction of a few hundred copies of a DNA construct into one of the two pronuclei of the fertilized egg. The injected DNA concantenates, generating tandem arrays, which integrate at what are assumed to be random sites in the genome in a small proportion of injected eggs. In mice, typically 10–20% of the pups that are born will carry the foreign DNA, and the majority of these founders will transmit the new DNA sequence to the next generation *(3)*.

Generally speaking, the efficiency of transgenic livestock production by pronuclear injection is somewhat lower than that obtained in the mouse. An important feature of this method for germline manipulation is that the investigator has no control over the site of integration or the copy number of the foreign DNA sequences. This has important consequences for the efficiency and reliability of transgene expression. Transgene loci may be subjected to position effects, which determine the level of transgene expression. These are owing to the effects of the flanking chromosomal sequence *(4)*, or the number of copies of the transgene present in the array *(5)*, as well as the nature of the expressed sequence *(6)*.

Until recently pronuclear injection was the only proven route for the genetic modification in livestock. In mice, embryonic stem (ES) cells can be used as a means to accomplish precise genetic changes by gene targeting by homologous recombination. Despite considerable efforts, the isolation of ES cells from livestock has not been proved *(7)*. Recently, however, Schnieke et al. *(8)* have

From: *Methods in Molecular Biology, vol. 180: Transgenesis Techniques, 2nd ed.: Principles and Protocols*
Edited by: A. R. Clarke © Humana Press Inc., Totowa, NJ

described the generation of transgenic sheep using a combination of cell transfection and nuclear transfer from cultured cells *(9)*. This approach will revolutionize the production of transgenic livestock and has opened up the possibility of accomplishing precise changes by gene targeting *(10–12)*.

Overall, pronuclear injection is rather an inefficient and imprecise method for engineering the germline, and it is limited to the introduction of new genetic material. Despite these drawbacks, it has been achieved in all the major species of livestock. In this chapter, I briefly review the major uses of transgenic farm animals, compare and contrast the techniques used for the genetic modification of these species, and provide detailed technical protocols for the production of transgenic sheep by this method.

## 2. Applications of Transgenic Livestock

The costs of working with livestock, in both time and money, far exceed those of laboratory animals such as mice. Therefore, the use of this technology in these species has focused on quite specific practical applications, in either agriculture or medicine *(13)*.

### 2.1. Agriculture

First attempts to exploit transgenic technology in livestock were directed at the manipulation of conventional production traits such as growth and feed efficiency. These approaches involved the introduction of transgenes encoding growth hormone (GH) *(14)*, GH-releasing factors, or other genes affecting body composition *(15)*. Overall, these approaches have not proved very successful. Thus, although transgenic pigs carrying human GH genes did exhibit a slightly enhanced growth rate and reduced carcass fat composition, these animals suffered from widespread deleterious effects, including susceptibility to stress, lameness, and reduced fertility *(14)*. One of the main problems in these experiments was the failure to regulate expression precisely, which led to uncontrolled systemic levels of potent biologic molecules, not surprisingly with deleterious side effects. However, transgenic sheep carrying an insulin-like growth factor-1 whose expression was targeted specifically to the hair follicle exhibited significantly enhanced wool growth with no apparent side effects *(16)*.

Another early target for transgenic technology was the enhancement of animal health; economic losses from animal disease have been estimated to comprise 10–20% of production costs *(17)*. In mice, transfer of the Mx1 gene was shown to confer resistance to influenza A viruses *(18)*, and similar experiments in pigs did produce a number of transgenics, although no expression was demonstrated *(19)*. Transgenic approaches to inactivate viruses at the cellular level also have been attempted. Sheep are widely infected with a lentivirus called mardi visna, which is similar to the human immunodeficiency virus. A fusion gene composed of the visna virus long tandem repeat fused to the gene encoding the envelope protein was introduced into sheep, and all three lambs produced expressed the foreign protein *(20)*. It is hypothesized that the envelope protein will bind the endogenous visna receptors and interfere with virus uptake when progeny from these animals are challenged with a virulent visna isolate. Milk, particularly colostrum, is a rich source of IgA, which provides the initial immunologic barrier against many pathogens that invade the body at mucosal surfaces. Transgenic animals secreting virus-neutralizing antibodies into their milk during pregnancy and lactation could thus confer additional immunity to their progeny. Recently, transgenic mice expressing high levels of a neutralizing antibody to porcine transmissible gastroenteritis virus (which causes a mortality close to 100% in infected pigs) have been produced *(21)*, and this may provide a strategy for protecting enteric infections in newborn piglets.

## 2.2. Medicine

Although there has been some progress in modifying livestock for agricultural purposes, it is probably fair to say that many of the early expectations have not been realized. This is owing to a number of factors, not least the difficulty of modifying traits that are controlled by a number of genes and the failure to engineer the germline precisely to achieve tightly regulated and controlled levels of gene expression. In fact, applications that develop entirely new uses of livestock, particularly for human medicine, have led the way and, in many respects, continue to advance this technology. One of the first biomedical applications proposed was to use transgenic animals for the production of valuable human therapeutic proteins. The most well-developed approach is to target the expression of the protein of interest to the mammary gland and harvest the product from the milk *(22)*. Over the last 10 yr, a wide variety of human proteins have been produced in milk with this approach using promoter elements from milk protein genes to target expression *(23)*; in some cases, very high levels of production have been achieved *(24)*. More important, the mammary gland is capable of carrying out the complex posttranslational modification these proteins require for biologic activity or stability. At least two human proteins that have been produced by this route, $\alpha$1-antitrypsin and antithrombin III, are well advanced in clinical trials.

A second and rapidly developing use for genetically modified animals is xenotransplantation. This application is greatly driven by the worldwide lack of donor organs for transplant surgery. Since the pig shares a number of anatomic and physiologic features with humans, this has been the species of choice for this application. Normal pig tissues are immunologically incompatible for transplantation into humans and provoke a rapid, complement-based, hyperacute rejection response that destroys the foreign tissue. Several transgenic approaches are being developed to overcome this rejection response. One strategy involves the production of transgenic pigs expressing human regulators of complement activity such as decay accelerating factor (DAF) *(25)*. Transplantation of pig hearts expressing human DAF into cynomolgus monkeys resulted in a dramatic improvement in graft survival time *(26)*. Modification of the hyperacute rejection response will not overcome T-cell-based mechanisms of rejection in the longer term, and, therefore, some form of immunosuppression will be required. In addition, recent concerns have been voiced as to the safety of this procedure given that porcine retroviruses can jump species and replicate in human cells with the possibility of unpredictable zoonoses *(27)*. Nevertheless, the demand for organs is so great that limited clinical trials are likely to occur in the near future, which will stimulate further the effort in this area.

Finally, transgenic animal models are being used widely in biomedical research. At present, these studies are almost exclusively confined to mice. Transgenic pigs carrying a defective rhodopsin gene have been described, however, and these are being used in the investigation of retinitis pigmentosa *(28)*. The pig retina is similar in size and anatomy to that of humans and is a far superior model to the mouse. Thus, although the ease and precision of germline manipulation of the mouse will ensure its continued use in biomedical research, the relevance of many of the models generated is questionable. The additional effort of working with livestock may be well justified in areas such as cardiovascular and respiratory systems, because pig and sheep organs are more similar to their human counterparts.

## 3. Techniques for Generating Transgenic Livestock

Successful production of transgenic livestock requires the provision and manipulation of fertilized embryos for microinjection, followed by their transfer and survival in recipient animals. This involves modulating the reproductive physiology of these species to obtain relatively large numbers of embryos, maintaining and culturing these embryos in vitro, and transferring them to recipients, in addition to adapting the pronuclear injection procedure. In principle, these requirements are the same for all the species of livestock, but there are important differences when it comes to realizing these in practice.

### 3.1. Pigs

Pigs were one of the first species of domestic livestock to be generated successfully by pronuclear injection *(1)*. As for all mammals, genetic manipulation in the pig is highly dependent on modulating the physiology of reproduction so that large numbers of fertilized one- and two-cell eggs are available for injection. Prepubertal females are injected with pregnant mare's serum gonadotropin (PMSG), or mature cycling females are fed progesterone for 2 to

3 wk *(29)* after which PMSG is injected to stimulate follicular development. Ovulation is induced by injection with human chorionic gonadotropin (hCG) about 3 d after administration of PMSG *(30)*. Up to 40 eggs can be generated by using the PMSG/hCG regimens, but in order to reduce the frequency of immature or degenerate eggs, the regimen should be adjusted to yield 25–30 ova. The superovulated eggs may be fertilized either by natural matings a day after the onset of estrus or by artificial insemination *(31)*.

The fertilized eggs are recovered surgically, by flushing them from the reproductive tract, usually using laparoscopic techniques. Both one- and two-cell stages have been used for injection *(32)*. As for most species of livestock, visualization of the pronuclei/nuclei is difficult because of the presence of opaque lipid droplets in the cytoplasm. This can be overcome by centrifuging the eggs at 15,000$g$ for 6 min *(33)*; this reveals the pronuclei in about 70% of the eggs. Injection generally involves the introduction of 1 to 2 pL of the DNA solution containing the DNA construct at a concentration of about 1 to 2 ng/mL.

Injected eggs are cultured for a short period prior to surgical transfer (midventral laparotomy) into one of the oviducts of the recipient females. Fresh females that have been hormonally synchronized with the donors may be employed, or, alternatively, transfer can take place in the oviducts of donor females immediately after embryo recovery *(34)*. Pigs have large litters, and a minimum of four developing fetuses is required to maintain pregnancy. Consequently, large numbers of injected eggs are transferred into an individual, and 35–40 microinjected embryos per recipient is quite standard *(30)*.

The efficiency of the production of transgenic pigs depends on a wide variety of factors, including the quality of the embryos and animals used, preparation of the DNA construct, and skill of the operator. The percentage of embryos that develops into transgenic founders has been reported to vary from 0.3 to 4.0% *(35)*. Mosaicism in the founder generation can prove to be a problem, and up to 20% of founders fail to transmit the transgene and a further 20–30% do so at lower frequency than the expected 50% *(36)*.

### 3.2. Sheep

As for pigs, in sheep the fertilized eggs are recovered from superovulated donor females (for details of the techniques associated with generating transgenic sheep, *see* **Subheading 4.**). Fertilization may be carried out by natural matings or artificial insemination. The fertilized eggs are recovered surgically and microinjected using differential interference contrast (DIC) microscopy to visualize the pronuclei (**Fig. 1A**). One limitation in sheep is that many of the commonly available breeds exhibit seasonal fertility, which limits the time during the year when eggs can be routinely recovered to the autumn

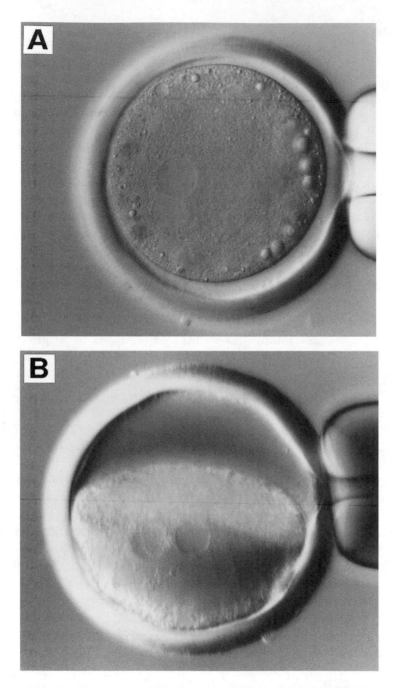

Fig. 1. Visualization of pronuclei for microinjection. (A) Fertilized one-cell sheep egg visualized by DIC microscopy. The holding pipet is shown on the right and the injection pipet on the left. The two pronuclei are visible at 2 to 3 o'clock. (B) Fertilized sheep egg after centrifugation; the two pronuclei are clearly visible on top of the sedimented cytoplasmic constituents.

and winter months; however, this problem can be overcome by using nonseasonal breeding breeds such as the Polled Dorset.

In contrast to pigs and cows (*see* **Subheading 3.4.**), centrifugation of the eggs is not usually employed for injection, although it does make the pronuclei easier to see (**Fig. 1B**). After injection, the eggs are transferred surgically to the oviducts of hormonally synchronized recipients. Sheep have small litters and rarely exceed twins without compromising the pregnancy or parturition. Consequently, this limits the number of injected eggs that can be reimplanted into the recipients to a maximum of four or five. Efficiencies for transgenic sheep production have been reported to vary from 0.1 to 2.0% *(37)*.

### 3.3. Goats

The procedures developed for the production of transgenic goats are similar to those described for sheep *(38,39)*. Thus, estrus is synchronized using a progesterone implant and superovulation induced by im injections of follicle-stimulating hormone (FSH). Fertilization is usually accomplished by natural matings, and females in estrus are mated to fertile males over a 2-d period. Goats are seasonal breeders, but to a much lesser extent than sheep, and minor changes to hormone doses can be used outside the breeding season to maintain the supply of eggs.

Embryos are recovered surgically on the second day following mating by means of a laparoscopic incision. The embryos are flushed out in phosphate-buffered saline (PBS). With standard protocols it has been reported that about five fertilized and injectable one-cell embryos are recovered *(39)*. The embryos are cultured short term in Hams F12 medium + 10% fetal calf serum (FCS) prior to injection. The fertilized eggs are visualized by DIC. As is the case for sheep, centrifugation is not required to see the pronuclei for injection. After injection, the embryos are cultured for a short time to evaluate survival prior to reimplantation into synchronized recipients. Recipients are synchronized by a combination of progesterone implants and injections of PMSG and are mated to vasectomized males to ensure synchrony. Three to four injected eggs are routinely transferred into the oviducts. Frequencies reported for the generation of transgenic goats are on a par with those reported for sheep, ranging from 6 to 10% of live offspring or 1 to 3% of injected and transferred embryos *(39)*.

### 3.4. Cattle

Cattle have proved to be the most difficult species of domestic livestock in which to accomplish genetic modification. In some ways this is surprising because many of the embryo manipulation and assisted reproduction techniques were developed in this species because of its overriding economic importance in agriculture. Nevertheless, the cost and timescale of working in cattle are much greater than in other species, and these factors have hindered progress.

Early work utilized zygotes recovered from superovulated donors using methods analogous to those just described for the other livestock species *(40)*. In these experiments, an average of approximately four injectable zygotes were recovered per donor. Given the small percentage of injected zygotes that eventually yields transgenics, this means that very large numbers of donor animals would be required to produce a single transgenic calf *(41)*.

A real advance in the procedures for generating transgenic cattle has been the use of zygotes produced by the in vitro maturation and fertilization of oocytes from slaughterhouse ovaries. Immature oocytes are aspirated from ovaries obtained immediately after slaughter; these are then matured and fertilized in vitro. Zygotes obtained this way are fertilized 16–24 h after insemination *(42)*. This is now the method of choice for the supply of fertilized donor eggs for injection and, it can provide, in effect, an unlimited quantity of material. Note, however, that the average quality of embryos produced by this route is low and large numbers of eggs need to be injected. A second downside to this approach is that it is difficult to ascertain the genetic or health status of eggs derived from slaughterhouse material. One solution to these problems has been to recover immature oocytes from defined animals by a nonsurgical procedure known as transvaginal echoscopy *(43)*, whereby large numbers of oocytes are recovered from an individual live animal and matured and fertilized in vitro to provide defined material for pronuclear injection.

As for pigs, cattle zygotes need to be centrifuged (6 min, 13,000$g$) to visualize the pronuclei. The microinjection procedure is essentially the same as that described for other species. In cattle, the conditions for in vitro embryo culture are well worked out *(43,44)*, and injected zygotes are cultured for 6 to 7 d prior to transfer to the recipients. This has a couple of advantages. One is that a rigorous developmental screen can be applied to the injected embryos, and, thus, only high-quality surviving embryos will be transferred. This is absolutely essential when working with slaughterhouse material since the quality of this material for injection is low and a large number of eggs must be injected; in vitro culture allows the majority of eggs that fail to develop to be weeded out. A second major advantage, dependent on the extended in vitro culture that is available for bovine eggs, is that after 6 to 7 d the embryos (morula/blastula stage) can be transferred nonsurgically into the uterus, avoiding the surgery required for embryo transfer into the oviduct.

Both pregnancy rates and calving rates are lower when using injected embryos as compared to nonmanipulated controls *(41)*. Overall, rates of transgenic cattle production are low. They are difficult to compare directly with figures from other livestock species because in vitro matured/in vitro fertilized (IVM/IVF) embryos are used, which have a much lower developmental potential. Thus, in a recently reported study by Eyestone *(45)*, 5717 injected eggs yielded five transgenic calves, or ~0.1%. This figure is considerably lower than those usually reported for other domestic species when in vivo–derived embryos are used. However, in that study the five transgenics represented 11% of the calves born, which is a favorable comparison with the figures obtained in other species *(46)*.

## 4. Procedures for Generating Transgenic Sheep

### *4.1. Superovulation and Fertilization*

Embryos are recovered from ewes that have been induced to superovulate at an estrus regulated with progestagen. Mature ewes of proved fertility are treated for 12–16 d with intravaginal sponges that are impregnated with progestagen (Intervet, Slough, UK). Superovulation is induced by administration of FSH. In the past, equine FSH was used and given in two im doses of between 2 and 4 mg 28 h before the end of progestagen treatment and at the time of sponge removal. Ewes are allowed to mate several times at the estrus, which occurs 20–72 h after sponge removal. The ewes are observed for the onset of heat at regular intervals, and embryos are recovered by surgery 36–72 h after the onset of estrus *(46)*.

Artificial insemination has been used to fertilize the eggs. In this protocol *(47)*, superovulation is induced by eight im injections of FSH (0.125 U/injection). PMSG (Intervet) is injected twice daily for 5 d and the ovulation synchronized by injection of a synthetic releasing hormone analog prior to insemination. Semen (0.2 mL/uterine horn) is administered by intrauterine laparoscopy.

### *4.2. Embryo Recovery*

Fertilized eggs are recovered surgically. Anesthesia is induced in the donor ewes by iv injection of 5% thiopentane sodium (Intraval, May and Baker ) at a dose rate of 3 mL/kg of body wt. Anesthesia is maintained by inhalation of 1 to 2% halothane. The reproductive tract is exposed through a midcentral incision and a nylon catheter inserted into the oviduct through the fimbria. Medium is introduced into the uterine lining through the uterotubal junction and along the oviduct. The embryos are recovered in PBS containing additional energy sources and protein (Ovum Culture Medium [OCM]; Advanced Protein Products, Brierly Hill, UK). Laparoscopic techniques may aid embryo recovery *(47)*. In this procedure, a laparoscopic incision is made, the uterus is externalized, and the eggs are flushed out with OCM. For culture and microinjection of the recovered eggs, the OCM is supplemented with 20% FCS.

### 4.3. Preparation of DNA for Microinjection

Prokaryotic vector sequences are thought to inhibit the expression of transgenes, and, therefore, it is common practice to remove these prior to microinjection. The vector and transgene sequences are separated by sucrose gradient centrifugation, or by agarose gel electrophoresis as described next.

The plasmid containing the transgene is digested with the appropriate enzymes overnight at 37°C. The digested DNA is run out in a wide well on a 1 to 2% agarose gel with relevant size markers. A small amount of the digest should be run in the lane next to the size marker. The gel should not contain ethidium bromide and can be either standard or low melting point agarose in TAE (0.04 $M$ Tris-acetate, 0.001 $M$ EDTA, pH 7.8). After running, the marker lanes and single lane containing the small amount of digest are removed and stained. The stained portion is realigned with the main body of the gel to ascertain the region to be excised. Using a clean, sharp scalpel, the area of the gel containing the fragment is removed. The DNA is purified on a QIAquick column (Qiagen, Gatwick Road, Crawley) using the QIAquick gel extraction protocol supplied by the manufacturer. This involves incubating the gel slice in buffer, applying to a QIAquick column, and then eluting the bound DNA from the column in double-distilled $H_2O$.

### 4.4. Microinjection of DNA

DNA (1 to 2 µg/mL) is injected into one pronucleus of a single-cell egg. The eggs are manipulated in a chamber filled with OCM. The chamber consists of a siliconized microscope slide with glass supports ($25 \times 2 \times 3$ mm) parallel to the long side of the slide. A cover slip is mounted on top of the supports, and the junction is sealed with silicone grease. The open ends of the chamber are filled with Dow Corning 200 fluid (BDH).

Eggs to be injected are held by suction on a blunt glass pipet that should have outer an inner diameters of 25–50 and 10–15 µm, respectively. Pronuclei are visualized using a Nikon Diaphet inverted microscope (Nikon, UK) preferably using ×40 DIC objectives (**Fig. 1**). The pronuclei are located by rotating the eggs and adjusting the holding pipet. DNA is injected into the pronucleus using a micropipet drawn from capillary tubing (borosilicate glass, 1 µm external diameter, thin wall with filament; Clark Electromedical) on a microelectrode puller (Campden, London, UK). The positions of the injection and holding pipets are controlled by micromanipulators (Leitz Mechanical Micromanipulators). The micropipet containing the DNA to be injected is connected via airtight tubing to a 100-mL glass syringe. Injection is performed by applying pressure using the syringe. Successful injection is indicated by visible swelling of the pronucleus to approximately twice its original volume, which represents an injected volume of approx 2 pL.

## 4.5. Transfer of Injected Embryos

Immediately after injection, eggs are cultured in OCM at 38.5°C in an atmosphere of 5% $CO_2$. Injected eggs can be returned to synchronized recipients on the same day or cultured overnight. Embryos judged to have survived injection are transferred to unmated recipient ewes whose estrus cycles are synchronized with those of the egg donors by treatment with progesterone. The sponge is removed 2 d prior to transfer, followed by an im injection of FSH. Recipients can be tested for the onset of estrus using a vasectomized ram. Embryos are transferred to the oviduct using fine-drawn mouth pipets. Up to four embryos are transferred to each ewe and are distributed between the two oviducts. Finally, the body wall is closed with soluble filament and the skin with Michel clips. Each ewe is given antibiotics at the time of surgery. More recently, laparoscopic techniques have been employed. The ewes are put under general anesthesia, the uterus is externalized with a laparoscopic excision, and the embryos are implanted into the oviducts.

## 4.6. Development and Growth

After recovery from anesthesia, the ewes are returned to the field and may be fed on supplements as required. During the third month of pregnancy, the number of fetuses is determined by ultrasonic scanning *(48)*. Management of pregnant ewes is then adjusted to take into account the number of fetuses.

## 4.7. Analysis of Transgenic Lambs

At least 2 wk after birth, 10-mL samples of blood are removed by venous puncture and collected into a heparinized tube. DNA is prepared from blood samples as follows. Thirty milliliters of lysis solution (155 m$M$ $NH_4Cl$, 10 m$M$ $KHCO_3$, 1 m$M$ EDTA) is added to a 10-mL blood sample, and the mixture is incubated for 15 min on ice. The white blood cells are spun at 1500$g$ for 10 min at 4°C, resuspended in 10 mL of SE (75 m$M$ NaCl, 2 nm of EDTA), and then washed once in SE. Proteinase K is added to 100 µg/mL followed by 1 mL of 20% sodium dodecyl sulfate, and the preparation is incubated for 4 h. Repeated phenol/chloroform extractions are performed until the preparation is completely deproteinized. One-thirtieth volume of 0.3 $M$ NaAc/vol of isopropanol is added to the aqueous phase to precipitate the DNA, which is hooked out, rinsed in 70% EtOH, and resuspended in TE.

DNA samples are analyzed by Southern blotting after digestion with the appropriate restriction enzyme and agarose gel electrophoresis. Alternatively, the polymerase chain reaction (PCR) can be used to screen for transgenic status, not only of the DNA but for partially purified DNA prepared from biopsy material. However, since PCR gives no information on the copy number of the transgenes or whether rearrangements have taken place and can be prone to detection errors, it is strongly recommended that Southern blotting be used to determine transgenic status unequivocaliy. Founder transgenic animals are selected and then used in the appropriate breeding programs.

## 5. Conclusion

It has been more than 15 yr since the first species of transgenic livestock were reported. In the intervening years, several applications in both agriculture and medicine have been developed. Whereas the full potential of this technology has yet to be realized in agriculture, medical products from genetically engineered livestock are now nearing the market. Until recently, the only proven route for the genetic manipulation of livestock was pronuclear injection. Whereas the reproductive physiology, embryology, and surgical procedures associated with the production of transgenic livestock have improved over the last decade, there really has not been any major improvement in the overall efficiency of the procedure except, perhaps, in cattle, in which the adoption of in-vitro matured/in vitro fertilization technology has made it a more realistic proposition. These low efficiencies, coupled with the expense in terms of both time and money, have constrained the application of the technology. The demonstration that cloning by nuclear transfer *(9)* offers a cell-based route for genetic modification, not only for the introduction of new genes *(9)* but also for gene targeting *(10)*, heralds a new dawn in the development of this technology in livestock.

## Acknowledgments

I wish to thank all my colleagues who have contributed to the transgenic program at Roslin Institute over the years and acknowledge support from the Biotechnology and Biological Sciences Research Council's core strategic grant to the institute.

## References

1. Hammer, R. E., Pursel, V. G., Rexroad, C., et al. (1985) Production of transgenic rabbits, sheep and pigs by microinjection. *Nature* **315,** 680–683.
2. Gordon, J. and Ruddle, F. (1981) Integration and stable germ-line transmission of genes injected into mouse pronuclei. *Science* **214,** 1244–1246.
3. Brinster, R. H., Chen, M., Trumbauer, A., Senear, R., Warren, R., and Palmiter, R. (1985) Factors affecting the efficiency of introducing foreign DNA into mice by microinjecting eggs. *Proc. Natl. Acad. Sci. USA* **82,** 239–241.
4. Clark, A. J., Bissinger, P., Bullock, D. W., et al. (1994) *Reprod. Fertil. Dev.* **6,** 589–598.
5. Garrick, D., Fiering, S., Martin, D. I. K., and Whitelaw, E. (1998) Repeat induced gene silencing in mammals. *Nat. Genet.* **18,** 56–59.

6. Clark, A. J., Harold, G., and Yull, F. E. (1997) Mammalian cDNA and CAT reporter constructs silence the expression of adjacent transgenes in transgenic mice. *Nucleic Acids Res.* **25,** 1009–1014.
7. Stice, S. L. (1998) Opportunities and challenges in domestic animal embryonic stem cell research, in *Animal Breeding: Technology for the 21st Century* (Clark, A. J., ed.), Harwood Academic, Switzerland, pp. 64–71.
8. Schnieke, A. E., Kind, A. J., Ritchie, W. A., et al. (1997) Human factor IX transgenic sheep produced by transfer of nuclei from transfected fetal fibroblasts. *Science* **278,** 2130–2133.
9. Wilmut, I., Schnieke, A. E., McWhir, J., Kind, A. J., and Campbell, K. H. S. (1997) Viable offspring derived from fetal and adult mammalian cells. *Nature* **385,** 810–813.
10. McCreath, K. J., Howcroft, J., Campbell, K. H. S., Colman, A., Schnieke, A. E., and Kind, A. J. (2000) Production of gene targeted sheep by nuclear transfer from cultured. *Nature* **405,** 1066–1068.
11. Clark, A. J., Burl, S., Denning, C., and Dickinson, P. (2000) Gene targeting in livestock: a preview. *Transgen. Res.* **9,** 263–275.
12. Campbell, K. H. S. (2001) Transgenic sheep from cultured cells (this volume).
13. Clark, A. J., Simons, J. P., and Wilmut, I. (1992) Germline manipulation: applications in agriculture and biotechnology, in *Transgenic Mice in Biology and Medicine* (Grosveld, F. and Kollias, G., eds.), Academic, London, pp. 247–269.
14. Pursel, V. G., Pinkert, C. A., Miller, K. F., et al. (1989) Genetic engineering of livestock. *Science* **244,** 1281–1288.
15. Pursel, V. G. (1998) Modification of production traits, in *Animal Breeding: Technology for the 21st Century* (Clark, A. J., ed.), Harwood Academic, Switzerland, pp. 183–200.
16. Damak, S., Su, H.-Y., Jay, N. P., and Bullock, D. W. (1996) Improved wool production in transgenic sheep expressing insulin-like growth factor I. *Bio/Technology* **14,** 185–188.
17. Muller, M. and Brem, G. (1991) Disease resistance in farm animals. *Experientia* **47,** 923–934.
18. Kolb, E., Laine, E., Strehler, D., and Staeheli, P. (1992) Resistance to influenza virus infection of Mx mice expressing Mx protein under the control of two constitutive promoters. *J. Virol.* **66,** 1709–1716.
19. Muller, M. and Brem, G. (1994) Transgenic strategies to increase disease resistance in livestock. *Reprod. Fertil. Dev.* **6,** 605–613.
20. Clements, J. E., Wall, R. J., Narayan, O., et al. (1994) Development of transgenic sheep that express the visna virus envelope gene. *Virology* **200,** 370–380.
21. Sola, I., Castilla, J., Pintado, B., Sanchez-Morgado, J. M., Whitelaw, C. B. A., Clark, A. J., and Enjuanes, L. (1998) Transgenic mice secreting coronavirus neutralising antibodies into the milk. *J. Virol.* **72,** 3762–3772.
22. Clark, A. J., Simons, J. P., Wilmut, I., and Lathe, R. (1987) Pharmaceuticals from transgenic livestock. *Trends Biotechnol.* **5,** 20–24.

23. Clark, A. J. (1998) The mammary gland as a bioreactor: expression, processing and production of recombinant proteins. *J. Mammary Gland Biol. Neoplasia* **3**, 337–350.

24. Wright, G., Carver, A., Cottom, D., et al. (1991) High level expression of active human alpha-1-antitrypsin in the milk of transgenic sheep. *Bio/Technology* **9**, 830–834.

25. Langford, G. A., Yannoutsos, N., Cozzi, E., et al. (1994) Production of pigs transgenic for human decay accelerating factor. *Transplant Proc.* **26**, 1400, 1401.

26. White, D. and Langford, D. (1998) Xenografts from livestock, in *Animal Breeding: Technology for the 21st Century* (Clark, A. J., ed.), Harwood Academic, Switzerland, pp. 229–242.

27. van der Laan, L. J., Lockey, C., Griffeth, B. C., Frasier, F. S., Wilson, C., Onions, D. E., Hering, B. J., Long, Z., Otto, E., Torbett, B. E., and Salomon, D. R. (2000) Infection by porcine endogenous retrovirus after islet xenotransplantation in SCID mice. *Nature* **407,** 90–94.

28. Petters, R. M. (1994) Transgenic livestock as genetic models of human disease. *Reprod. Fertil. Dev.* **6,** 643–645.

29. Davis, D. L., Knight, J. W., Killian, D. B., and Day, D. N. (1979) Control of estrus in gilts with a progestagen. *J. Anim. Sci.* **49,** 1506–1509.

30. Martin, M. J. and Pinkert, C. A. (1994) Production of transgenic swine, in *Transgenic Animal Technology* (Pinkert, C. A., ed.), Academic, San Diego, pp. 315–338.

31. Diehl, J. R., Danion, J. R., and Thompson, L. H. (1990) Artificial insemination in swine, in *Iowa State Univ. Pork Ind. Handbook*, no. 64, Iowa State University Press, Ames, IA.

32. Rexroad, C. E., Jr., Pursel, V. G., Hammer, R. E., Bolt, D. J., Miller, K. F., Mayo, K. E., Palmiter, R. D., and Brinster, R. L. (1988) Gene insertion: role and limitations of technique in farm animals as a key to growth, in *Biomechanisms Regulating Growth and Development*, vol. 12 (Steffens, G. L. and Rumsey, T. S., eds.), Kluwer, Dordrecht, The Netherlands, pp. 87–97.

33. Wall, R., Pursel, V., Hammer, R., and Brinster, R. (1985) Development of porcine ova that were centrifuged to permit visualisation of pronuclei and nuclei. *Biol. Reprod.* **32,** 645–651.

34. Brem, G., Springmann, K., Meier, E., Krausslich, H., Brenig, B., Muller, M., and Winnacker, E. L. (1989) Factors in the success of transgenic pig programmes, in *Transgenic Model in Medicine and Agriculture* (Church, R. B., ed.), Wiley-Liss, NY, pp. 61–72.

35. Pursel, V. G. and Rexroad, C. E., Jr. (1993) Status of research with transgenic farm animals. *J. Anim. Sci.* **71(Suppl. 3),** 10–19.

36. Pursel, V. G., Hammer, R. E., Bolt, D. J., Palmiter, R. D., and Brinster, R. L. (1990) Genetic engineering of swine: integration expression and germline transmission of growth-related genes. *J. Reprod. Fertil.* **41(Suppl.),** 77–87.

37. Clark, A. J., Archibald, A. L., McClenaghan, M., Simons, J. P., Whitelaw, C. B. A., and Wilmut, I. (1990) The germline manipulation of livestock. *Proc. NZ Anim. Prod.* **50,** 167–179.
38. Ebert, K., Selgrath, J., DiTullio, J., et al. (1991) Transgenic production of a variant of human tissue plasminogen activator in goat milk: generation of transgenic goats and the analysis of expression. *Bio/Technology* **9,** 835–838.
39. Gavin, W. G. (1997) Gene transfer into goat embryos, in *Transgenic Animals— Generation and Use* (Houdebine, L.-M., ed.), Harwood Academic, Switzerland, pp. 19–21.
40. McEvoy, T. and Sreenan, J. (1990) The efficiency of production, centrifugation, microinjection and transfer of one- and two- cell bovine ova in a gene transfer programme. *Theriogenology* **33,** 819– 828.
41. Eyestone, W. H. (1994) Challenges and progress in the production of transgenic cattle. *Reprod. Fertil. Dev.* **6,** 647–652.
42. Krimpenfort, P., Rademakers, A., Eyestone, W., et al. (1991) Generation of transgenic dairy cattle by *in vitro* embryo production. *Bio/Technology* **9,** 844–847.
43. de Loos, F. S., Hengst, F., Pieper, F., and Saladdine, M. (1996) Trans-vaginal oocyte recovery used for generation of bovine embryos for DNA microinjection. *Theriogenology* **45,** 349.
44. Gardner, D. K. (1998) Embryo development and culture techniques, in *Animal Breeding: Technology for the 21st Century* (Clark, A. J., ed.), Harwood Academic, Switzerland, pp. 13–46.
45. Eyestone, W. H. (1998) Techniques for the production of transgenic livestock, in *Animal Breeding: Technology for the 21st Century* (Clark, A. J., ed.), Harwood Academic, Switzerland, pp. 167–181.
46. Simons, J. P., Wilmut, I., Clark, A. J., et al. (1988) Gene transfer into sheep. *Bio/Technology* **6,** 179–183.
47. Gibson, Y. and Colman, A. (1997) The generation of transgenic sheep by pronuclear injection, in *Transgenic Animals—Generation and Use* (Houdebine, L.-M., ed.), Harwood Academic, Switzerland, pp. 23–25.
48. White, I. R., Russel, A. S. F., and Fowler, D. G. (1984) Realtime ultrasonic scanning in the diagnosis of pregnancy and determination of fetal numbers in sheep. *Vet. Rec.* **115,** 140–143.

# 14

## Transgenic Sheep from Cultured Cells

### Keith H. S. Campbell

## 1. Introduction

Recent progress in the production of mammals by nuclear transfer using donor nuclei from cultured cell populations has provided a novel route for genetic manipulation. The technique of nuclear transfer allows the production of offspring by the reconstruction of an embryo. Genetic material from a donor cell (karyoplast) is transferred to a suitable recipient cell from which the nuclear or genomic genetic material has been removed. In the first demonstrations of this technique, successful development was obtained only when the donor genetic material was taken from apparently undifferentiated cells or blastomeres from early embryos (for a review *see* **ref.** *1*). Subsequently, development has been obtained using donor genetic material from differentiated cells maintained in culture and isolated from embryonic *(2)* fetal and adult animals *(3)*. More recently, live offspring have been obtained in the mouse using quiescent cell populations derived directly ex vivo as nuclear donors *(4)*. The successful use of differentiated cells has now been demonstrated in sheep, cattle, and mice.

A number of animals may be produced by nuclear transfer using donor nuclei from either a single embryo or cells derived from a single tissue source. However, these animals will not be true clones because differences will exist in the mitochondrial DNA of the reconstructed embryos unless oocytes or zygotes from the maternal line of the cell donor are used as cytoplast recipients. In addition, point mutations or rearrangements may arise in the genomic DNA either during culture or development before or after embryo reconstruction.

### 1.1. Nuclear Transfer from Cultured Cell Populations

The use of nuclear transfer technology has many benefits and uses in the production of mammalian embryo fetuses and offspring, including the following:

From: *Methods in Molecular Biology, vol. 180: Transgenesis Techniques, 2nd ed.: Principles and Protocols*
Edited by: A. R. Clarke © Humana Press Inc., Totowa, NJ

1. To carry out precise genetic modification of cultured cells to be used as nuclear donors prior to embryo reconstruction. Such modifications include random gene addition, addition of multiple copies of a transgene, addition of a transgene at a precise location (targeted addition or knockin), gene removal (knockout), gene inactivation by targeted insertion, gene replacement, modification of any gene or its control sequences, and gene multiplication.

2. To carry out multiple genetic modifications in a single animal either by multiple genetic modifications of a cell population in culture or by sequential genetic modification, nuclear transfer, and reisolation of a cell population from the embryo, fetus, or animal so produced.

3. To increase the lifespan of cultured cell populations to be used for genetic modification by nuclear transfer and reisolation of a cell population from the embryo, fetus, juvenile, or adult animal so produced.

4. To produce multiple copies of an animal from a genetically modified selected and cloned cell population.

5. To produce multiple copies of any embryo, fetus, juvenile, or adult animal by nuclear transfer from cells taken directly ex vivo or cell populations derived from any tissues taken from any of these stages with or without culture in vitro.

6. To produce true clones by utilizing oocytes from the maternal line of the cell donor as cytoplast recipients for embryo reconstruction.

7. To store intact genomes for long periods (e.g., by freezing cell populations in liquid $N_2$) and subsequently to use these stored cells for the production of offspring by nuclear transfer.

8. To dedifferentiate somatic nuclei and to produce undifferentiated cells that may be used for the production of chimeric embryos, fetuses, and adult animals by embryo aggregation or injection or to produce embryonic stem or embryonic germ cell populations.

Genetic modification of animals and the production of stem cell and differentiated cell populations by nuclear transfer technology have numerous uses in the fields of human medicine, agriculture, genetic preservation, and research. These include the production of human therapeutic proteins in the bodily fluids, disease prevention, increasing required production traits, cell-based therapies, cell-based delivery systems for genetic therapy, as well as tissue and organ transplantation (for a review *see* Chapter 1).

Preliminary experiments have demonstrated that cultured cell populations may be used for genetic modification prior to embryo production in both sheep *(5)* and cattle *(6)*. The use of nuclear transfer coupled with genetic modification of cells in culture and their selection prior to animal production has several advantages:

1. The production of nonmosaic animals ensuring germline transmission of the genetic modifications.

2. An increased efficiency in the production of such genetically modified animals.

3. The production of multiple copies of the offspring, thereby reducing the generation interval to produce flocks or herds for production purposes or increasing the numbers of animals for dissemination of genetic modification into the population as a whole.
4. The production of animals containing multiple genetic modifications.
5. The production of transgenic animals with superior expression characteristics by utilizing the preselection of the integration site of the transgene.

## 1.2. Embryo Production by Nuclear Transfer

The process of embryo reconstruction and production of viable offspring by nuclear transfer is a multistep procedure. Each of these steps is described in the following sections.

### 1.2.1. Recipient Cell or Cytoplast

Oocytes, fertilized zygotes, and two-cell embryos have been used as cytoplast recipients for nuclear transfer. In general, oocytes arrested at metaphase of the second meiosis have become the cytoplast of choice. At this point in oocyte development, the genetic material is arranged on the meiotic spindle and is easily removed using mechanical means. Several reports have demonstrated that during maturation—i.e., between the germinal vesicle stage (prophase of the first meiotic division) and arrest at metaphase of the second meiotic division—genomic DNA can be removed and the resulting cytoplast used for nuclear transfer *(7)*. The use of fertilized zygotes as cytoplast recipients has been reported in the mouse *(8,9)* and pig *(10)*. In cattle and pigs, development of embryos reconstructed using zygotes as cytoplast recipients is low and, on the whole, restricted to the exchange of pronuclei, suggesting that factors essential for successful development are removed with the pronuclei.

### 1.2.2. Preparation of Cytoplast Recipient by Removal of Genomic Genetic Material

The process of preparing a cytoplast recipient by removing the genomic genetic material has, in general, been termed *enucleation*. However, this is a misleading description because in matured oocytes (arrested at MII), the genomic DNA is not enclosed within a nuclear membrane at the time of removal. Removal of the genetic material is possible by physical and or chemical means. In the early reports of nuclear transfer, MII oocytes were simply cut in half on the basis that one half would contain the genetic material and the other would not. Modifications to this approach have been made in order to reduce the volume of cytoplasm, which was removed. This may be achieved by aspirating a small amount of cytoplasm from directly beneath the first polar body using glass micropipets or by using a knife to cut away that part of the oocyte beneath the polar body. To facilitate plasticity of the oocyte, it may be pretreated with the microtubule inhibitor cytochalasin B or other such agent that disrupts the cytoskeleton. In contrast to physical aspiration to achieve enucleation, other treatments have been demonstrated to cause complete removal of the genetic material. In the mouse, treatment of maturing oocytes with the topoisomerase inhibitor ectoposide results in the expulsion of all genomic material with the first polar body *(11)*. However, no development to term has been described using cytoplast recipients produced by this method, and there are no reports of this procedure in other species. Centrifugation of MII oocytes combined with cytochalasin B treatment has been reported to cause enucleation in cattle oocytes *(12)*. The development of embryos reconstructed from such cytoplasts has been reported in cattle; however, the frequency of development is low.

### 1.2.3. Introduction of Genetic Material (Embryo Reconstruction)

Once a suitable recipient cell or cytoplast has been prepared, the donor genetic material must be introduced. Various techniques have been reported including the following:

1. Cell fusion induced by chemical, viral, or electrical means.
2. Injection of an intact cell.
3. Injection of a lysed or damaged cell.
4. Injection of a nucleus.

Any of these methods may be used in any species with some modifications of individual protocols.

### 1.2.4. Activation of Reconstructed Embryo

In addition to the transfer of donor genetic material from the karyoplast to the cytoplast, the cytoplast must be stimulated to initiate development. When using a fertilized zygote as a cytoplast recipient, development has already been initiated by sperm entry at fertilization, whereas when using MII oocytes, the oocyte must be activated by other stimuli. Various treatments have been reported to induce oocyte activation and promote early embryonic development, including, but not limited to, application of a direct current (DC) electric stimulus; treatment with ethanol *(13)*, ionomycin *(14,15)*, calcium ionophore A23187 *(16)*, or extracts of sperm *(17,18)*; or any other treatment that induces calcium entry into the oocyte or release of internal calcium stores and results in initiation of development. In addition, any of these treatments in combination, with their application at the same or different times, or in combination with inhibitors of protein synthesis (i.e., cycloheximide or puromycin) *(18)* or inhibitors of serine threonine protein kinases (i.e., 6-DMAP) *(14)* may be applied.

### 1.2.5. Culture of Reconstructed Embryos

Nuclear transfer–reconstructed embryos may be cultured in vitro to a stage suitable for transfer to a final recipient using any suitable culture medium or culture process. Alternatively, embryos may be cultured in vivo in the ligated oviduct of a suitable host animal (in general sheep) until a stage suitable for transfer to a final surrogate recipient is reached. Embryos from cattle, sheep, and other species may be cultured in a transspecies recipient; for simplicity, a sheep provides a suitable recipient for bovine, ovine, and porcine species. To prevent mechanical damage or attack by macrophages to the reconstructed embryos while in the oviduct of the temporary recipient, it is usual to embed the embryos in a protective layer of agar or similar material.

### 1.2.6. Cell-Cycle Coordination of Reconstructed Embryo

The development of embryos reconstructed by nuclear transfer is dependent on many factors. Central to obtaining development is the maintenance of correct ploidy in the reconstructed embryo. In this chapter, methods are described for the production of transgenic lambs by nuclear transfer from quiescent fetal fibroblasts utilizing MII-arrested oocytes as cytoplast recipients. In this situation, correct ploidy of the reconstructed embryos is maintained. A more detailed review of the effects of cell-cycle coordination can be found in **ref. *19***.

### 1.2.7. Donor Cell or Karyoplast

Successful development to term has now been obtained from a variety of cultured cell populations isolated from embryo, fetal, and adult tissues *(2–4,6,20)*. The selection of a suitable donor cell type is dependent on the required outcome of the nuclear transfer procedure. For example, if the production of a genomic copy of an adult animal is required, then an adult cell type is required. In the context of this chapter, the outcome is to produce transgenic offspring by nuclear transfer. The cell type used for this is dictated only by the properties of the cells; to produce a transgenic animal, the cultured cell population must be amenable to transfection and selection in culture and still retain the ability to produce viable offspring when utilized as donors of genetic material for embryo reconstruction by nuclear transfer. In addition, for precise genetic modification, the cells must also have the ability to form single-cell clones. At the present time, the properties of individual cell types for these manipulations are under study. However, the use of fetal-derived cell populations offers several advantages. Large numbers of cells can be obtained from a single fetus, and the fetus can be recovered at a stage of gestation when the probability of development to term is high. The suitability of cells as successful nuclear donors is only evidenced by the production of viable offspring; however, the karyotype of individual populations can be used as a measure of suitability. Cell populations with a high frequency of aneuploidy have a lesser chance of successful development, but we must realize that a high frequency of cells having a normal modal number of chromosomes is not predictive of development to term.

The isolation, culture, and genetic manipulation of cells in culture have been dealt with in depth by several other reviews and, therefore, are not discussed here. However, for use as donors of genetic material, the transfected selected cell population must be amenable to the induction of quiescence or synchronization in the $G_0$-phase of the cell cycle by reduction in the levels of serum or other nutrient in the culture medium or by the addition of specific chemicals or growth factors depending on the cell type.

### 1.3. Discussion

This chapter summarizes a method for the production of transgenic lambs by nuclear transfer from cultured cell populations. The advantages of nuclear transfer technology in the genetic manipulation of farm animal species is discussed here and in Chapter 1. The application of this technology is in its infancy, and further studies are required to reveal its true potential. In particular, new methodologies for the precise genetic manipulation of cells in culture (i.e., homologous recombination) or the identification of cell types more amenable to such modifications are needed.

## 2. Materials

1. For embryo manipulation, a Nikon Diaphot TDM inverted microscope with differential interference contrast (DIC) and epifluorescence is used. This is fitted with 2× Narishige MO-188 Hydraulic Joystick Micromanipulators and 2× Narishige IM-188 microinjectors that are modified to accept gastight syringes. A 500-µL Hamilton syringe (S 0142; Sigma, St. Louis, MO) is mounted on the right-hand side of the microscope in order to control the holding pipet on the left-hand side of the chamber. A similar syringe, but with a capacity of 250 µL, is mounted on the left-hand side to control the enucleation pipet on the right-hand side of the chamber. The syringes are connected to tubing with a two-way stopcock (Vigon VG1) that allows a 5-mL syringe to be used as a hydraulic reservoir, and the system is filled with Fluorinert FC 77 (F4758; Sigma).
2. Enucleation pipets, which can be purchased from Cook Veterinary (cat. no. V-NTP-1800).
3. For the preparation of holding pipets, glass capillaries can be purchased from Clark (cat. no. GC10-100).
4. A Campden Instruments Moving Coil Microelectrode puller (Model 753) is used in conjunction with a microforge (Research Instruments MF 1).

5. Wild M8 and M3Z dissecting microscopes.
6. Drummond model 105, 5-μL and model 110, 10-μL pipets.
7. M2 medium: Dissolve 0.356 g KCl (cat. no. P5405; Sigma), 0.252 g $CaCl_2 \cdot 2H_2O$ (cat. no. C7902; Sigma), 0.162 g $KH_2PO_4$ (cat. no. P5655; Sigma), 0.293 g $MgSO_4 \cdot 7H_2O$ (cat. no. M1880; Sigma), 0.349 g $NaHCO_3$ (cat. no. S5761; Sigma), 4.969 g HEPES (cat. no. H9136; Sigma), 2.610 g sodium lactate (cat. no. L7900; Sigma), 0.036 g sodium pyruvate (cat. no. P5280; Sigma), 1.0 g of glucose (cat. no. G6138; Sigma), 4.0 g bovine serum albumin (cat. no. A6003; Sigma), 0.060 g penicillin G potassium salt (cat. no. P4687; Sigma), 0.050 g streptomycin sulfate (cat. no. S1277; Sigma), 0.010 g phenol red (cat. no. P0290; Sigma) in 800 mL of Milli-Q water. Adjust to pH 7.4 and 280 mosM, and make up the volume to 1 L with Milli-Q water.
8. M2 medium without calcium and magnesium: As in **item 7** but omit $CaCl_2 \cdot 2H_2O$ and $MgSO_4 \cdot 7H_2O$.
9. Hyaluronidase stock solution: Dissolve hyaluronidase IV-S (cat. no. H3884; Sigma) in phosphate-buffered saline to give 1500 IU/mL. Dispense into 20-μL aliquots and store at −20°C.
10. Cytochalasin B stock solution: Dissolve cytochalasin B (cat. no. C6762; Sigma) in dimethyl sulfoxide (cat. no. D4540; Sigma) to give a 1.0 mg/mL solution. Aliquot and store at −20°C.
11. Bisbenzimide stock solution: Dissolve P1754bis-benzimide (Hoechst 33342) (cat. no. B2261; Sigma) in Milli-Q water to give 1.0 mg/mL. Aliquot in 20-μL volumes and store at −20°C.
12. Dow Corning Silicone Fluid 200/50 cs (cat. no. 63006 4V) from BDH.
13. Fluorinert FC 77.
14. NaCl (cat. no. S5886; Sigma).
15. Sigmacote (cat no. Sl2; Sigma).
16. Mannitol (cat. no. M9546; Sigma).
17. L-Glutamine (cat. no. G7513; Sigma).
18. Agar (Difco).
19. PBS Dulbecco "A" (BR14a) Oxoid.
20. Electronic thermometer (PTM1) Petracourt.
21. Fusion machine (BTX).
22. Fusion chamber, homemade, consisting of two electrodes of 100-μ platinum wire glued to the bottom of a 4.5-cm glass Petri dish with a 200-μm gap.
23. TC 199 (cat. no. M0148; Gibco-BRL).
24. Ovine follicle-stimulating hormone (oFSH) (Ovagen; Immuno-chemicals, NZ).
25. Vaginal sponges (Veramix; Upjohn).
26. Gonadotropin-releasing hormone (GnRH) (Receptal; Hoechst, UK).
27. Pregnant mare's serum gonadotropin (PMSG) (Folligon; Intervet).
28. Suture (Dexon; Davis and Geck).
29. Scottish Blackface ewes.

## 3. Methods

### 3.1. Superovulation of Ewes to Act as Oocyte Donors

1. On d 0, insert vaginal sponges (Veramix) in the ewes.
2. On d 10–13 inclusive, inject ewes subcutaneously with 2 mL of oFSH (1 U in 16 mL of $H_2O$) twice daily (8:00 AM and 5:00 PM).
3. On d 14, inject ewes with 2 mL of GnRH Receptal intramuscularly at 7:00 AM.
4. Start oocyte recovery 26 h post-GnRH injection.

### 3.2. Preparation of Ewes to Act as Temporary Recipients for Embryo Culture

1. On d 0, insert vaginal sponges (Veramix) in the ewes.
2. On d 14, inject ewes with PMSG Folligon (500 U 0.5 mL/ewe intramuscularly).
3. Check ewes d 15 for heat, starve them, and use them as temporary recipients.

### 3.3. Preparation of Ewes to Act as Final Recipients for Development to Term

1. On d 0, insert vaginal sponges (Veramix) in the ewes.
2. On d 13, remove the sponges at 8:00 AM.
3. On d 15, check the ewes for heat.
4. Select the ewes in heat. On d 21, starve the selected ewes, and on d 22, use them as recipients.

### 3.4. Recovery of Oocytes from Superstimulated Ewes and Preparation for Nuclear Transfer

1. At 26–30 h following GnRH injection, anesthetize superovulated donor ewes and perform midline laparotomy.
2. Exteriorize the uterus to expose the ovaries and oviducts. Place a small catheter into the fimbria. Using a blunt 18-gage needle introduced close to the uterotubal junction, flush 20 mL of warm (37°C) PBS/fetal calf serum (FCS) through each oviduct. Collect the flushings in an embryo culture or Petri dish.
3. Examine the flushings under the dissecting microscope; transfer cumulus oocyte complexes to fresh calcium-free M2; and keep at 37°C, 5% $CO_2$ in air.
4. Remove the cumulus cells from the oocytes by placing them into calcium-free M2 without FCS but with 600 U/mL of hyaluronidase added. Incubate for 10 min at 37°C, and then repeatedly pipet with an automatic pipet (Gilson 200 µL) until the cumulus cells are removed.
5. Wash the oocytes in calcium-free M2 10% FCS until all the cumulus cells are removed.
6. Place batches of 10 oocytes into calcium-free M2 10% FCS containing 7.5 µg/mL of cytochalasin B plus 5 µg/mL of bis-benzimide, and culture at 37°C for 15 min prior to enucleation.

## 3.5. Preparation of Holding Pipet

1. Pull glass capillaries (GC10-100) by hand over a very small flame to give a diameter of 100–150 μ.
2. Cut the pipet using a diamond pencil to mark the glass making sure that the break is at right angles to the pipet.
3. Place the broken end of the pipet over the filament on the microforge. Apply heat until the open tip is almost closed ensuring a smooth end with a diameter of approx 20 μ.

## 3.6. Enucleation of MII Oocytes

1. Place a clean depression slide onto a microscope stage, pipet 300 μL of manipulation medium into the depression, and overlay with mineral oil.
2. Attach the holding pipet to the left-hand side tool holder ensuring that all air is removed from the hydraulic system. Move the holding pipet into the manipulation chamber, and draw a small volume of manipulation medium into it.
3. Wash the inside of the enucleation pipet thoroughly in a solution of 1.25% Tween-80 water, by aspirating the solution through the pipet several times. Then coat the inside of the pipet with FCS.
4. Mount the pipet on the right side of the chamber, and ensure that the system is free from air bubbles. A slight angle, of approx 5° from horizontal, allows the pipet to pick up cells from the bottom of the chamber.
5. Transfer a group of 10 oocytes into the manipulation chamber.
6. Using ×40 magnification, pick up and attach a single oocyte to the holding pipet using negative pressure. Change the magnification to ×200 DIC, and focus on the oocyte held by the pipet.
7. Using the manipulator control, bring the enucleation pipet into focus. Using the enucleation pipet, rotate the oocyte into a position where the polar body plus the area of cytoplasm adjacent to it can be aspirated into the pipet.
8. Insert the enucleation pipet through the zona pellucida at a point opposite the holding pipet, and manipulate into a position next to the polar body. Apply a small amount of negative pressure, and aspirate the polar body and a small amount of cytoplasm directly beneath into the pipet.
9. Withdraw the pipet from the oocyte, and then remove the oocyte from the field of view. Turn off the transmitted light source, change to ultraviolet illumination (blue light), and examine the aspirated karyoplast with fluorescence using filter block UV-2A. If the metaphase has been removed, it will fluoresce with a blue color. The metaphase plate fluoresces with less intensity than the polar body.
10. Remove completed batches of oocytes from the manipulation chamber and place into a microdrop or dish containing calcium-free M2 10% FCS. Maintain at 37°C until all the oocytes have been enucleated.

### 3.7. Preparation of Donor Cells

1. Subculture the selected cell population into fresh growth medium containing 10% FCS, and culture for 24 h under suitable conditions to allow the cells to enter log growth.
2. Wash the cell monolayer three times in warm culture medium containing 0.5% FCS, and then replace the medium with low-serum medium. Continue incubation at 37°C, 5% $CO_2$ for at least 48 h (*see* **Note 1**).
3. Harvest the cells by trypsinization *(21)* and resuspend in growth medium containing 10% FCS.
4. Incubate the cell suspension at 37°C, 5% $CO_2$.
5. Use for embryo reconstruction.

### 3.8. Preparation of Oocyte Cell Couplets

1. Using a hand-drawn capillary mouth pipet, place the nuclear donor cells and a batch of 10 enucleated oocytes into the manipulation chamber.
2. Focus the microscope onto the bottom of the chamber. Move the enucleation pipet to the bottom of the chamber, maneuver it to a suitable cell, and gently aspirate it into the pipet.
3. Refocus on the enucleated oocyte and move the enucleation pipet until it is in focus.
4. Insert the enucleation pipet through the hole previously made in the zona pellucida. While holding the pipet against the cytoplasm, expel the donor cell into the perivitelline space.
5. On completion of each batch, carry out electrofusion as soon as possible.

### 3.9. Transfer of Donor Nucleus by Electrofusion of Cytoplast/Karyoplast Couplet and Activation of MII Oocyte

1. Place the cytoplast/karyoplast couplets in 500 µL of warm (37°C) fusion medium and allow to sink.
2. Place 200 µL of fusion medium, spanning the electrodes, into the fusion chamber.
3. Pipet the batch of cytoplast/karyoplast couplets into the fusion medium but outside the electrodes, and allow to settle.
4. Place a couplet between the electrodes, using a hand-drawn capillary mouth pipet, and move the couplet until the plane of contact between the cytoplast and karyoplast is parallel to the electrodes.
5. Apply the fusion pulse, which consists of a 5-s alternating current pulse of 3 V followed by three DC pulses of 1.25 kV/cm for 80 µs each.
6. Remove the couplets from the chamber wash in TC199 10% FCS, and then place each batch into a 20-µL drop of TC 199 10% FCS containing 7.5 µg/mL of cytochalasin B. Incubate at 37°C for 1 h, wash in TC 199 10% FCS without cytochalasin, transfer to M2 medium, and examine for cell fusion.
7. Transfer fused couplets to 20-µL drops of TC199 10% FCS under oil, and culture overnight at 37°C.

### 3.10. Preparation of Reconstructed Embryos (Fused Couplets) and Culture in Ligated Oviduct of Temporary Recipient Ewe

1. Place molten agar (1.2% [w/v]) into embryo culture dishes, and cool and maintain at 37°C.
2. Place one or two of the fused embryos into the agar. Aspirate a small amount of agar containing the embryos into a Drummond model 105, 5-μL pipet.
3. Remove the pipet from the agar and allow to cool.
4. Place the tip of the pipet into cool M2 (at room temperature), and expel the agar cylinder formed in the medium.
5. Trim the agar cylinders containing the embryos using two 25-gage needles attached to two 1-mL syringe barrels.
6. Place the agar-embedded embryos in M2 medium at 37°C.
7. Anesthetize a pseudopregnant ewe using a short-acting barbiturate, intubate, and perform a midventral laparotomy.
8. Using Dexon suture, double ligate each oviduct near the uterotubal junction.
9. Pick up agar cylinders containing the nuclear transfer embryos in a small volume of medium using a 10-μL Drummond pipet.
10. Carefully insert the tip of the pipet into the fimbria end of the ligated oviduct. Push the end of the pipet as far down the oviduct as possible using a pair of fine, nontoothed forceps to grasp the oviduct. Then carefully expel the agar cylinders into the oviduct while withdrawing the pipet. Suture the incision and allow the ewe to recover.

### 3.11. Recovery of Embryos and Transfer to Final Recipient Ewes for Development to Term

1. At d 7 (from the time of embryo reconstruction), euthanize the temporary recipient and recover the oviducts.
2. Dissect excess tissue from the oviducts; remove the ligatures; and flush each oviduct, toward the uterotubule junction, with 5 mL of M2 medium using a syringe and blunt 18-gage needle.
3. Collect the flushings, examine under the dissecting microscope, and locate the agar cylinders. Wash the agar cylinders in M2 and assess embryo development.
4. Dissect morula/blastocyst-stage embryos from the agar using two 25-gage hypodermic needles attached to 1-mL syringe barrels.
5. Anesthetize a d-7 pseudopregnant ewe and perform a midline laparotomy.
6. Pick up two or three of the recovered morula/blastocyst-stage embryos in a small volume of M2 using a long-form Pasteur pipet with a flame-polished end.
7. Make a small hole in the wall of the uterus using a blunt 18-gage needle. Insert the end of the pipet carrying the embryos into this hole, and expel the embryos into the uterus.
8. Suture the incision and allow the ewe to recover.
9. Ultrasound scan the recipient after 45 d to determine pregnancy.

## 4. Note

1. The precise time required for exit from the growth cycle is dependent on the growth characteristics of individual cell populations. For each population, the time to exit the cycle should be determined experimentally.

## References

1. Wolf, E., Zakhartchenko, V., and Brem, G. (1998) Nuclear transfer in mammals: recent developments and future perspectives. *J. Biotechnol.* **65,** 99–110.
2. Campbell, K. H., McWhir, J., Ritchie, W. A., and Wilmut, I. (1996) Sheep cloned by nuclear transfer from a cultured cell line. *Nature* **380,** 64–66 (see comments).
3. Wilmut, I., Schnieke, A. E., McWhir, J., Kind, A. J., and Campbell, K. H. (1997) Viable offspring derived from fetal and adult mammalian cells. *Nature* **385,** 810–813 (see comments) (published erratum appears in *Nature* 1997 Mar 13;386[6621]:200).
4. Wakayama, T., Perry, A. C., Zuccotti, M., Johnson, K. R., and Yanagimachi, R. (1998) Full-term development of mice from enucleated oocytes injected with cumulus cell nuclei. *Nature* **394,** 369–374 (see comments).
5. Schnieke, A. E., Kind, A. J., Ritchie, W. A., Mycock, K., Scott, A. R., Ritchie, M., Wilmut, I., Colman, A., and Campbell, K. H. (1997) Human factor IX transgenic sheep produced by transfer of nuclei from transfected fetal fibroblasts. *Science* **278,** 2130–2133 (see comments).
6. Cibelli, J. B., Stice, S. L., Golueke, P. J., Kane, J. J., Jerry, J., Blackwell, C., Ponce de Leon, F. A., and Robl, J. M. (1998) Transgenic bovine chimeric offspring produced from somatic cell-derived stem-like cells. *Nat. Biotechnol.* **16,** 642–646 (see comments).
7. Kato, Y. and Tsunoda, Y. (1993) Totipotency and pluripotency of embryonic nuclei in the mouse. *Mol. Reprod. Dev.* **36,** 276–278.
8. Kwon, O. Y. and Kono, T. (1996) Production of identical sextuplet mice by transferring metaphase nuclei from four-cell embryos. *Proc. Natl. Acad. Sci. USA* **93,** 13,010–13,013.
9. Prather, R. S. and First, N. L. (1990) Cloning embryos by nuclear transfer. *J. Reprod. Fertil. Suppl.* **41,** 125–134.
10. Prather, R. S., Sims, M. M., and First, N. L. (1989) Nuclear transplantation in early pig embryos. *Biol. Reprod.* **41,** 414–418.
11. Elsheikh, A. S., Takahashi, Y., Katagiri, S., and Kanagawa, H. (1998) Functional enucleation of mouse metaphase II oocytes with etoposide. *Jpn. J. Vet. Res.* **45,** 217–220.
12. Tatham, B. G., Sathananthan, A. H., Dharmawardena, V., Munesinghe, D. Y., Lewis, I., and Trounson, A. O. (1996) Centrifugation of bovine oocytes for nuclear micromanipulation and sperm microinjection. *Hum. Reprod.* **11,** 1499–1503.
13. Bordignon, V. and Smith, L. C. (1998) Telophase enucleation: an improved method to prepare recipient cytoplasts for use in bovine nuclear transfer. *Biol. Reprod.* **49,** 29–36.
14. Susko-Parrish, J. L., Leibfried-Rutledge, M. L., Northey, D. L., Schutzkus, V., and First, N. L. (1994) Inhibition of protein kinases after an induced calcium tran-

sient causes transition of bovine oocytes to embryonic cycles without meiotic completion. *Dev. Biol.* **166**, 729–739.

15. Zakhartchenko, V., Stojkovic, M., Brem, G., and Wolf, E. (1997) Karyoplast-cytoplast volume ratio in bovine nuclear transfer embryos: effect on developmental potential. *Biol. Reprod.* **48**, 332–338.

16. Soloy, E., Kanka, J., Viuff, D., Smith, S. D., Callesen, H., and Greve, T. (1997) Time course of pronuclear deoxyribonucleic acid synthesis in parthenogenetically activated bovine oocytes. *Biol. Reprod.* **57**, 27–35.

17. Swann, K. (1990) A cytosolic sperm factor stimulates repetitive calcium increases and mimics fertilization in hamster eggs. *Development* **110**, 1295–1302.

18. Tanaka, H. and Kanagawa, H. (1997) Influence of combined activation treatments on the success of bovine nuclear transfer using young or aged oocytes. *Anim. Reprod. Sci.* **49**, 113–123.

19. Campbell, K. H., Loi, P., Otaegui, P. J., and Wilmut, I. (1996) Cell cycle co-ordination in embryo cloning by nuclear transfer. *Rev. Reprod.* **1**, 40–46.

20. Wells, D. N., Misica, P. M., Day, T. A., and Tervit, H. R. (1997) Production of cloned lambs from an established embryonic cell line: a comparison between in vivo- and in vitro-matured cytoplasts. *Biol. Reprod.* **57**, 385–393.

21. Campbell, K. H. S., Ritchie, W. A., and Ferrier, P. M. (1998) Nuclear transfer from an established cell line, in *Cell Biology: A Laboratory Handbook* (Celis, J. E., ed.), Academic, pp. 487–501.

# IX

## Characterization and Analysis of Transgenic Strains

# 15

## Analysis of Transgenic Mice

### Stefan Selbert and Dominic Rannie

## 1. Introduction

The first step in the analysis of potentially transgenic mice is the characterization of their altered genotype. A strategy must be devised that allows easy and reliable discrimination between wild-type animals and those carrying additional transgenes or introduced targeted mutations. It is also necessary to demonstrate that the characteristics of the alteration are as desired.

Transgenic animals fall into two general categories: those that have been generated by gene insertion in a randomly integrating fashion, as is usually the case with animals harboring additional transgenes; and those produced by targeting specific gene sequences within their genome via homologous or site-specific recombination, e.g., knockout animals. Although the analysis of all transgenics utilizes many of the same molecular techniques, the investigator must devise a screening strategy that will allow appropriate characterization of each individual transgenic line.

Animals that have been generated by random gene integration must be described in terms of modification in genetic structure, spatial and/or temporal distribution of expression, and assurance that the recombinant protein is biologically active. Genetically, it is usually important to determine the number of transgene integration sites while estimating the overall number of copies successfully integrated. Expression studies should ascertain the spatial and temporal distribution of expression throughout the animal at an RNA and/or protein level. To characterize transgene expression properly, it is important to examine as many tissues as possible, including those in which no transcripts are expected, because expression patterns are not always truly predictable by the DNA elements employed to regulate the transgene. Frequently expected patterns are given, but many animals often exhibit no or "ectopic" expression,

From: *Methods in Molecular Biology, vol. 180: Transgenesis Techniques, 2nd ed.: Principles and Protocols*
Edited by: A. R. Clarke © Humana Press Inc., Totowa, NJ

i.e., expression in tissues or at developmental stages that would not be predicted by the regulatory DNA elements used. The integration site of the transgene, e.g., integration adjacent to regulatory elements such as enhancers or silencers, or into transcriptionally inactive heterochromatin, can usually explain these phenomena.

When characterizing animals generated by gene targeting, it is important to demonstrate that the introduced genetic alterations result in the desired modifications of transcription. With knockouts, perhaps the most common form of targeted animal, it is essential to establish that transcription of the gene and translation of the resulting mRNA have been abolished. Truncated mRNAs may still give rise to protein domains that contain biologic activity, and it is therefore vital to exclude the possibility that any remaining protein fragments are functional.

This chapter describes appropriate methods frequently used to address these areas of question. It does not, however, attempt to be a comprehensive guide to all such techniques. Because no two laboratories are the same, decisions made about the exact approaches adopted will depend on factors such as the nature of the transgene and the equipment available.

In the following sections, we explore these areas:

1. *Genotyping of transgenes*: the isolation of good-quality genomic DNA from tail biopsies and subsequent analysis by polymerase chain reaction (PCR) and Southern blotting.
2. *RNA analysis*: extraction of total or mRNA and their analysis by reverse transcriptase polymerase chain reaction (RT-PCR) and Northern blotting. Also, the relatively new field of RT-*in situ* hybridization is introduced as a valuable method for determining expression patterns on a cellular level.
3. *Protein analysis*: the detection of protein expression by Western blot analysis and immunohistochemistry. Three widely used reporter enzyme assays (CAT, β-Gal, and luciferase) are presented, which allow the fast characterization of introduced transgenic regulatory DNA elements.

## *1.1. Genotyping*

Perhaps the simplest way to identify transgenic mice is to assess their genotype via DNA extracted from a tail tip, taken from an animal over 3 wk old. This easy, quick healing, noninvasive biopsy is the recommended source of genomic DNA, which, once purified, can be analyzed for the presence of transgenic alterations.

Initial screening for transgenics is generally performed by PCR. This technique is very rapid and allows relatively easy and cost-effective analysis of a large number of samples. The sensitivity of this technique, however, dictates that considerable caution must be taken to prevent the contamination of samples, either with plasmid DNA or with genomic DNA from other samples. The generation of false positive results is an inherent hazard, and it is recommended that PCRs be done in duplicates, with positive results confirmed by Southern blot analysis. A well-designed Southern blot strategy will also generate valuable additional information, including transgene copy number and the number of transgene integration sites, or in the case of gene targeting, conformation of correct targeting events.

### 1.1.1. Isolation of Genomic DNA from Rodent Tail Biopsies

Here we describe two procedures for the isolation of genomic DNA. First is the isolation of high-quality genomic DNA by phenol/chloroform extraction. This method is relatively time-consuming but results in DNA of a quality good enough for all common downstream applications, including PCR and restriction digests. The second, and more rapid, method is less labor-intensive and is therefore recommended when screening a large number of samples. DNA generated by this method is generally suitable for PCR but can often be resistant to digestion by restriction endonucleases, owing to the relatively high levels of contaminating protein, sodium dodecyl sulfate (SDS), and salts. Both methods are used regularly when isolating genomic DNA from rodent tail biopsies, but they can be equally well employed for the extraction of DNA from almost any other biologic tissue.

### 1.1.2. Genotyping by PCR Analysis

Before embarking on the screening of potential transgenics by PCR, it is advisable to establish that the PCR strategy and reagents to be used are able to detect a single copy of the transgene within the context of a complete genome *(1)*. The easiest way to simulate the actual screening procedure is to dilute the cloned transgene with wild-type mouse genomic DNA. The chosen PCR strategy should be able to detect a single transgene or recombination event in $3 \times 10^9$ bp of genomic DNA (approximately the size of the murine genome). The PCR, after optimization, should allow the detection of the target DNA sequence without background amplification.

### 1.1.3. Genotyping by Southern Blot Analysis

In contrast to the PCR, Southern blot analysis can supply detailed information about the genetic structure of the transgenic alteration. Information derived from Southern blot analysis can describe conclusively a homologous recombination event or assess transgene copy number and the number of integration sites within the genome. This procedure is less sensitive to cross contamination than PCR but is more labor-intensive and time-consuming.

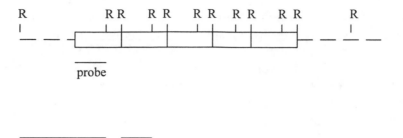

Fig. 1. Strategy for analyzing genomic digests to estimate transgene copy number. The open box represents five copies of the transgene in head-to-tail orientation, each exhibiting two restriction sites for the enzyme R. The broken line indicates that the distances between restriction sites are unknown and vary between hereditary lines. To estimate the transgene copy number, the genomic DNA is digested with R, then separated on an agarose gel, transferred, and fixed to a nylon membrane. The probe used to screen the membrane and its localization is indicated. On the autoradiograph, two types of signal are expected (A and B). B always has the same predicted size, independent of the integration site because it is formed by restriction sites *within* the transgene. The size of the type A bands vary with each different site of integration, with the number of A bands representing the number of integration sites. Quantifying the intensity of band B in comparison with band A can be used to determine the copy number of the transgene. In this example, band B should be five times as intense as band A. Note, however, that if the transgene has only one copy, and therefore only one integration site, only a single band A will appear with no band B.

Southern analysis involves three steps *(2)*:

1. The genomic DNA is digested with appropriate restriction enzymes. The fragments are then size separated by electrophoresis in an agarose gel.
2. DNA is transferred from the gel and covalently bound to a nylon or nitrocellulose membrane.
3. A labeled probe is hybridized to the bound DNA.

Before starting a Southern blot, it is important to have an accurate restriction enzyme map of the transgene required. Nucleotide sequences and restriction maps can be obtained using computerized DNA databanks, such as the EMBL nucleotide sequence database and mapping programs from UWGCG (University of Wisconsin Genetics Computer Group *[3]*; see also http://www.hgmp.mrc.ac.uk). An introduction to databases, computer networks, and molecular biology can be found in **ref. 4**.

The aim is to find restriction sites that are only present once or twice in the transgene. One should also be aware that some of the CpG-containing enzymes, which might be useful for the detection of the transgene, are methylation sensitive (e.g., *Sfi*I, *Sac*II, *Sal*I, or *Xho*I) and therefore result in only partial digestion of the genomic DNA *(5)*. As salts often co-precipitate with DNA, it is advisable to choose enzymes that cut in high-salt conditions—precipitated salts may inhibit those enzymes dependent on low-salt concentrations. **Figure 1** shows a typical example of a screening strategy.

### 1.1.3.1. PROBE LABELING AND PURIFICATION

The probe chosen to screen the genomic DNA should be labeled using one of a variety of commercially available labeling kits, following the manufacturer's protocol. Two of the more commonly used methods for oligonucleotide probe labeling are random priming *(6)* and nick translation *(7)*. Roche offers one of the fastest protocols with a random priming kit called High Prime. It contains premixed solutions for 50 labeling reactions.

## 1.2. RNA Analysis

The analysis of RNA is often the quickest and easiest way to study the expression patterns of a transgene or manipulated genomic sequence within a transgenic animal. RNA is, however, extremely sensitive to degradation by RNases; thus, considerable caution must be taken when purifying and manipulating these molecules. The use of disposable plasticware or designated glassware baked at 200°C overnight and diethylpyrocarbonate (DEPC)-treated reagents are essential when working with RNA. Gloves must be worn at all times, and long hair should be tied back.

### 1.2.1. Isolation of RNA

RNA may be isolated as either total RNA or mRNA, depending on the investigator's requirements. Total RNA is simpler to isolate and is suitable for most downstream applications, including RT-PCR and Northern blots of moderate to highly expressed genes. From this total RNA, mRNA—consisting of about 1 to 2% of the total RNA pool—can be further purified using poly(T) cellulose. This uncomplicated step allows greatly enhanced signals when blotted, because up to 100 times as much RNA can be loaded on an equivalent gel, thus increasing the relative abundance of the message to be investigated accordingly.

Ready-to-use reagents for RNA isolation are available from several suppliers (TRIzol, Gibco; RNAzol, Ambion). These utilize the strong denaturing activity of guanidinium thiocyanate, thus allowing an RNase-free environment from which the nucleotides can be isolated. It is possible to extract good-quality RNA from a large number of samples in less than 2 h using these reagents.

To isolate RNA free from degradation, it is essential that tissue be either snap-frozen or processed *immediately* after it has been harvested. Homogenizing fresh tissue directly into guanidinium thiocyanate–containing reagents usually gives better results than tissue that has been subjected to a freeze/thaw cycle. Rapid and complete homogenization in the presence of RNase-inhibiting agents is essential to achieve optimum results.

### 1.2.2. Northern Blot

Northern Blot analysis is used to determine qualitatively the relative level of gene expression within the specific tissue from which the RNA was isolated. In this technique, RNA is size separated in a denaturing formaldehyde containing agarose gel, transferred to a nitrocellulose or nylon membrane, and subsequently hybridized to a specific labeled DNA or RNA probe *(8)*.

### 1.2.3. Reverse Transcriptase Polymerase Chain Reaction

RT-PCR allows the specific detection of mRNA transcripts even if they exist at very low levels; as few as one copy in 1000 cells is said to be detectable *(9)*. Transcripts that even the most sensitive Northern blots fail to detect are often easily detectable by RT-PCR. An RT-PCR strategy also can be designed in such a way that even the most closely related genes (e.g., genes from the same family) can be distinguished *(10)*. Carefully designed primers can allow transcripts that differ in as little as one nucleotide to be discerned, permitting, e.g., a mutant transgene to be preferentially amplified over the endogenous wild-type allele *(10)*.

RT-PCR is essentially composed of two phases. In the first phase, mRNA is reverse transcribed to form complementary (c)DNA. Priming for this reaction is achieved either by gene-specific primers, random hexamers, or oligo-$(dT)_{15}$ primers. During the second phase, specific primers allow the amplification of the target cDNA by thermocycling as in standard PCR.

The protocol described herein is based on the RT-PCR reaction using Superscript II RT from Gibco-BRL.

### 1.2.4. In Situ RT-PCR and In Situ Hybridization

*In situ* hybridization (ISH) and *in situ* PCR are the methods of choice to investigate the tissue distribution and cellular localization of mRNA molecules or DNA rearrangements at a cellular level. When high levels of the target molecule are present, ISH is sufficient for detection without the need for any PCR amplification. For single-copy DNA sequences or low-copy mRNAs, it is usually necessary to amplify the target molecules before performing the hybridization step, as is the case with direct or indirect *in situ* PCR. Single-stranded complementary RNA, cloned cDNA sequences, and synthetic oligonucleotides are the most commonly used probes for the detection of specific mRNA or DNA molecules. ISH or PCR can be performed on cell suspensions *(11)*, cytospins *(12)* and tissue sections *(13)*. With direct *in situ* PCR, labeled nucleotides (digoxigenin-11-dUTP [DIG-11-dUTP], fluorescin-dUTP, or [$^{35}$S]-UTP) are incorporated during the *in situ* amplification, and the mRNA or DNA of interest is then detected immunohistochemically (DIG-11-dUTP and fluorescin-dUTP) or by using autoradiography ([$^{35}$S]-UTP) *(14)*. Immunohistochemistry is performed using anti-DIG (DIG-11-dUTP) or antifluorescin (fluorescin-dUTP) alkaline phosphatase (AP)–conjugated antibodies or gold-labeled antibodies followed by protein A–gold and silver enhancement. It is also possible to detect fluorescin-dUTP-labeled probes directly using fluorescent microscopy. Immunohistochemistry, however, allows for signal amplification and thus gives better results if signals are weak. During indirect *in situ* PCR, the *in situ* amplification is performed without labeled nucleotides, with the PCR product being detected by the hybridization of a labeled probe *(15)*.

Direct *in situ* PCR is less time-consuming but can result in high levels of background signal caused by the incorporation of labeled nucleotides into damaged DNA sequences by the *Taq* DNA polymerase. Because the DNA quality is dramatically reduced during the preparation of cytospins or tissue sections, the use of direct *in situ* PCR is best reserved for cell suspensions. Indirect *in situ* PCR, in general, gives much greater specificity and is more reproducible *(14)*.

### 1.2.4.1. PREPARATION OF SPECIMEN

Fixation of tissue specimen in 4% paraformaldehyde/phosphate-buffered saline (PBS) is recommended for the demonstration of DNA and mRNA targets *(13)*. For optimal fixing, the tissue is immersed for at least 2 h but not longer than 24 h in the fixation solution. Extensive fixation times can lead to protein-nucleotide crosslinking, thus masking nucleic acid sequences and reducing hybridization and PCR signals. For cryostat or cell preparations, formaldehyde fixation is recommended.

### 1.2.4.2. INDIRECT *IN SITU* PCR

The basic protocol for *in situ* PCR is much the same as standard "in-tube" PCR (*see* **Subheading 3.1.2.**), except that the cycle number must remain low (20–25 cycles) in order to maintain tissue morphology and reduce nonspecific amplification. PCR reactions to be performed *in situ* should first be optimized on a blank slide with cDNA or genomic DNA as a template. The amplified product can then be analyzed by gel electrophoresis. It is crucial for the outcome of the *in situ* experiment that the only product generated by the PCR is that of interest. Conditions *in situ* should remain essentially unchanged, allowing the investigator some degree of confidence that the amplification is specific, as seen on the blank slide.

## 1.3. Western Blot Analysis

Western Blot analysis is a commonly applied method that permits the detection of a specific protein within a given tissue or cell culture. Using this method, the investigator can directly describe and quantify the presence of endogenous or transgenic proteins, even allowing (assuming that an appropriate antibody is available) the detection of truncated protein products. Proteins, present in crude tissue or cell homogenate, are size separated by electrophoresis on an acrylamide gel and then transferred to a nitrocellulose membrane and probed with antibodies against the protein of interest *(16)*. An excess of negatively charged SDS in the acrylamide gel guarantees that all proteins are completely denatured, with secondary structures destroyed. As a result, protein migration is relative to molecular mass, irrespective of structural morphology. This also dictates that any primary antibody used in Western blot analysis must recognize the protein that it was raised against in its denatured state. Once the primary antibody has bound to its target, the blot is incubated with an enzyme-labeled secondary antibody capable of recognizing the $F_C$ region of the first antibody. Multiple secondary antibodies will bind to one primary antibody, thereby amplifying the detectable signal generated by the attached enzyme, often horseradish peroxidase (HRP) or AP.

### 1.3.1. Cell Lysis

The most common method for preparing samples for immunoblots is to lyse the entire tissue mass in sample buffer. If, however, the protein of interest is of very low abundance, it is recommended that steps be taken to enrich for the protein of interest with either subcellular fractionation or immunoprecipitation prior to SDS polyacrylamide gel electrophoresis. Methods relating subcellular fractionation and immunoprecipitation can be found in **refs.** *17* and *18*.

### 1.3.2. Protein Transfer

Protein transfer can be accomplished either in a tank filled with transfer buffer or in a semidry blotter *(19)*. Both methods are effective, although the latter gives faster transfer times.

### 1.3.3. Immunodetection of Western Blot

The variety of commercially obtainable primary and secondary antibodies has greatly expanded. Suppliers offer catalogs with available primary antibodies listed according to the protein class that they recognize (e.g., signal transduction proteins, cell adhesion proteins, receptor proteins). Secondary antibodies raised in a variety of host species and labeled with a wide range of marker enzymes are readily available. **Subheading 3.3.4.** describes a general protocol using an HRP-labeled secondary antibody. This protocol can be applied to other secondary antibodies with, if a different marker enzyme conjugate is used, the exception of the final color reaction (*see* **Notes 52–56**).

## 1.4. Analysis of Transgene Expression by Immunohistochemistry

Immunohistochemistry can be used to examine the distribution and localization of a transgenic, or indeed any, protein within tissue sections or single cells. A variety of protocols are available for the detection of specific proteins *(20)*. In general, the fixed and permeabilized tissue is incubated with a primary antibody that was raised against the protein to be investigated. Only in rare cases, when the epitope is very abundant, can the primary antibody be directly labeled (e.g., anti BrdU-antibody *[21]*). In most cases, the primary binding event will require amplification in order to provide signals sufficient for detection. These secondary, or in some cases even tertiary, levels of amplification are provided by the use of secondary antibodies and enzyme conjugates. Secondary antibodies are linked either to a fluorochrome, which is then observed using fluorescent microscopy *(22)*, or to enzymes such as AP or HRP, which results in a permanent stain that can be examined under a bright-field microscope *(23)*. For low-abundance proteins, detection can be further enhanced by the use of heavily biotinylated secondary antibodies. These are able to bind multiple avidin molecules that themselves are linked to enzyme reporter molecules such as AP or HRP. It is also possible to boost weak signals by applying methods based on three-enzyme complexes. These include peroxidase antiperoxidase complexes and AP antialkaline phosphatase complexes *(24)*. Such methods are highly sensitive and usually give excellent results on fixed, paraffin-processed material or cryostat sections.

## 1.5. Reporter Enzyme Assays

The following section describes techniques that can be employed to assess the transcriptional activity of regulatory DNA elements such as promoters, enhancers, and silencers. Potential regulatory sequences of the chosen mammalian gene are fused to the coding region of one of several, mostly bacterial, reporter genes. The most commonly used reporter enzymes are chloramphenicol acetyl transferase (CAT) *(25)*, β-galactosidase *(26)*, and luciferase *(27)*. Each of these enzymes carries a unique enzymatic activity that can be detected readily without interference from endogenous homologs. Sensitive assays are available that allow quantitative measurements of the transgene expression (for a review, *see* **ref. 28**).

### 1.5.1. CAT Assay

*CAT* gene is derived from transposon 9 of *Escherichia coli (29)* and encodes for a trimeric protein consisting of three 25-kDa subunits *(30)*. The enzyme confers resistance to the antibiotic chloramphenicol by allowing bacteria to acetylate the drug *(31)*. This enzyme is particularly useful as a reporter because it lacks any mammalian equivalent. The coding region of *CAT* is also relatively small (660 bp), which means that vectors carrying this gene can be easily modified. The CAT assay is based on the transfer of a labeled acetyl group from a water-soluble acetyl-CoA donor to a water-insoluble chloramphenicol acceptor. The labeled end product can then be extracted from its substrates and quantitatively assayed using liquid scintillation spectroscopy.

### 1.5.2. β-Galactosidase Assay

β-Galactosidase is a tetrameric bacterial protein encoded for by the *E. coli lacZ* gene *(32)* and is responsible for the hydrolysis of β-galactosides. Mammalian tissues contain a β-galactosidase activity; however, since the bacterial β-galactosidase is active at pH 7.3, whereas its mammalian counterpart is inactive unless in acidic conditions—having an optimum pH of 3.5—the two proteins can be easily distinguished *(26)*. The enzyme activity is determined by a photometric assay that measures the hydrolysis of the substrate *o*-nitrophenyl β-D-galactopyranoside (ONPG), with the hydrolyzed product causing an increase in absorbance at 420 nm *(26)*. The main advantage of using this reporter system is that it allows easy *in situ* analysis of expression with histochemical staining.

### 1.5.3. Luciferase Assay

The enzyme luciferase most commonly used is derived from the firefly *Photinus pyralis (33,34)*. The 60.7-kDa monomeric protein is encoded for by the *luc* gene. Luciferase is capable of catalyzing a reaction involving D-luciferin and adenosine triphosphate (ATP) in the presence of oxygen and $Mg^{2+}$ that results in the production of light *(35)*. Such emissions can be measured and quantified using a luminometer, giving sensitive and immediate results. The assay's main advantages are that it has a greatly improved sensitivity over the CAT assay and that results can be obtained in a fraction of the time. Luciferase also has a shorter half-life than CAT or β-galactosides, making it particularly useful for transient and inducibility studies. Because no equivalent exists in mammalian systems, endogenous activity is not an issue.

## 2. Materials

### 2.1. Genotyping

### 2.1.1. Preparation of High-Quality Genomic DNA (*see* **Notes 1–3**)

1. Sterile scissors.
2. Water bath or incubator.

3. Digestion buffer: 50 m*M* Tris-HCl (pH 8.0), 100 m*M* EDTA, 100 m*M* NaCl, 1% (w/v) SDS.
4. Proteinase K (20 mg/mL) (cat. no. P0390; Sigma, St. Louis, MO).
5. RNase A (1 mg/mL).
6. Tris-buffered phenol.
7. Chloroform/isoamyl alcohol (24:1).
8. 100 and 70% Ethanol.

### 2.1.1.1. FAST PROTOCOL FOR ISOLATION OF GENOMIC DNA

1. Homogenization buffer: 0.3 *M* NaOAc, 10 m*M* Tris-HCl (pH 7.9), 1 m*M* EDTA, 1% SDS; before use add 0.2 mg/mL of proteinase K.

## 2.1.2. Genotyping by PCR Analysis

1. Thermocycler (e.g., Hybaid, Perkin-Elmer).
2. Microcentrifuge tubes (0.5 mL).
3. Primers (*see* **Notes 4** and **5**).
4. 10X PCR buffer: 200 m*M* Tris-HCl (pH 8.3), 20 m*M* MgCl$_2$, 250 m*M* KCl, 0.5% (v/v) Tween-20, and 1 mg/mL of bovine serum albumin (BSA).
5. 10 m*M* solutions of dATP, dCTP, dGTP, dTTP (100 m*M* dNTP-Set, cat. no. 84205520-53; Roche).
6. *Taq* DNA polymerase (Gibco-BRL or Roche).
7. Mineral oil (cat. no. M3516; Sigma).
8. Transilluminator.

## 2.1.3. Genotyping by Southern Blot Analysis

1. Genomic DNA (10 μg/digest).
2. Restriction enzyme of choice (approx 30 U/digest) and appropriate buffer.
3. 37°C Water bath.
4. DNA sample buffer: 40% sucrose, 0.5% SDS, 0.25% bromophenol blue in TBE.
5. Molecular biology grade agarose (Seakem LE® agarose; Flowgen, UK).
6. Ethidium bromide (10 mg/mL).
7. DNA size standards (e.g., Roche DNA Molecular Weight Marker X, 0.07–12.2 kb; cat. no. 84103121).
8. 10X TBE; 108 g Tris, 55 g boric acid, 7.4 g EDTA; add up to 1 L with water.
9. Gel-casting tray and running tank (10 × 15 cm).
10. Electrophoresis power pack.
11. Ultraviolet (UV) light transilluminator (312 nm).
12. UV crosslinker or oven.
13. Nylon hybridization membrane (e.g., Hybond N+; Amersham, UK).
14. 3MM Whatman paper.
15. Transfer buffer: 20X saline sodium citrate (SSC) (3.0 *M* NaCl, 0.3 *M* sodium citrate).
16. 0.25 *M* HCl.

17. Denaturing buffer: 0.5 $M$ NaOH, 1.5 $M$ NaCl.
18. Neutralizing buffer: 1.0 $M$ Tris-HCl (pH 7.4), 1.5 $M$ NaCl.

### 2.1.3.1. PROBE LABELING AND PURIFICATION

1. Water bath or heating block.
2. Random labeling kit (e.g., High Prime, cat. no. 1585592; Roche).
3. $\alpha^{32}$P-ATP or $\alpha^{32}$P-CTP, 3000 Ci/mmol, aqueous solution (e.g., cat. no. 39011X; ICN).
4. Denatured DNA (25–200 ng).
5. 0.2 $M$ EDTA.
6. Sephadex G-50 columns (cat. no. 17-0855-02; Pharmacia Biotech).
7. TE buffer: 10 m$M$ Tris-HCl, pH 7.5, 1 m$M$ EDTA.
8. Sonicated salmon sperm DNA (10 mg/mL) (cat. no. D-9156; Sigma).

### 2.1.3.2. HYBRIDIZATION

1. Hybridization buffer: 15 mL of 20X SSC, 5 mL of 10% (w/v) SDS, 30 mL of 16.7% (w/v) dextransulfate, 0.1 mg/mL of sonicated salmon sperm DNA.
2. Washing solutions 1–5:
   a. Solution 1: 2X SSC + 1% SDS.
   b. Solution 2: 1X SSC + 1% SDS.
   c. Solution 3: 0.5X SSC + 1% SDS.
   d. Solution 4: 0.25X SSC + 0.1% SDS.
   e. Solution 5: 0.1X SSC + 0.1% SDS.
3. Hybridization oven.
4. Autoradiography films (e.g., Biomax MS-1, Kodak) and exposure holders preferentially with amplifier screens.
5. –80°C Freezer.

## 2.2. RNA Analysis

### 2.2.1. Isolation of RNA

1. Desktop centrifuge.
2. TRIzol (Gibco-BRL).
3. Liquid nitrogen.
4. DEPC-treated water: Stir 0.01% (v/v) DEPC in water overnight and autoclave.
5. Autoclaved Eppendorf tubes.
6. Chloroform.
7. Glass-Teflon or Polytron power homogenizer.
8. Isopropanol.
9. 70% Ethanol.

### 2.2.2. Northern Blot

1. Gel-casting tray and running tank 10 × 15 cm.
2. DEPC-treated water: 0.01% (v/v) DEPC in water overnight and autoclave.
3. Molecular biology grade agarose (cat. no. A-9539; Sigma).

4. 10X MOPS running buffer: Dissolve 41.8 g of MOPS in 800 mL of DEPC-treated water. Adjust the pH to 7.0 by adding NaOH or acetic acid, and add 16.6 mL of 3 $M$ DEPC-treated sodium acetate and 20 mL of 0.5 $M$ DEPC-treated EDTA. Bring to 1 L with DEPC-treated water. MOPS solutions should be wrapped with aluminum foil.
5. RNA sample buffer: 50 µL of 10X MOPS running buffer, 87.5 µL of formaldehyde, 250 µL of formamide.
6. 37% Formaldehyde (pH > 4.0).
7. Deionized formamide.
8. Loading buffer: 1 m$M$ EDTA, 0.25% bromophenol blue, 0.25% xylene cyanol, 50% glycerol.
9. 10X SSC.
10. Ethidium bromide (10 mg/mL).

## 2.2.3. Reverse Transcriptase Polymerase Chain Reaction

1. Water bath or heating block.
2. Total RNA (1–10 µg).
3. DNase I (RNase-free, cat. no. 18047-019; Gibco-BRL).
4. RNase inhibitor (RNasin, cat. no. N2111; Promega).
5. Oligo (dT)$_{15}$ primer (cat. no. 83718423; Roche).
6. RNA-dependent DNA polymerase (Superscript II RT, cat. no. 18064-014; Gibco-BRL).
7. 10 m$M$ dNTP Mix diluted in DEPC-treated water (cat. no. 84205520; Roche).
8. 0.1 $M$ Dithiothreitol (DTT).
9. DEPC-treated water: Stir 0.01% (v/v) DEPC in water overnight and autoclave.

## 2.2.4. In Situ RT-PCR and ISH

### 2.2.4.1. PREPARATION OF DIG-LABELED PROBES

1. Thermocycler (e.g., Hybaid).
2. *Taq* DNA polymerase (Gibco-BRL), 10X buffer, and 50 m$M$ MgCl$_2$.
3. 10 m$M$ dATP, dCTP, dGTP, dTTP (cat. no. 83718423; Roche).
4. DIG-11-dUTP (cat. no. 1093088; Roche).

### 2.2.4.2. PRETREATMENT OF PARAFFIN BLOCKS

1. Poly-L-lysine (cat. no. P1274; Sigma): 10 mg/mL in 10 m$M$ Tris-HCl, pH 8.0.
2. Glass slides.
3. Absolute ethanol.
4. Xylene.
5. 0.25 $M$ Ammonium acetate.
6. 0.2 $N$ HCl.
7. DEPC-treated water: 1 mL of DEPC/L of water, stirred overnight and autoclaved.
8. PBS (in DEPC-treated water, 1 L): 2.35 g Na$_2$HPO$_4$, 0.45 g NaH$_2$PO$_4$, 8.76 g NaCl, 0.2 g NaAz.

9. 0.01% Triton X-100 in PBS.
10. Proteinase K stock solution (20 mg/mL) (cat. no. 84010727-62; Roche).
11. DNase I (cat. no. 18047-019; Gibco-BRL).
12. Slide thermal cycler (OmniSlide, Hybaid).

### 2.2.4.3. RT REACTION

1. Slide thermal cycler (OmniSlide, Hybaid).
2. RNase inhibitor (RNasin; Promega).
3. Oligo (dT)$_{15}$ primer (cat. no. 83718425-21; Roche).
4. RNA-dependent DNA polymerase (Superscript II RT, Gibco-BRL).
5. 10 m$M$ dNTP Mix (in DEPC-treated water).
6. 0.1 $M$ DTT.
7. DEPC-treated water.
8. 20X SSC: 3 $M$ NaCl, 0.3 $M$ sodium citrate.
9. 0.4% Paraformaldehyde.
10. PBS: 2.35 g/L Na$_2$HPO$_4$, 0.45 g/L NaH$_2$PO$_4$, 8.76 g/L NaCl, 0.2 g/L NaNa$_3$.

### 2.2.4.4. INDIRECT *IN SITU* PCR

1. Slide thermal cycler (OmniSlide, Hybaid).
2. Sure seal (Hybaid).
3. Primers (40 pmol/µL).
4. 10X PCR buffer: 200 m$M$ Tris-HCl, pH 8.3, 20 m$M$ MgCl$_2$, 250 m$M$ KCl, 0.5% (v/v) Tween-20, and 1 mg/mL of BSA.
5. 10 m$M$ solutions of dATP, dCTP, dGTP, dTTP (100 m$M$ dNTP-Set, cat. no. 84205520-53; Roche).
6. *Taq* DNA polymerase (Gibco-BRL or Roche).

### 2.2.4.5. HYBRIDIZATION AND DETECTION

1. Hybridization buffer (500 µL): 20 µL of 50X Denhardt's (cat. no. D2532; Sigma), 100 µL of 50% dextran sulfate (cat. no. D8906; Sigma), 20 µL of denatured salmon sperm DNA, 200 µL of 20X SSC, 100 µL of TE, 1 µL of labeled probe (50 ng/µL). Make up to 500 µL with dH$_2$O.
2. Buffer 1: 100 m$M$ Tris-HCl, pH 7.5, 150 m$M$ NaCl, 2 m$M$ MgCl$_2$.
3. Buffer 2: buffer 1 + 3% BSA (cat. no. B-2518; Sigma).
4. Buffer 3: 100 m$M$ Tris-HCl, pH 9.5, 100 m$M$ NaCl, 50 m$M$ Mg Cl$_2$.
5. Nitroblue tetrazoliumchloride (NBT)/5-Bromo-4-chlor-indolylphosphate (BCIP) (cat. no. 1681451; Roche).

## *2.3. Western Blot Analysis*

### *2.3.1. Cell Lysis*

1. 1X Sample buffer: 1 mL 0.5 $M$ Tris-HCl, pH 6.8, 1 mL glycerol, 1 mL 10% SDS, 4.9 mL distilled water, 120 mg DTT, 1 mg bromophenol blue.
2. Glass-Teflon homogenizer.
3. Needles (20 and 26 gage) and 1- to 5-mL syringes.

## 2.3.2. Sodium Dodecyl Sulfate Polyacrylamide Gel Electrophoresis

1. Electrophoresis tank (e.g., Bio-Rad Mini-Gel Apparatus), glass plates (e.g., 7.5 × 10 × 0.05 cm minigels), 0.75 mm spacers, and combs.
2. Constant-current power supply.
3. 4X Stacking gel buffer: 0.5 $M$ Tris-HCl, pH 6.8, 0.4% SDS.
4. 4X Separating gel buffer: 1.5 $M$ Tris-HCl, pH 8.8, 0.4% SDS.
5. 10X SDS running buffer: 250 m$M$ Tris, 1.9 $M$ glycine, 1.0% SDS (adjust to pH 8.3).
6. 40% Acrylamide/bis solution (37.5:1) (cat. no. 161-0148; Bio-Rad).
7. Ammonium persulfate (cat. no. A-7460; Sigma).
8. TEMED (cat. no. T-8133; Sigma).
9. $H_2O$-saturated isobutyl alcohol.
10. Protein molecular weight standard.

## 2.3.3. Protein Transfer

1. Nitrocellulose membrane or polyvinyl difluoride (PVDF) membrane: 0.2-μm pore size (Schleicher & Schuell or Millipore).
2. Whatman 3MM paper.
3. Ponceau-S (0.2% in 3% trichloroacetic acid, available from Sigma).
4. Constant-current power supply.
5. Semidry blotter (e.g., Millipore) or tank system (e.g., Bio-Rad) with two foam pads (e.g., Scotch Brite pads).
6. Blotting buffer for semidry transfer: Add the following and make up to 1 L with d$H_2O$: 5.82 g of Tris, 2.93 g of glycine, 200 mL of methanol, 3.75 mL of 10% SDS; do not adjust the pH.
7. Transfer buffer for tank blotting: Add the following and make up to 3.5 L with d$H_2O$: 10.6 g of Tris, 50.4 g of glycine, 35 mL of 10% SDS, 700 mL of methanol; do not adjust the pH.

## 2.3.4. Immunodetection of Western Blot

1. TBS buffer: 20 m$M$ Tris-HCl, pH 8.0, 150 m$M$ NaCl.
2. Blocking solution: TBS, 1% BSA (Sigma).
3. Antibody solution: TBS, 1% BSA, 0.1% Tween-20.
4. Washing solution: TBS, 0.1% Tween-20.
5. HRP-conjugated secondary antibody (Amersham).
6. Electrochemiluminescence (ECL) Western Blotting Detection Reagents (Amersham).
7. Hyperfilm ECL (Amersham).

## 2.4. Analysis of Transgene Expression by Immunohistochemistry

### 2.4.1. Avidin-Biotin Technique for Paraffin Sections

1. Xylene.
2. Absolute alcohol, 70 and 65% alcohol.
3. Trypsin (cat. no. T-8128; Sigma).

4. Tris-buffered saline (TBS).
5. Normal serum (from the species used when raising the secondary antibody).
6. Primary antibody (either monoclonal or polyclonal).
7. Polyclonal biotinylated secondary antibody (biotinylated swine antirabbit [cat. no. E0353] or biotinylated rabbit antimouse [cat. no. E0354]; Dako).
8. Avidin biotin complex/AP (cat. no. K0376; Dako).
9. Vector Red Alkaline Phosphatase Substrate Kit I (cat. no. SK-5100; Vector) or AP substrate solution: Dissolve 2 mg of Naphtol AS MX phosphate in 0.2 mL of dimethylformamide in a glass beaker. Dilute to 10 mL with Tris-HCl, pH 8.2. To block endogenous AP, add 10 µL of levamisole. Immediately before use, dissolve 10 mg of Fast red TR in the solution and filter onto the sections.
10. 3,3-Diaminobenzidene (DAB) solution: Dissolve 7.5 g of DAB tetrahydro-chloride in 300 mL of Tris (pH 7.6), and store in 1-mL aliquots at –20°C. Care must be taken when handling DAB; it is a suspected carcinogen.
11. Hematoxylin (cat. no. 9249; Merck).
12. Mounting medium such as DPX mountant (cat. no. 360292F; BDH) or Crystal mount (Biomeda, Foster City, CA).

## 2.5. Reporter Enzyme Assays

### 2.5.1. CAT Assay

1. Homogenization buffer: 250 m$M$ Tris-HCl, pH 7.8.
2. Teflon homogenizer (1–5 mL).
3. Water bath or heating block.
4. Reaction mixture: 30 µL of 250 m$M$ Tris-HCl, pH 7.8, 5 m$M$ EDTA; 20 µL of 8 m$M$ chloramphenicol (Sigma); 20 µL of diluted acetyl CoA solution. Prepare the diluted acetyl CoA solution by diluting 20 µL of $^{14}$C acetyl CoA (1 mCi) (Amersham) with 0.5 m$M$ unlabeled acetyl CoA (cat. no. A2056; Sigma) to a final concentration of 5 mCi/mL.
5. Ethyl acetate (4°C).
6. Liquid scintillation vials and fluid.
7. Liquid scintillation counter.

### 2.5.2. β-Galactosidase Assay

1. ONPG (cat. no. N1127; Sigma). Prepare a 4 mg/mL solution in 0.1 $M$ sodium phosphate buffer (pH 7.5).
2. 1 $M$ Na$_2$CO$_3$.
3. 0.1 MgCl$_2$, 4.5 $M$ β-mercaptoethanol.

### 2.5.3. Luciferase Assay

1. Homogenization buffer: 1% Triton X-100, 25 m$M$ glycylglycine (pH 7.8), 15 m$M$ MgSO$_4$, 4 m$M$ EGTA, and 1 m$M$ DTT.
2. Incubation buffer: 25 m$M$ glycylglycine (pH 7.8), 15 m$M$ MgSO$_4$, 4 m$M$ EGTA, 15 m$M$ potassium phosphate (pH 7.8), 1 m$M$ DTT, and 2 m$M$ ATP.

3. Luciferin solution: 0.2 m$M$ luciferin (cat. no. L6882; Sigma). Dissolve in 25 m$M$ glycylglycine (pH 7.8), 15 m$M$ MgSO$_4$, 4 m$M$ EGTA, and 2 m$M$ DTT.
4. Luminometer with autoinjection system (Lumac Biocounter M2500; Landgraaf, The Netherlands).

## 3. Methods

### 3.1. Genotyping

### 3.1.1. Preparation of High-Quality Genomic DNA (*see* **Notes 1–3**)

1. Anesthetize the animals by an approved procedure (e.g., halothane).
2. Restrain the animal with one hand, and with the other, remove approx 1 cm of the tail with a sharp pair of sterile scissors.
3. Place the tail biopsy in 500 µL of digestion buffer and add Proteinase K to a final concentration of 100 µg/mL.
4. Incubate at 55°C for 3 to 4 h or overnight, preferentially on an orbital shaker.
5. Add 5 µL of RNase A and incubate for 1 to 2 h at 37°C.
6. Add an equal volume of Tris-buffered phenol (500 µL), and mix thoroughly but avoid vortexing.
7. Microfuge the sample for 10 min at room temperature.
8. Carefully transfer the upper aqueous layer to a clean Eppendorf tube.
9. Add an equal volume of chloroform/isoamyl alcohol (24:1) and mix thoroughly.
10. Microfuge the sample for 10 min at room temperature.
11. Transfer the upper aqueous layer to a clean Eppendorf tube, and add 800 µL of 100% ethanol.
12. Microfuge the precipitated DNA for 2 min and discard the supernatant.
13. Wash the pellet in 700 µL of 70% ethanol.
14. Air-dry and redissolve the pellet in 200 µL of distilled water or TE. DNA yield is assayed at 260 nm. An OD$_{260}$ of 1 equals 50 µg/mL. Genomic DNA is stored at 4°C.

#### 3.1.1.1. FAST PROTOCOL FOR ISOLATION OF GENOMIC DNA (*SEE* **NOTES 6** AND **7**)

1. Prepare tissue as described in **Subheading 3.1.1.**
2. Incubate about 100 mg of tissue in 700 µL of SDS containing homogenization buffer overnight at 37°C.
3. Freeze samples for 30 min to precipitate the SDS.
4. Centrifuge the frozen samples at 4°C for 30 min at 14,000$g$.
5. Carefully transfer the supernatant into a fresh Eppendorf tube using a pipet. Avoid transferring any of the SDS containing the pellet.
6. Add an equal volume of isopropanol, and invert the tube until a white precipitate is seen.
7. Centrifuge at 14,000$g$ for 5 min at room temperature.
8. Wash the DNA pellet with 500 µL of 70% ethanol and recentrifuge.
9. Dry the pellet and resuspend in 150 µL of water or TE.

## 3.1.2. Genotyping by PCR Analysis (see **Notes 4, 5,** and **8–14**)

1. Add the following to a 0.5-mL Eppendorf tube on ice: 5 µL of 10X buffer, 1 µL of 10 m$M$ dNTPs, 1 µL (40 pmol/µL) of primer 1, 1 µL (40 pmol/µL) of primer 2, 1 µg of genomic DNA, and 1 U of *Taq* DNA polymerase. Make up to 50 µL with dH$_2$O.
2. Overlay with 50 µL of mineral oil to prevent evaporation.
3. Place the reactions directly from ice to a preheated (94°C) thermocycler block (hot start). One cycle at 94°C for 1 min will denature the template DNA.
4. Perform 35 PCR cycles, e.g., 45 s at 94°C (denaturing), 1 min at 50–70°C (annealing), and 1–5 min at 72°C (elongation).
5. Perform one cycle at 72°C for 5–10 min to allow all DNA strands to be fully synthesized.
6. Visualize PCR products by performing agarose gel electrophoresis with 10–20 µL of the reaction (for agarose gel electrophoresis, *see* **Subheading 3.1.3.**, which refers to the Southern transfer protocol).

## 3.1.3. Genotyping by Southern Blot Analysis

### 3.1.3.1. SOUTHERN TRANSFER (*SEE* NOTES 15–21)

1. Digest 10 µg of genomic DNA with the chosen restriction enzyme in a water bath overnight at the appropriate temperature. The total volume of the digest should not exceed the maximum volume held by a single well of an agarose gel. (If required, digests can be precipitated and resuspended in a reduced volume before loading.)
2. Add 1/10 vol of sample buffer to the digest and load on a large (at least 10- to 15-cm running length is recommended) TBE or TAE agarose gel containing 1 µg/mL of ethidium bromide and run at 4 V/cm. Generally, 0.7% (w/v) agarose gels are convenient, giving good resolution between 500 bp and 15 kb. Stop the run as the blue dye (bromophenol blue) contained within the sample buffer reaches the end of the gel. A DNA size marker should be loaded in a lane adjacent to the digests.
3. Photograph the gel on a UV light transilluminator with a ruler placed alongside the edge of the gel next to the DNA size standard.
4. Incubate the gel in 0.25 $M$ HCl for 10 min (only required if fragments larger than 15 kb are expected).
5. Incubate the gel in denaturing solution for 40 min.
6. Put the gel into neutralizing buffer two times for 20 min each time.
7. Build up the blot (*see* **Fig. 2**): Place the gel on a 3MM piece of paper soaked with 20X SSC and in contact with the reservoir filled with 20X SSC. On top of the gel put one piece of nylon membrane prewetted in water and then 20X SSC for 5 min, three layers of 3MM Whatman paper soaked with water, and on top a stack of dry paper towels. Weigh down the sandwich with an ~1-kg weight.
8. Leave the transfer overnight.
9. The next day, dismantle the blot, wash the membrane briefly in 2X SSC, and UV-crosslink the DNA to the membrane. Alternatively, the DNA can be crosslinked by placing the membrane in an oven and baking it at 70°C for 2 h.

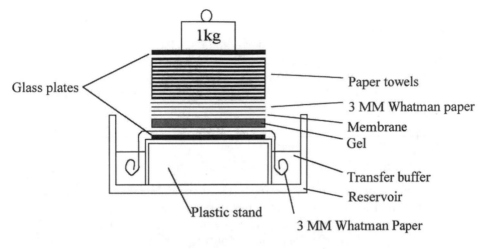

Fig. 2. Southern transfer system.

### 3.1.3.2. PROBE LABELING AND PURIFICATION (*SEE* **NOTES 22–24**)

1. Denature 25–200 ng of DNA in a total volume of 13 μL by boiling it for 5 min.
2. Collect the contents at the bottom of the tube by briefly centrifuging, and transfer 11 μL into a fresh Eppendorf tube.
3. Add 4 μL of High Prime solution.
4. Add 5 μL (=50 μCi) of $\alpha^{32}$P-ATP or $\alpha^{32}$P-CTP.
5. Incubate for 10 min at 37°C.
6. Stop the reaction by adding 2 μL of 0.2 $M$ EDTA (pH 8.0).
7. Add 20 μL of sonicated salmon sperm DNA.
8. Drain the Sephadex G-50 column and equilibrate with 1 mL of TE.
9. Load the stopped labeling reaction onto the column.
10. Wash the column with 400 μL of TE.
11. Elute with 400 μL of TE and collect the flowthrough, containing the labeled probe, in a clean tube.
12. Compare the radioactivity of the column with that of the labeled probe. The radioactivity counted in the probe should be at least that of the column. In a good labeling, the majority of the isotope should have been incorporated into the DNA probe.
13. Boil the purified probe for 5 min to denature the DNA before adding it to the hybridization.

### 3.1.3.3. HYBRIDIZATION (*SEE* **NOTES 25** AND **26**)

1. Soak the membrane briefly in 2X SSC and then prehybridize in hybridization buffer supplemented with 100 μg/mL of salmon sperm DNA.

2. After at least 30 min at 64°C, add the labeled and denatured probe to the prehybridization solution.
3. Allow the hybridization to proceed overnight at 64°C.
4. Discard the hybridization solution.
5. Wash for 20 min in each washing solution in turn *(1–5)*.
6. Wrap the membrane in cling film and expose to an autoradiography film at −80°C.

## 3.2. RNA Analysis

### 3.2.1. Isolation of RNA (see **Notes 27–33**)

1. Completely homogenize 50–100 mg of tissue in not less than 1 mL of TRIzol.
2. Let the samples stand for 5 min at room temperature.
3. Add 0.2 mL of chloroform/mL of TRIzol and shake vigorously.
4. Centrifuge samples at 10,000$g$ in a table centrifuge for 15 min at room temperature.
5. Transfer the colorless aqueous phase to a fresh, RNase-free Eppendorf tube.
6. Add an equal volume of isopropanol, and leave at room temperature for 10 min to precipitate the RNA.
7. Centrifuge at 10,000$g$ in a table centrifuge for 15 min at 4°C.
8. Carefully discard the supernatant, wash the pellet with 1 mL of 70% ethanol, and centrifuge again for 10 min.
9. Discard the wash and dry the pellet. Dissolve the pellet in 50 μL of DEPC-treated water.
10. Measure the quantity of RNA at an OD of 260 nm. One $OD_{260}$ is equivalent to an RNA concentration of 40 μg/mL. The solubilized RNA should have a 260: 280 ratio higher than 1.9. A lower value indicates protein or phenol contamination.

### 3.2.2. Northern Blot (see **Notes 34–36**)

1. For a 1.2% gel, boil 3.6 g of agarose in 213 mL of distilled water. Leave the agarose to cool to 60°C in a heated water bath.
2. To the cooled agarose, add 30 mL of 10X MOPS running buffer and 54 mL of formaldehyde. After adding ethidium bromide to a final concentration of 0.1 μg/mL, the gel is ready to be poured (**caution:** formaldehyde is toxic and the gel should be poured and run in a fume hood).
3. Up to 10 μg of RNA should be added to 3 vol of RNA sample buffer.
4. Incubate the probes at 65°C for 10 min.
5. Add 1/10 vol of loading buffer and load the samples on the gel.
6. Run the gel in 1X MOPS running buffer at a constant voltage of 5 V/cm until the bromophenol blue has migrated halfway down the gel.
7. Destain the gel in 1X MOPS running buffer for 2 to 3 h or until the RNA is clearly seen.
8. Photograph the gel on a UV light transilluminator (312 nm) with a ruler placed alongside the gel.

9. Transfer of RNA to a nylon membrane should be performed immediately after electrophoresis, without any further denaturing or neutralizing steps. The construction of the capillary (or vacuum) blot and the hybridization procedure of the membrane are equivalent to the Southern Blot procedure (*see* **Fig. 2**).

### 3.2.3. Reverse Transcriptase Polymerase Chain Reaction (*see* **Notes 37–41**)

1. Dissolve between 1 and 10 μg of total RNA in a maximum of 9 μL of DEPC-treated water. Add 4 μL of 5X first-strand buffer (supplied with the polymerase), 2 μL of 0.1 *M* DTT, 1 μL of 10 m*M* dNTPs, 1 μL of DNase I (10 U), and 1 μL of RNasin (40 U).
2. Incubate the reaction at 37°C for 30 min to allow degradation of genomic DNA.
3. Incubate at 75°C for 5 min to inactivate the DNase I.
4. Chill the samples on ice and centrifuge briefly.
5. Add 1 μL of oligo-(dT)$_{15}$ primer (500 μg/mL) and 1 μL of Superscript II RT.
6. Incubate at 42°C for 50 min.
7. Heat-inactivate the RT at 70°C for 15 min. The reverse-transcribed cDNA can now be handled like other DNA and should be stored at either +4 or –20°C.
8. Use 2 μL of the generated cDNA pool for a standard PCR reaction.

### 3.2.4. In Situ *RT-PCR and ISH*

#### 3.2.4.1. PREPARATION OF DIG-LABELED PROBES (*SEE* **NOTES 42–46**)

The PCR conditions for this step must be optimized for every pair of primers (*see* **Subheading 1.1.2.**). The following protocol, therefore, only serves as a general guideline.

1. Mix the following in a 0.5-mL Eppendorf tube: 1 μg of template DNA, 10 μL of 10X PCR buffer, 5 μL of 50 m*M* MgCl$_2$, 2 μL of 10 m*M* dNTPs, 8 μL of 0.1 m*M* DIG-11-dUTP, 1 μL of primer 1 (40 pmol/μL), 1 μL of primer 2 (40 pmol/μL), and 0.5 μL of *Taq* DNA polymerase (Gibco-BRL). Make up to 100 μL with distilled water.
2. Perform PCR as described in **Subheading 3.1.2.**
3. Purify labeled probe using DNA purification columns or gel extraction kits (Qiatex II, Qiagen).

#### 3.2.4.2. PRETREATMENT OF PARAFFIN BLOCKS (*SEE* **NOTES 47–49**)

1. To coat poly-L-lysine slides, dip alcohol-cleaned microscope slides in 0.25 *M* ammonium acetate solution, oven dry, soak in 50 μg/mL of poly-L-lysine made up in 10 m*M* Tris (pH 8.0) for 30 min, and oven dry again.
2. Cut 5-μm paraffin sections and float onto precoated slides. Incubate at 50°C for 2 to 3 d to allow the sections to adhere.
3. Dewax in xylene at room temperature for 30 min.

4. For rehydration, wash the sections for 10 min in 100% ethanol, 5 min each in 70% ethanol, 2 min in water, and finally 10 min in PBS.
5. Permeabilize with 0.01–0.1 mg/mL of proteinase K, 100 m$M$ Tris (pH 7.6), and 50 m$M$ EDTA at 37°C for 10 min. This step can be performed on the slide thermal cycler.
6. Stop the reaction by immersing the slides in 0.1 $M$ glycine/PBS for 5 min.
7. Wash the slides two times for 3 min each in PBS.
8. For DNase I treatment, prepare in a microcentrifuge tube 1 μL of RNasin, 8 μL of 10X DNase buffer, 8 μL of DNase I (1 U/μL), and 63 μL of DEPC-treated water.
9. Place the slides in a humid chamber, and add 80 μL of the solution in **step 8** to each section. Keep at room temperature overnight.
10. Stop the DNase reaction by adding 5 μL of 20 m$M$ EDTA. Incubate for 10 min at 65°C on the slide thermal cycler.
11. Wash three times for 10 min each in 50 m$M$ Tris (pH 7.5).

### 3.2.4.3. RT REACTION

1. To each of the DNase-treated sections, add 65 μL of DEPC-treated water and 5 μL of oligo-(dT)$_{15}$. Incubate for 10 min at 70°C on the slide thermal cycler.
2. Incubate for 1 min on ice.
3. Add to each slide 10 μL of 10X RT buffer, 5 μL of 10 m$M$ dNTPs, 10 μL of 100 m$M$ DTT, and 5 μL of RT (200 U/μL).
4. Incubate the sections for 10 min at room temperature, 50 min at 42°C, and 15 min at 70°C on the slide thermal cycler.
5. Chill the sections for 1 min on ice.

### 3.2.4.4. INDIRECT *IN SITU* PCR (*SEE* NOTES 50 AND 51)

1. Perform the *in situ* PCR according to a standard PCR protocol (*see* **Subheading 3.1.2.**) but attaching Sure Seal adhesive strips to form sealed chambers around the sections. A minimum of 5 U of *Taq* polymerase must be used for each reaction.
2. Wash the sections twice for 5 min each time in 2X SSC at room temperature and twice for 5 min each time at 37°C.
3. Fix the slides in 0.4% paraformaldehyde for 15 min on ice.
4. Wash the slides at room temperature for 5 min in PBS.

### 3.2.4.5. HYBRIDIZATION AND DETECTION

1. Add 10 μL of hybridization buffer to each section and incubate at 95°C for 6 min.
2. Chill the slides on ice for 1 min.
3. Hybridize overnight at 42°C in a humid chamber.
4. Wash three times in 2X SSC at room temperature for 5 min, then once in 0.1% SSC at 42°C.
5. Wash briefly in buffer 1.
6. Block for 15 min in 60–80 μL of buffer 2.

**Table 1**
**Preparation of Resolving Gel Solution**

| | Volume for different percentages of gels (mL) | | | | |
|---|---|---|---|---|---|
| Resolving gel components | 8% | 10% | 12% | 15% | 20% |
| 4X Separating gel buffer | 2.5 | 2.5 | 2.5 | 2.5 | 2.5 |
| Water | 5.4 | 4.9 | 4.4 | 3.65 | 2.4 |
| 40% Acrylamide:bis stock | 2.0 | 2.5 | 3.0 | 3.75 | 5.0 |
| 10% (w/v) Ammonium persulfate | 0.05 | 0.05 | 0.05 | 0.05 | 0.05 |
| TEMED | 0.005 | 0.005 | 0.005 | 0.005 | 0.005 |

7. Incubate for 1 h with 70 µL of anti-DIG AP-conjugated antibodies (diluted 1:500) in buffer 2.
8. Wash two times for 10 min in buffer 1.
9. Equilibrate the sections for 5 min in buffer 3.
10. Add 20 µL of NBT/BCIP to the sections, and incubate in the dark for several hours or until blue/purple precipitate appears.
11. Rinse the sections briefly in water and then mount in water-soluble mounting medium (e.g., Crystal mount; Biomeda).

## 3.3. Western Blot Analysis

### 3.3.1. Cell Lysis

1. Homogenize 100 mg of tissue in 10 vol (~1 mL) of 1X sample buffer.
2. To shear the chromosomal DNA, pass the samples several times through 20- and 26-gage needles.
3. For full protein denaturation, boil the samples for 5 min.
4. Spin the samples for 10 min at 10,000$g$. The supernatant is now ready for electrophoresis.
5. On minigels (7 × 10 cm), load approx 20 µg of protein per lane in a maximum total volume of 20 µL.

### 3.3.2. Sodium Dodecyl Sulfate Polyacrylamide Gel Electrophoresis (see *Notes 57–61*)

1. Assemble the glass-plate sandwich using 0.75-mm spacers and the casting rig supplied.
2. Prepare the resolving gel solution as directed in **Table 1** adding the ammonium persulfate and TEMED last. Immediately pour the gel using a pipet or syringe and leave ~2 cm for the stacking gel and comb.
3. Cover the resolving gel with $H_2O$-saturated isobutyl alcohol being careful not to disturb the gel surface.
4. Leave for 30–60 min, allowing the gel to polymerize completely.

**Table 2**
**Preparation of Stacking Gel Solution**

| Stacking gel components | Volume added (mL) |
|---|---|
| 4X Stacking gel buffer | 2.5 |
| Water | 6.6 |
| 40% Acrylamide:bis stock | 0.8 |
| 10% (w/v) Ammonium persulfate | 0.01 |
| TEMED | 0.01 |

5. Pour off the $H_2O$-saturated isobutyl alcohol layer and wash with 1X separating buffer taking care to completely remove any residual alcohol.
6. Prepare the stacking gel solution as directed in **Table 2** and again add the ammonium persulfate and TEMED last. Immediately pour the gel using a pipet or syringe and leave enough space for the comb to be introduced.
7. Insert a 0.75-mm comb into the stacking gel while being careful not to introduce air bubbles. Top up with stacking gel solution if needed.
8. Allow the stacking gel to polymerize for approx 30 min.
9. Remove the comb and run in 1X SDS running buffer with the gel rig provided.

## 3.3.3. Protein Transfer (see **Notes 62–65**)

### 3.3.3.1. SEMIDRY

1. Soak three sheets of 3MM paper cut to a size slightly larger than the gel in transfer buffer.
2. Put the gel onto the 3MM paper taking care to exclude all air bubbles.
3. Wet a sheet of nitrocellulose cut to the size of the gel in transfer buffer and place on top of the gel.
4. Place three more prewetted sheets of 3MM paper on top.
5. Connect the blotting apparatus to the power supply. Check that the membrane is facing the anode and that the gel is facing the cathode.
6. Run at 2.5 mA/cm$^2$ for 30–45 min.
7. At the end of the transfer, remove the membrane and rinse in deionized water.

### 3.3.3.2. TANK BLOTTING

1. Soak two Scotch Brite pads (or two sheets of sponge, 1 to 2 cm thick) in transfer buffer.
2. Make a sandwich consisting of one Scotch Brite pad, three sheets of 3MM paper cut to size just larger than the gel and wetted in buffer, the gel, one sheet of nitrocellulose wetted in transfer buffer, three sheets of prewetted 3MM paper, and finished with another Scotch Brite pad. Remove all air bubbles by rolling a pipet over the sandwich after each layer is applied.
3. Close the cassette and fit into the electrophoresis tank with the membrane facing the anode and the gel facing the cathode.

4. Fill the electrophoresis tank with buffer, and run for 1 to 2 h at 6–8 V/cm interelectrode. You may need to cool and stir the buffer during the transfer to prevent overheating (i.e., run on a magnetic stirrer in a 4°C room).

### 3.3.4. Immunodetection of Western Blot (see **Notes 52–56**)

1. Block the membrane carrying the immobilized antigens for 30 min (or overnight) in blocking solution.
2. Incubate the blocked membrane for 1 h (or overnight) with an appropriate dilution of the primary antibody in antibody solution.
3. Discard the antibody solution and wash the blot three times for 5 min each in washing solution.
4. Incubate the blot for 1 h with a 1:2000 dilution of HRP-conjugated secondary antibody (Amersham) in antibody solution.
5. Wash the membrane three times for 5 min each in washing solution.
6. Initiate the chemiluminescence reaction by incubating the membrane for 1 min in a 1:1 mixture of the ECL detection reagents, wrap it in cling film, and expose an autoradiography film (Hyperfilm ECL; Amersham) to the membrane for between 30 s and 2 h.

## 3.4. Analysis of Transgene Expression by Immunohistochemistry

### 3.4.1. Avidin-Biotin Technique for Paraffin Sections (see **Notes 66–68**)

1. Deparaffinize the sections for 10 min in xylene and rehydrate in descending grades of alcohol to water (incubate for approx 2 min in each grade).
2. For antigen retrieval, *see* **Note 66**.
3. Use the following protocol for microwaving:
   a. Transfer the sections to a microwaveable container, and immerse the sections in excess citrate buffer (1.05 g of citric acid in 500 mL of distilled water; adjust to pH 6.0 with 2 *N* NaOH).
   b. Place Cling film over the top of the container to prevent evaporation. Puncture the Cling film in two places before placing in the microwave.
   c. Microwave on high power (650–700 W) three times for 5 min each time. Tap the slides after each heating step to dislodge any bubbles formed. (Top up with citrate buffer if the level falls.)
   d. With great care remove the container from the microwave while wearing heat protective gloves.
   e. Leave for 20 min in the buffer to cool.
   f. Wash in running tap water.
   g. Wash in TBS for 5 min.
4. Use the following protocol for trypsinization:
   a. Dissolve 1 mg of trypsin in 100 mL of 0.1% calcium chloride. Adjust to pH 7.6 using 0.2 *M* Tris.
   b. Incubate the sections for 10–20 min in the trypsin solution.
5. Wash in water and then rinse in TBS for 5 min.

6. For blocking, cover the sections with normal serum (from the animal species used to produce the secondary antibody) diluted 1:5 with TBS for 20 min.
7. Drain off the serum and incubate for 1 h or overnight with the primary antibody (see manufacturer's data sheet for instructions, how to dilute the antibody/serum).
8. Drain and wash twice in TBS for 5 min each.
9. Incubate in biotinylated secondary antibody 1:400 diluted in TBS (e.g., rabbit antimouse, swine antirabbit) for 1 h.
10. Drain and wash twice in TBS for 5 min each.
11. Incubate the sections for 30 min in avidin biotin–AP complex (Dako) in Tris-HCl buffer (pH 7.6). The complex must be made up at least 30 min before use. Do not use if it is more than 2 d old.
12. Drain and wash twice in TBS for 5 min each.
13. In the dark, incubate the AP substrate solution at room temperature (10–20 min) or use the Vector Red Kit (cat. no. SK-5100). Structures that are positively stained will be red, whereas nuclei will be light blue.
14. Rinse in distilled water.
15. Counterstain lightly in hematoxylin.
16. Allow the sections to dry before mounting in crystal mount; alternatively, mount in Aquamount or glycerine jelly.

## 3.5. Reporter Enzyme Assays

### 3.5.1. CAT Assay (see **Note 69**)

1. Prepare tissue extracts by homogenizing tissue samples in 0.2 mL of homogenization buffer, followed by three freeze/thaw cycles. Immerse the extracts in liquid nitrogen (3 min) and then thaw in a 37°C water bath (3 min) (repeated three times). Then vortex and spin the homogenates for 10 min at 10,000$g$ in a table centrifuge in a microcentrifuge at 4°C.
2. Retain the supernatant and assay directly or store at –20°C.
3. Destroy endogenous CAT inhibitors by heating the samples at 65°C for 10 min.
4. To initiate acetylation, add 30 µL of tissue extract to 70 µL of the reaction mixture and incubate at 37°C for 1 h. Stop the reaction by placing the tubes on ice.
5. Add 100 µL of cold ethyl acetate to the labeled tissue extract, mix the two layers vigorously, and separate by centrifuging at 10,000$g$ in a table centrifuge for 3 min at 4°C. Collect 80 µL of the organic upper phase containing the labeled reaction product, and repeat the extraction once more. Remove 100 µL of the organic phase and combine the two.
6. Combine the organic extracts and add to 2 mL of scintillation fluid. Use liquid scintillation spectroscopy to ascertain the level of radioactivity. The cycles per minute measured represents the quantity of acetylated chloramphenicol products within each sample.

### 3.5.2. β-Galactosidase Assay (see **Note 70**)

1. Prepare tissue extracts by homogenizing tissue samples in 0.2 mL of homogenization buffer, followed by three freeze/thaw cycles. Immerse extracts in liquid

nitrogen (3 min) and then thaw in a 37°C water bath (3 min) (repeated three times). Then vortex and spin the homogenates for 10 min at 10,000*g* in a table centrifuge in a microcentrifuge at 4°C.

2. Retain the supernatant and assay directly or store at –20°C.
3. Add 3 µL of 0.1 $MgCl_2$, 4.5 *M* β-mercaptoethanol, 66 µL of ONPG, and 201 µL of 0.1 *M* sodium phosphate buffer (pH 7.5) to 30 µL of tissue extract.
4. Incubate the reaction at 37°C for 30 min or until a faint yellow color appears.
5. Stop the reaction by adding 0.5 mL of 1 *M* $Na_2CO_3$.
6. Measure the absorbance of each sample at 420 nm.

### 3.5.3. Luciferase Assay (see **Note 71**)

1. Homogenize the tissue in homogenization buffer.
2. Centrifuge the lysates at 14,000*g* for 5 min at 4°C and transfer the supernatant to a new tube.
3. Mix 0.1 mL of tissue extract with 0.36 mL of incubation buffer.
4. Initiate the reaction by injecting 0.2 mL of luciferin solution into each sample using an automatic injection system. Measure the integrated light output for 5 s. The total light output is rapid and proportional to the amount of luciferase over a wide range of enzyme concentrations when assayed in the presence of excess substrate.

## 4. Notes

1. This method routinely results in 50–100 µg of DNA. Increased yields can be achieved by mincing the tail tissue prior to digestion, but this also increases the risk of cross contamination unless care is taken to effectively clean instruments between samples.
2. Vortexing should be avoided at all stages of genomic DNA isolation because this can cause shearing and will consequently give poor results, especially during Southern blot analysis.
3. Resuspending genomic DNA is best done by heating to 55°C for 1 to 2 h while flicking the tube occasionally; alternatively, the tube can be left overnight at 4°C. Genomic DNA should be redissolved before the pellet has completely dried. Vacuum or extended air-drying is not recommended. Avoid attempting to resuspend by repeated pipetting because this can cause DNA shearing.
4. For best results, the PCR primers should be designed to result in a product of between 150 bp and 2 kb after amplification. Larger products may be harder to amplify and will result in reduced product quantities, which are harder to detect. Roche offers PCR kits for the amplification of very long templates (>5 kbp).
5. Best PCR reliability is usually achieved with primers bearing Gs and Cs at their 3' terminus.
6. When analysis is to be performed by PCR, it is often possible to omit the alcohol precipitation step completely, using 1 µL of crude tail digest supernatant as the template. Care must be taken, however, not to introduce SDS (which should

remain pelleted in the digest) into the PCR reaction, because high SDS concentrations will inhibit *Taq* polymerase.

7. DNA isolated using this rapid method, while usually sufficient for PCR, may not be suitable for restriction digests. If restriction digests fail, subsequent phenol/chloroform extraction is recommended.

8. The annealing temperature $(T_a)$ of each of the two opposing primers should be approximately the same and between 50 and 70°C. $T_a$ can be calculated as follows: $T_a [°C] = 2 \times \#(A + T) + 4 \times \#(G + C)$ *(36)*. Subtracting 5–10°C from $T_a$ can improve results, and temperatures can be raised again if increased stringency is required, i.e. if nonspecific amplification products are produced.

9. When calculating elongation times, allow 1 min/1 kb of product *(37)*.

10. The PCR conditions are likely to vary depending on the DNA template, primers, magnesium concentration, and DNA polymerase used. The supplier usually provides a general protocol describing how best to use its individual polymerase. In some cases, to achieve best results, slight variations to this basic protocol may be necessary, such as adding dimethylsulfoxide, Triton X-100, or glycerol *(38)*.

11. A variety of *Taq* DNA polymerases and PCR kits are on the market (e.g., Roche, Gibco-BRL, and Promega). Since, in general, several hundred PCR reactions will be performed for any transgenic line, it may be advisable to try the cheapest polymerases first. In most cases, they perform equally well, and because fidelity is of no concern during screening, they confer no real disadvantage over more costly products.

12. It is recommended that a second primer pair be included, if possible, in the same PCR reaction, allowing the amplification of an endogenous gene as a positive control, thus confirming the integrity of the reaction. Various primer sets that amplify endogenous genes are commercially available (e.g. β-actin and G3PDH from Clontech, UK). Be aware, however, that second primer pairs can compete with the primary primers, resulting in poor amplification of one or the other of the targets.

13. For the detection of homologous recombination in embryonic stem cells and in the resulting mice, one PCR primer is usually complementary to sequence in the introduced selectable marker, while the second primer anneals to a DNA sequence outside the targeting construct. Positive PCR signals, therefore, only result from a successful homologous recombination event.

14. More detailed information about PCR and its potentials can be found in **ref. *39***.

15. To prepare agarose gels, dissolve molecular biology grade agarose in TBE buffer by boiling the solution briefly in a microwave or on a hot plate. Cool to 50°C and add ethidium bromide to give a final concentration of 1 µg/mL. Mix well and pour into a gel mold and allow to set in a fume hood.

16. Digests should be performed in volumes >10 µL; with smaller volumes, pipetting errors become too dramatic.

17. For restriction digests, the enzyme volume added should not exceed 1/10 of the total reaction volume. Glycerol in the storage buffer may inhibit digestion and cause star activity *(40)*.

18. Transfer of DNA molecules from agarose gels to nitrocellulose, or more typically nylon membranes, can be accomplished using a variety of methods. Transfer can be achieved under either high salt, as described in **Subheading 3.1.3.1.**, or alkaline conditions *(41)*. In our hands, both work equally well. Alkaline transfer saves time as the denaturing and neutralizing steps can be avoided, with the gel being blotted immediately after electrophoresis in 0.4 *N* NaOH. A variety of membranes are commercially available both neutrally (Amersham Hybond-N) or positively charged (Amersham Hybond N+, Bio-Rad Zetaprobe). The chosen transfer and crosslinking method (UV or baking) should be compatible with the exact membrane used; see manufacturer's instructions.

19. Most researchers still perform the well-established capillary transfer. However, several companies (e.g., Pharmacia LKB or Hybaid) now offer vacuum blotting systems that reduce the transfer time to as little as 1 h while retaining the same transfer efficiencies.

20. Strategies for the screening of homologously recombined sequences should encompass any additional or lost restriction sites caused by the introduction of a selectable marker, deletion, or mutation. The labeled probe for screening genomic digests by Southern blot should always be located outside the targeting construct, allowing differentiation of homologous and random recombination events.

21. Additional information about the Southern Blot technology can be found in **ref. *42***.

22. Although several companies now offer nonradioactive labeling kits (e.g., DIG from Roche and ECL from Amersham), radioactively labeling DNA probes is still recommended when attempting to detect single-copy genes. They are still more reliable than their nonradioactive counterparts and tend to require less optimization when minimizing background.

23. Purifying labeled DNA from unincorporated nucleotides using Sephadex G-50 columns is recommended because it not only indicates the labeling efficiency but also can greatly reduce background.

24. Labeling efficiencies of distinct DNA fragments can vary. Contaminants or factors such as incompatible DNA secondary structures may cause poor labeling. The problem sometimes can be overcome by truncating the probe and always ensuring that the DNA to be labeled is as free from contaminating protein and salts as possible.

25. The signal from a single-copy gene usually requires an exposure time of up to 1 wk at –80°C. However, newly developed BIOMAX MS-1 film (Kodak) can shorten exposure times to one-fourth of that needed by standard film.

26. To estimate the molecular size of the appearing bands, compare the migration distance measured from the autoradiograph with that of the molecular size markers as seen on the photograph of the ethidium bromide–stained gel.

27. All reagents and plasticware coming into contact with samples after extraction from guanidinium thiocyanate–containing solutions must be guaranteed RNase-free.

28. Tissue volumes should never exceed 10% of the isolation reagent (e.g., TRIzol) because overloading can cause incomplete inactivation of RNases.

29. The quality of a total RNA preparation can be ascertained by visualizing rRNA bands on an ethidium bromide–stained RNA-formaldehyde agarose gel (*see* **Subheading 3.2.2.**). RNA free from degradation will resolve two distinct bands representing the 18S and 28S rRNA species. The heavier 28S rRNA should be double the intensity of the 18S band.

30. With the described method, 0.6–1 mg of total RNA can be expected from 100 mg of liver or spleen, with kidney yielding 0.3–0.4 mg, and skeletal muscle and brain 0.1–0.15 mg.

31. Oligo(dT) cellulose spin columns provide a simple method for mRNA isolation and are available from several suppliers (e.g., Pharmacia's mRNA purification kit, cat. no. 27-9258-01; or Ambion's Poly[A]Pure, cat. no. 1915).

32. RNA should be stored in DEPC-treated water at –80°C.

33. A good technical library covering many aspects of RNA work is available online (http://www.ambion.com/techlib).

34. If the mRNA to be studied is expressed at a high level (e.g., when a strong promotor is used to regulate a transgene) and is expected to consist of >0.05% of the total mRNA species within a cell, then total cellular RNA can be used for analysis. If, however, the mRNA of interest is of low abundance, then it is advisable to purify and load mRNA instead of total RNA *(43)*. Tools such as poly(dT) cellulose columns that enable mRNA purification in a single step are available from several suppliers (e.g., Roche).

35. To examine RNAs >1.5 kb, agarose concentrations of 1 to 1.2% are appropriate. Smaller RNAs (1.5–0.5 kb) are separated on 1.8% agarose gels.

36. When the gel is photographed, two bands should appear, which represent the 28S and 18S rRNA subunits. Without any degradation, both bands are sharp and the 28S band should have double the intensity of the 18S band. Both rRNA subunits can be used as standards to estimate the size of hybridization signals. The mouse 28S rRNA contains 4712 nucleotides, and the 18S rRNA 1869 nucleotides. In addition, RNA molecular weight markers are available (e.g., cat. no. 15620-016; Gibco-BRL). DNA size markers should not be used because DNA and RNA migrations through agarose gels are not equivalent.

37. PCR conditions to amplify specific cDNA sequences are dependent on the primers as well as the chosen polymerase and buffer system (*see* **Subheading 3.1.2.**). Appropriate protocols are generally supplied by the manufacturer.

38. RT-PCR is a sensitive technique that is prone to false positive results. As with PCR, contaminating DNA from other samples is a major concern. Care must be taken at all stages to avoid the introduction of contaminating DNA/RNA.

39. Primers should be designed, where possible, such that sequence spanning an intron is amplified, thus allowing cDNA and genomic DNA to be distinguished by size separation on an agarose gel. When employed, this strategy allows the omission of the DNase I treatment (in **Subheading 3.2.3.**, **steps 2–4** and DNase I being substituted for DEPC H$_2$O). This can be of significant benefit because the incubation of RNA samples at 37 and 75°C may cause unnecessary RNA degradation owing to incomplete RNase inhibition.

40. When intron-spanning primers are not available, the complete digestion of genomic DNA is vital, since any intact molecules will give rise to amplification products indistinguishable from those derived from cDNA. To control for the complete digestion of genomic DNA, duplicate the procedure, substituting DEPC-treated dH$_2$O for Superscript II RT. A second pair of primers amplifying an intronic or other nontranscribed region (e.g., promoter) also can be included in the final PCR reaction, allowing the detection of genomic DNA.

41. Always use filtered pipet tips and wash dissection tools and homogenizer with RNA-destroying cleaning solutions (e.g., Absorb, DuPont) between samples to minimize the risk of cross contamination.

42. The incorporation of DIG-dUTP can be monitored on an agarose gel; the DIG gives the labeled probe a slightly higher molecular weight, causing a visible band shift. Alternatively, the proportion of labeled DNA can be quantified using DIG quantification test strips from Roche (cat. no. 1669966).

43. Before labeling a probe, it should be purified, either by using DNA purification columns (e.g., Promega or Qiagen) or by gel extraction (Qiagen).

44. Alternative methods of labeling include the incorporation of [$^{35}$S]dATP, which can be subsequently detected by autoradiography. $^{35}$S-labeled probes provide a similar degree of sensitivity as the DIG system *(15)*. Biotin/streptavidin-based detection systems are extremely sensitive but have a widespread endogenous distribution.

45. Although DNA probes are easily obtained and handled, in comparison to riboprobes, they often give reduced hybridization efficiencies. DNA:RNA duplexes are less stable than their RNA:RNA counterparts, and double-stranded DNA probe sequences tend to reanneal, thus reducing probe availability. RNA probes, on the other hand, hybridize well but are more difficult to produce and handle. Oligonucleotides hybridize efficiently but are insensitive unless "cocktails" of several labeled sequences are used *(44)*.

46. More detailed information about ISH can be obtained from Roche (http://Roche.com) or Hybaid *(45)*.

47. The DNase I step can be omitted if primers for amplification are specific for cDNA sequences and are not able to amplify intron-containing genomic sequences.

48. Fixation of specimens prior to paraffin embedding results in reduced sensitivity of endogenous mRNA molecules to "normal" nuclease activities so that only the hybridization solutions need to be scrupulously free of RNase enzymes.

49. The fixation process closes many of the natural pores within the membrane of the cell. Proteinase K treatment (step 5) enables the entry of amplification reagents into the cell and thus access to the target nucleic acid. It is important to optimize this step for each different tissue sample, because while it allows entry of reagents, it can also potentially allow exit of small amplification reagents *(46)*.

50. In standard PCR, specificity can be increased by adding the *Taq* DNA polymerase to the reaction at 94°C, during the so-called hot start. Since this is impossible in the sealed chambers used for *in situ* PCR, the use of anti-*Taq* antibodies (TaqStart

antibodies, Clontech) can be employed as an alternative. Such antibodies guarantee that the polymerase is not initiated until the reaction reaches 94°C.

51. Because samples are heated repeatedly to 94°C, the chambers formed by the Sure Seal must be completely airtight to prevent loss of moisture by evaporation.

52. Depending on the source of the primary antibody, the type of the secondary antibody varies: If the primary antibody is monoclonal, the secondary antibody will be either rabbit, goat, or sheep antimouse immunoglobulin. If the primary antibody is a polyclonal antibody raised, e.g., in rabbits, the secondary antibody must be an antirabbit immunoglobulin raised in another animal (e.g., goat, sheep, or swine).

53. The concentration of the primary or secondary antibody has to be optimized for each study. General information is usually given by the manufacturer and should be used as a guideline. When buying primary antibodies, it is necessary to confirm that the antibody is able to recognize the denatured protein and preferably has been shown to work in Western blots. The secondary antibody should be either the whole molecule or just the $Fab_2$ fragment.

54. The ECL detection system is extremely sensitive, and film exposure times should initially be kept short (≤5 min). Extending this time may result in an overexposed film, which makes reading difficult and increases the risk of mistakenly identifying "background" bands as positive results.

55. The ECL system, including the film needed to monitor the signal, is a relatively expensive method, and if a large number Western-Blots are to be done, a cheaper alternative may be favored. AP-labeled secondary antibodies and the reagents required are considerably cheaper than ECL. With AP, a color reaction takes place using the substrates NBT (0.66%) and BCIP (0.33%) (Promega) in a 0.1 $M$ sodium carbonate buffer (pH 10.2). After several minutes, purple positive bands appear that are stable on the membrane. The developed blots can be stored at room temperature in a dry, dark place for up to 1 yr. The main disadvantage of the AP over the ECL system is that it generates signals on fragile membranes, which are more difficult to handle than autoradiography films.

56. A variety of blocking agents can be used, but to achieve optimal blocking, thus reducing background to a minimum, the addition of serum extracted from the host species in which the secondary antibody was raised is recommended. For example, if the secondary antibody is a rabbit antimouse antibody, the blocking solution should contain rabbit serum diluted in TBS. Commonly used reagents also include BSA, milk powder, and Tween-20.

57. If more than one gel is needed, it is possible to prepare numerous gels at the same time. Gels can be stored for up to 1 wk at 4°C when wrapped in a moistened paper towel and Cling film.

58. Running times are about 1 h. The starting voltage should be 100 V until the dye front has reached the separating gel, and then it can be increased to 200 V.

59. If gels fail to polymerize, it usually indicates a problem with either the ammonium persulfate or the TEMED. Always use freshly made ammonium persulfate.

60. For the separation of proteins with molecular weights between 10 and 40 kDa, 12% acrylamide gels are recommended, 10% for 40- to 100-kDa proteins and 8% for between 100- and 200-kDa proteins.

61. Molecular weight standards always should be loaded in parallel to the protein samples. They can be purchased from suppliers such as Promega or Pharmacia (low-range markers: 10–40 kDa; midrange: 21–100 kDa; high-range: 40–200 kDa). Prestained markers (Amersham) allow the separation of proteins to be followed during electrophoresis and also can be easily detected on nitrocellulose membranes after blotting.

62. To confirm efficient transfer of protein, the gel can be stained with Coomassie brilliant blue postblotting. Incubate the gels for 10 min in staining solution (0.1% Coomassie brilliant blue R, 50% methanol, 10% glacial acetic acid). Nonbound stain can be washed off in 10% ethanol, 7% glacial acetic acid.

63. A second check of transfer efficiency, which allows the visualization of transfer imperfections caused by trapped air bubbles in the transfer sandwich, is the reversible Ponceau-S membrane stain. This stain, unlike Coomassie blue, does not interfere with immunodetection. After Ponceau-S staining, it may be useful to mark the position of the molecular weight markers on the membrane with a pencil. Incubate the membrane in Ponceau-S (0.5% in 1% glacial acetic acid) (cat. no. P7170; Sigma) for 5 min, and rinse briefly in $dH_2O$.

64. If the filters are not going to be used immediately, they can be stored dry at 4°C or wrapped in Cling film.

65. PVDF membranes (e.g., Immobilon-P; Millipore) have to be incubated briefly in methanol prior to their introduction into transfer buffer. PVDF membranes are more rigid than nitrocellulose membranes and thus ease repeated probing.

66. The retrieval of antigens is optional and has to be tested for every antibody. Treatment with proteolytic enzymes, such as trypsin, results in the breaking of protein crosslinks formed by formalin fixation and thus may increase the availability of antigen epitopes that were previously masked. Extended treatment should be avoided because it can result in an increase in false positive signals. Alternatively, microwaving tissue for 5–20 min, in either citrate buffer or 10 m$M$ EDTA (pH 8.0), may increase exposure of hidden antigens.

67. Avidin-biotin-based methods are very powerful; they do, however, have disadvantages, which should be kept in mind. Because of avidin's high isoelectric point (~10), it may bind nonspecifically to negatively charged structures such as the nucleus. Furthermore, avidin may interact with glycoprotein binding molecules, such as lectins, resulting in false positive signals. Both problems may be overcome with the substitution of streptavidin for avidin. In addition, blocking kits are available that may be used to block the nonspecific binding sites of biotin/avidin system reagents (cat. no. SP-2001; Vector).

68. A wide range of secondary antibody tags are commercially available, including avidin/biotin; AP; HRP; and fluorochromes such as fluorescein isothiocyanate (FITC), Rhodamine, Texas Red, and Cy3.

a. AP: Visualized in an alkaline substrate buffer (*see* **Subheading 2.4.**). Tissues such as the gut contain high levels of endogenous AP and therefore are unsuitable for this system.

b. HRP: To develop the color reaction, add 3 mL of DAB solution and 12 drops of hydrogen peroxide (30% [w/v]) to 400 mL of Tris (pH 7.6). Incubate with sections for 10 min or until staining appears. Macrophages and red blood cells contain high levels of endogenous peroxidases that require blocking prior to primary antibody incubation. Blocking can be achieved by incubating the sections in 3% (v/v) $H_2O_2$ for 20 min.

c. Fluorochromes: Laboratories that have access to a fluorescence microscope may prefer to employ fluorochrome-labeled secondary antibodies such as FITC, Rhodamine, Texas Red, or Cy3. Fluorescence labeling, however, is generally only applied to single-cell fixations rather than tissue sections. Fluorescence microscopy often gives good resolution at a cellular level but does not allow for counterstaining.

69. To quantify CAT activity, a standard curve using purified CAT (cat. no. E1051; Promega) has to be prepared. As a guideline, the standard curve should be established from 0.1 to 0.00625 U of enzyme. Dilute the CAT enzyme to 0.02 U/µL with Tris-HCl (pH 7.8). Prepare four serial twofold dilutions of the 0.02 U/µL solution (1:2–1:16). Five-microliter aliquots of these solutions contain 0.05, 0.025, 0.0125, and 0.00625 U.

70. To estimate enzyme activity, a standard curve using purified β-galactosidase (cat. no. G-6512; Sigma) has to be developed. For baseline activity, 30 µL of normal tissue extract should be measured.

71. Promega is a good source of luciferase vectors and assay kits (cat. no. E1500; Promega).

## References

1. Mason, I. J., Murphy, D., Munke, M., Francke, U., Elliot, R. W., and Hogan, B. L. M. (1986) Developmental and transformation-sensitive expression of the Sparc gene on chromosome 11. *EMBO J.* **5,** 1831–1837.

2. Southern, E. M. (1975) Detection of specific sequences among DNA fragments separated by gel electrophoresis. *J. Mol. Biol.* **98,** 503–517.

3. Devereux, J., Haeberli, P., and Smithies, O. (1984) A comprehensive set of sequence analysis programs for the VAX. *Nucleic Acid Res.* **12,** 387–395.

4. Fuchs, R. and Cameron, G. N. (1995) Databases, computer networks, and molecular biology, in *DNA Cloning 3, A Practical Approach* (Glover, D. M. and Hames, B. D., eds.), IRL, NY, pp. 151–172.

5. Bickmore, B. (1995) Long-range restriction mapping, in *DNA Cloning 3, A Practical Approach* (Glover, D. M. and Hames, B. D., eds.), IRL, NY, pp. 182–188.

6. Rigby, P. W. J., Dieckmann, M., Rhodes, C., and Berg, P. (1977) Labelling desoxyribonucleic acid to high specific activity *in vitro* by nick translation with DNA polymerase I. *J. Mol. Biol.* **113,** 237–251.

7. Feinberg, A. P. and Vogelstein, B. (1984) A technique for radiolabelling DNA restriction endonuclease fragments to high specific activity. *Anal. Biochem.* **137,** 266, 267.

8. Lehrbach, H., Diamond, D., Wozney, J. M., and Boedtker, H. (1977) RNA molecular weight determinations by gel electrophoresis under denaturing conditions: a critical reexamination. *Biochemistry* **16,** 4743–4751.

9. Chelly, J., Kaplan, J.-C., Maire, P., Gautron, S., and Kahn, A. (1988) Transcription of the dystrophin gene in human muscle and non-muscle tissues. *Nature* **333,** 858–860.

10. Newton, C. R. (1995) Mutational analysis: known mutations, in *PCR 2, A Practical Approach* (McPherson, M. J., Hames, B. D., and Taylor, G. R., eds.), IRL, NY, pp. 219–253.

11. Haase, A. T., Retzel, E. F., and Staskus, K. A. (1990) Amplification and detection of lentiviral DNA inside cells. *Proc. Natl. Acad. Sci. USA* **87,** 4971–4975.

12. Chen, R. H. and Fuggle, S. V. (1993) In situ cDNA polymerase chain reaction: a novel technique for detecting mRNA expression. *Am. J. Pathol.* **143,** 1527–1534.

13. Komminoth, P. and Long, A. A. (1993) In situ polymerase chain reaction: an overview of methods, applications and limitations of a new molecular technique. *Virchows Archiv B Cell Pathol.* **64,** 67–73.

14. Long, A. A., Komminoth, P., Lee, E., and Wolfe, H. J. (1993) Comparison of indirect and direct in-situ polymerase chain reaction in cell preparations and tissue sections. *Histochemistry* **99,** 151–162.

15. Komminoth, P., Merk, F. B., Leav, I., Wolfe, H. J., and Roth, J. (1992) Comparison of $^{35}$S- and digoxigenin-labelled RNA and oligonucleotide probes for in situ hybridization. *Histochemistry* **98,** 217–228.

16. Laemmli, U. K. (1970) Cleavage of structural proteins during the assembly of the head of bacteriophage T4. *Nature* **227,** 680–685.

17. Spector, D. L., Goldman, R. D., and Leinwand, L. A. (1997) *Cells, A Laboratory Manual*, Vol. 1, Culture and biochemical analysis of cells. Cold Spring Harbor Laboratory Press, Cold Spring Harbor, NY, pp. 34.1–55.11.

18. Harlow, E. and Lane, D. (1988) *Antibodies, A laboratory manual*, Vol. 1, Cold Spring Harbor Laboratory Press, Cold Spring Harbor, NY, pp. 421–470.

19. Towbin, H., Staehelin, T., and Gordon, J. (1979) Electrophoretic transfer of proteins from polyacrylamide gels to nitrocellulose sheets: procedure and some applications. *Proc. Natl. Acad. Sci. USA* **76,** 4350–4354.

20. Beesley, J. E., ed. (1993) *Immunocytochemistry, A Practical Approach*, IRL, New York.

21. Gratzner, A. G. (1982) Monoclonal-antibody to 5-bromodeoxyuridine and 5-iododeoxyuridine—a new reagent for detection of DNA-replication. *Science* **218,** 474, 475.

22. Selbert, S., Fischer, P., Pongratz, D., Stewart, M., and Noegel, A. A. (1995) Expression and localization of annexin VII (synexin) in muscle cells. *J. Cell Sci.* **108,** 85–95.

23. Midgeley, C. A., White, S., Howitt, R., Save, V., Dunlop, M. G., Hall, P. A., Lane, D. P., Wyllie, A. H., and Bubb, V. J. (1997) APC expression in normal human tissues. *J. Pathol.* **181,** 426–433.

24. Mason, L. Y. and Sammons, R. (1978) Alkaline phosphatase and peroxidase for double immunoenzymatic labelling of cellular constituents. *J. Clin. Pathol.* **31,** 454–460.

25. Gorman, C. M., Moffat, L. F., and Howard, B. H. (1982) Recombinant genomes which express chloramphenicol acetyltransferase in mammalian cells. *Mol. Cell. Biol.* **2,** 1044–1051.

26. Rosenthal, N. (1987) Identification of regulatory elements of cloned genes with functional assays, in *Methods in Enzymology*, vol. 152 (Berger, S. L. and Kimmel, A. R., eds.), Academic, London, pp. 704–720.

27. Stanley, P. E. and Kricka, L. J., eds. (1991) *Bioluminescence and Chemiluminescence: Current Status*, John Wiley & Sons, NY.

28. Wood, K. V. (1995) Marker proteins for gene expression. *Curr. Opin. Biotechnol.* **6,** 50–58.

29. Alton, N. K. and Vapnek, D. (1979) Nucleotide sequence analysis of the chloramphenicol resistance transposon Tn9. Nature **282,** 864–869.

30. Leslie, A. G. W., Moody, P. C. E., and Shaw, W. V. (1988) Structure of chloramphenicol acetyltransferase at 1.75 A resolution. *Proc. Natl. Acad. Sci. USA* **85,** 4133–4137.

31. Shaw, W. V. (1975) Chloramphenicol acetyltransferase from chloramphenicol-resistant bacteria, in *Methods in Enzymology*, vol. 43 (Hash, J. H., ed.), Academic, London, pp. 737–755.

32. Folwer, A. V. and Zabin, I. (1983) Purification, structure and properties of hybrid β galactosidase proteins. *J. Biol. Chem.* **258,** 14,354–14,358.

33. DeWet, J. R., Wood, K. V., Helinski, D. R., and Deluca, M. (1985) Cloning of firefly luciferase cDNA and the expression of active luciferase in *Escherichia coli*. *Proc. Natl. Acad. Sci. USA* **82,** 7870–7873.

34. DeWet, J. R., Wood, K. V., Deluca, M., Helinski, D. R., and Subramain, S. (1987) Firefly luciferase gene—structure and expression in mammalian cells. *Mol. Cell. Biol.* **7,** 725–737.

35. Brasier, A. R., Tate, J. E., and Habener, J. F. (1989) Optimized use of the firefly luciferase assay as a reporter gene in mammalian cell lines. *BioTechniques* **7,** 1116–1122.

36. Wallace, R. B. and Miyada, C. G. (1987) Oligonucleotide probes for the screening of recombinant DNA libraries, in *Methods in Enzymology*, vol. 152 (Berger, S. L. and Kimmel, A. R., eds.), Academic, London, pp. 432–442.

37. Innis, M. A., Myambo, K. B., Gelfand, D. H., and Brow, M. A. D. (1989) DNA sequencing with Thermus aquaticus DNA polymerase and direct sequencing of polymerase chain reaction-amplified DNA. *Proc. Natl. Acad. Sci. USA* **85,** 9436–9440.

38. Kidd, K. K. and Ruano, G. (1995) Optimizing PCR, in *PCR 2, A Practical Approach* (McPherson, M. J., Hames, B. D., and Taylor, G. R., eds.), IRL, NY, p. 15.

39. Kidd, K. K. and Ruano, G. (1995) Optimising PCR, in *PCR2, A Practical Approach* (McPherson, M. J., Hames, B. D., and Taylor, G. R., eds.), IRL, NY, pp. 1–22.
40. Titus, D. E., ed. (1991) *Promega, Protocols and Application Guide*, Promega, Madison, WI.
41. Chomczynski, P. and Qasba, P. K. (1984) Alkaline transfer of DNA to plastic membrane. *Biochem. Biophys. Res. Commun.* **122,** 340–344.
42. Selden, R. F. (1991) Analysis of DNA sequences by blotting and hybridization, in *Current Protocols in Molecular Biology* (Ausubel, F. M., Brent, R., Kingston, R. E., Moore, D. D., Seidman, J. G., Smith, J. A., and Struhl, K., eds.), vol. 1, Green Publishing Associates and Wiley-Interscience, New York, Section IV, Unit 2.9., pp. 2.9.1–2.9.17.
43. Walker, J. M., ed. (1984) *Methods in Molecular Biology, Vol. 2: Nucleic Acids*, Humana, Clifton, NJ.
44. Starling, J. (1994) *A Guide to* In Situ, Hybaid, Middlesex, UK.
45. Staecker, H., Cammer, M., Rubinstein, R., and Van De Water, T. R. (1994) A procedure for RT-PCR amplification of mRNAs on histological specimens. *Biotechniques* **16,** 76–80.
46. Lewis, M. E., Sherman, T. G., and Watson, S. J. (1985) *In Situ* hybridization histochemistry with synthetic oligonucleotides: strategies and methods. *Peptides* **6(Suppl. 2),** 75–87.

# Index

Adenovirus-mediated gene transfer,
    mouse,
  cassette cosmids, 74
  conditional transgenesis, 81
  Cre expressing virus, 81
  embryo transfer, 80
  fertilization in TYH medium, 77
  infection of eggs,
    dose and integration efficiency, 78, 79
    incubation conditions, 78, 80
    X-gal staining, 80
  materials, 74, 75, 77
  overview, 44, 73, 74
  recombinant virus generation, 74, 81
  Southern blot analysis, 80
  zona pellucida removal, 78

Biopharming, 6–8
Blotting, see Northern blot; Southern blot;
    Western blot

Casein, transgenic animal production, 9
CAT, see Chloramphenicol
    acetyltransferase
Cattle, transgenesis, 279–281, 284
Chimera, mouse,
  cell markers, 129, 146
  donor mice, 130, 146
  embryo culture media,
    blastocysts, 130
    cleavage-stage embryos, 130
    decompaction and aggregation, 131
    embryonic stem cells, 131
    zona pellucida removal, 131

Chimera, mouse (cont.),
  embryo transfer, 142–144
  embryonic stem cell aggregation with
      morulae,
    advantages, 139
    clump aggregation, 140, 141
    coculture, 141
    compaction extent, 140
    embryonic stem cell preparation,
        137
    media, 131
    tetraploid embryo aggregation
        with embryonic stem cells,
        141, 142
    zona pellucida removal, 131, 140
  embryonic stem cell injection into
      embryos,
    blastocyst injection, 137–139
    embryonic stem cell preparation,
        injection preparation, 137
    karyotype analysis, 147
    injection chamber, 131
    micromanipulator assembly,
      manipulators and optics, 135,
          136
    microsuction, 136
    orientation within injection
        chamber, 136
    morula injection, 139
    pipette preparation,
      bend angle, 135
      holding, 134, 135
      injection, 132, 134

From: *Methods in Molecular Biology, vol. 180, Transgenesis Techniques, 2nd ed., Principles and Protocols*
Edited by: A. R. Clarke © Humana Press, Inc., Totowa, NJ